GENE EXPRESSION
General and
Cell-Type-Specific

Progress in Gene Expression

Series Editor:

Michael Karin
Department of Pharmacology
School of Medicine
University of California, San Diego
La Jolla, CA 92093-0636

Books in the Series:

Gene Expression: General and Cell-Type-Specific
M. Karin, editor
ISBN 0-8176-3605-6

Forthcoming:

Inducible Transcription, I: Control of Gene Expression
Through Cell Surface Receptors
P. Baeuerle, editor

Inducible Transcription, II: Control of Gene Expression
Through Nuclear Receptors
P. Baeuerle, editor

GENE EXPRESSION
General and Cell-Type-Specific

Michael Karin

Editor

Birkhäuser

Boston • Basel • Berlin

Michael Karin
Department of Pharmacology
School of Medicine
University of California, San Diego
La Jolla, CA 92093

Library of Congress Cataloging In-Publication Data
Gene expression : general and cell-type-specific / Michael Karin,
    editor.
        p.   cm. -- (Progress in gene expression)
        Includes bibliographical references and index.
        ISBN 0-8176-3605-6 (acid-free paper). -- ISBN 3-7643-3605-6 (acid
    -free paper)
        1. Gene expression.   2. Genetic transcription.   I.   Karin,
    Michael.   II. Series.
    QH450.G4619   1993   93-3907
    574.87'322--dc20   CIP

© Birkhäuser Boston 1993

Printed on acid-free paper   *Birkhäuser* 

ISBN 0-8176-3605-6
ISBN 3-7643-3605-6

Typset by TEXniques, Inc., Newton, MA.
Printed and bound by Edwards Brothers, Ann Arbor, MI.
Printed in the U.S.A.

9 8 7 6 5 4 3 2 1

Contents

Preface

This book is the first volume in a new series *Progress in Gene Expression*. The control of gene expression is a central-most topic in molecular biology as it deals with the utilization and regulation of gene information. As we see huge efforts mounting all over the developed world to understand the structure and organization of several complex eukaryotic genomes in the form of Gene Projects and Genome Centers, we have to remember that without understanding the basic mechanisms that govern the use of genetic information, much of this effort will not be very productive. Fortunately, however, research during the past seven years on the mechanisms that control gene expression in eukaryotes has been extremely successful in generating a wealth of information on the basic strategies of transcriptional control. (Although regulation of gene expression is exerted at many different levels, much of the emphasis in this series will be on transcriptional control. A future volume, however, will deal with other levels of regulation).

The progress in understanding the control of eukaryotic transcription can only be appreciated by realizing that seven years ago we did not know the primary structure of a single sequence specific transcriptional activator, and those whose primary structures were available (e.g., homeodomain proteins) were not yet recognized to function in this capacity. Also, seven years ago transcription was thought to be carried out by an abstract assembly of transcription factors and RNA polymerases referred to as the "transcriptional machinery," while nowadays many of these basic components have been purified to homogeneity and are available as molecular clones. While the progress in this field has been incredible, it is far from reaching a plateau and it is likely that the next seven years will result in an even greater and faster increase in our understanding of gene regulation. However, we have reached a point at which some generalizations can be made, recurrent themes can be identified, and unifying hypotheses formulated. The purpose of this series is to summarize this overwhelming amount of information in a small number of volumes, each containing chapters written by well-recognized experts dealing with highly related topics. By studying the progress made in a select number of model systems, it is hoped that the reader will be able to apply this knowledge to his own favorite experimental system.

The topic chosen for the first and introductory volume is general and cell-type-specific gene expression. Although the major emphasis is placed on various approaches to the central problem of cell-type-specific gene transcription, it is impossible to discuss this problem without discussing the structure and composition of the basic transcriptional machinery because this complex assembly of proteins is the target for the action of various cell-type-specific regulators. The basic transcriptional machinery, its composition, and the mechanisms that regulate its activity are discussed in the first chapter by Stefan Roberts and Michael Green. The second chapter by Helen Blau presents a general discussion of the various mechanisms used to achieve differential gene expression and the regulatory strategies involved in maintenance of the differentiated state. Since differential gene expression involves an intricate interplay between cell-type-specific and more generally distributed transcriptional regulators, some of which are responsive to a variety of inductive signals, the next two chapters by Miguel Beato and Marc Montminy deal with transcription factors whose activity is regulated by two classes of hormonal signals: steroid hormones and polypeptide hormones, respectively. The factors described by these authors serve as prototypes for many other transcriptional factors whose activity is regulated by a variety of hormonal and environmental signals. More extensive discussions of such factors will be presented in an upcoming volume in this series.

Following these more general chapters, the remaining four chapters deal with a small number of model systems that illustrate the regulatory principles used to determine and regulate cell-type-specific gene expression. These systems were chosen because they have been extensively studied as generally applicable models in many different laboratories, they are relatively well understood, and the major transcription factors responsible for selective gene activation have been identified. While these chapters present an excellent and extensive summary of each system, the wealth of information regarding the regulation of many other individual genes justify the publication of future volumes, each dedicated to one of the organ systems discussed here (liver, muscle, hematopoietic cells, and neuroendocrine cells) and those not covered in this volume (the neuronal system, bone and connective tissue, kidney, lung, etc.).

John Schwartz, James Martin, and Eric Olson present an extensive discussion of the various mechanisms that operate to achieve differential and selective gene expression in various types of muscle cells. They chose to focus their discussion on several muscle-specific transcription factors responsible for these processes and the mechanisms that regulate their activity. Gennaro Ciliberto, Riccardo Cortese, and their colleagues present a comprehensive review of gene expression in another specialized organ system, the liver. The liver is a rather complex organ consisting of several cell types including those that specialize in the production of various serum proteins. These authors discuss the mechanisms involved in the transcriptional control of liver-specific genes, including both signal-regulated and cell-type-specific gene expression. Gene expression in an even more heterogenous organ system, the hematopoietic system, is discussed by Beverly Emerson, who chose to focus her discussion on a single gene family — the globin genes. This is an excellent choice because more is known about the structure and regulation of both the avian and mammalian globin loci than any other gene.

In the last chapter, Lars Theill presents an extensive review of gene regulation in neuroendocrine cells. As with the previous author, Theill has chosen to focus his review on a single well-studied model gene — the growth hormone gene. His discussion deals with the mechanisms responsible for both the selective expression of the growth hormone gene in specialized cells of the anterior pituitary, and for regulating this expression by a variety of hormonal signals.

It is our goal that the *Progress in Gene Expression* series serves as an important resource for graduate students and experienced researchers alike, in the fields of molecular biology, cell biology, biochemistry, biotechnology, cell physiology, endocrinology, and related fields. Other exciting volumes are in the planning stages, including *Inducible Transcription*, volumes 1 and 2, and *Gene Expression and Environmental Stress*. Suggestions for future volumes are appreciated, and should be directed to the Series Editor.

Michael Karin
La Jolla, CA
April 1993

List of Contributors

Miguel Beato, Institut für Molekularbiologie und Tumorforschung, IMT, Universität Marburg, Emil-Mannkopff-Straße 2, D-3550 Marburg, Germany

Helen M. Blau, Department of Pharmacology, Stanford University School of Medicine, 300 Pasteur Drive, Stanford, California 94305-5332, USA

Gennaro Ciliberto, Istituto di Ricerche di Biologia Molecolare, Via Pontina Km. 30, 600, 00040 Pomezia (Roma), Italy

Vittorio Colantuoni, Dipartimento di Biochimica e Biotecnologia, Università degli Studi di Napoli, Via Pansini 5, 80131 Naples, Italy

Riccardo Cortese, Istituto di Ricerche di Biologia Molecolare, Via Pontina Km. 30, 600, 00040 Pomezia (Roma), Italy

Raffaele De Francesco, Istituto di Ricerche di Biologia Molecolare, Via Pontina Km. 30, 600, 00040 Pomezia (Roma), Italy

Vincenzo De Simone, CEINGE, Via Pansini 5, 80131 Naples, Italy

Beverly M. Emerson, The Salk Institute, 10010 North Torrey Pines Road, La Jolla, California 92037, USA

Michael R. Green, Program in Molecular Medicine, University of Massachusetts Medical Center, Worcester, Massachusetts 01605, USA

James F. Martin, Department of Biochemistry and Molecular Biology, The University of Texas M.D. Anderson Cancer Center, Houston, 1515 Holcombe Boulevard, Houston, Texas 77030, USA

Paolo Monaci, Istituto di Ricerche di Biologia Molecolare, Via Pontina Km. 30, 600, 00040 Pomezia (Roma), Italy

Marc R. Montminy, The Clayton Foundation Laboratories for Peptide Biology, The Salk Institute, 10010 North Torrey Pines Road, La Jolla, California 92037, USA

Alfredo Nicosia, Istituto di Ricerche di Biologia Moleculare, Via Pontina Km. 30, 600, 00040 Pomezia (Roma), Italy

Eric N. Olson, Department of Biochemistry and Molecular Biology, The University of Texas M.D. Anderson Cancer Center, Houston, 1515 Holcombe Boulevard, Houston, Texas 77030, USA

Dipak P. Ramji, Istituto di Ricerche Biologia Molecolare, Via Pontina Km. 30, 600, 00040 Pomezia (Roma), Italy. Current affiliation: Department of Biochemistry, University of Wales, College of Cardiff, P.O. Box 903, Museum Avenue, Cardiff CR1 1ST, United Kingdom

Stefan G. E. Roberts, Program in Molecular Medicine, University of Massachusetts Medical Center, Worcester, Massachusetts 01605, USA

John J. Schwarz, Department of Biochemistry and Molecular Biology, The University of Texas M.D. Anderson Cancer Center, Houston, 1515 Holcombe Boulevard, Houston, Texas 77030, USA

Lars Eyde Theill, Department of Cell Biology, Amgen, 1840 DeHavilland Drive, Thousand Oaks, California 91320-1789, USA

Carlo Toniatti, Istituto di Ricerche Biologia Molecolare, Via Pontina Km. 30, 600, 00040 Pomezia (Roma), Italy

Chapter 1

The Basic Transcriptional Machinery

Stefan G. E. Roberts and Michael R. Green

Recent years have seen a vast increase in our understanding of the mechanisms by which gene transcription is initiated. Indeed, several of the components of the transcriptional machinery have now been cloned, and current work is concentrating on the function of these proteins and the way in which they assemble to form the transcriptionally competent preinitiation complex. The preinitiation complex is an obvious target for the action of sequence specific transcription factors including those involved in cell-type-specific gene expression.

Introduction

In contrast to the cells of prokaryotes, eukaryotic cells contain three RNA polymerases (termed pol I, II, and III), each involved in the transcription of a subset of genes. Pol I is responsible for the transcription of ribosomal RNA, and pol III transcribes small RNAs, which include transfer RNAs and the 5S subunit of ribosomal RNA. Pol II, which is required for the transcription of all protein-coding genes, is the focus of this chapter. The finding that eukaryotic cells contain three RNA polymerases, each involved in the transcription of a specific group of genes, and the observation that these classes of genes contained different control elements, indicated that the protein complexes involved in these systems may be mutually exclusive. Although this dogma appears to be mostly true, recent work has shown that at least one protein may be utilized in the transcription of all three classes of eukaryotic genes.

GENE EXPRESSION: GENERAL AND CELL-TYPE-SPECIFIC
Michael Karin, Editor
© 1993 Birkhäuser Boston

Transcription by RNA Polymerase II

RNA polymerase II, while fully capable of catalyzing mRNA synthesis from a DNA template, is unable to initiate transcription at specific sites (Manley et al., 1980; Weil et al., 1979). The task of directing pol II to the promoter is accomplished by a group of proteins termed the general transcription factors (GTFs). The initiation sites of pol II genes contain DNA sequences that are recognized by the GTFs as targets for the formation of a transcriptional complex. These recognition sequences include the TATA motif and the less well studied initiator element (INR). Binding of specific GTFs to these sites nucleates the formation, on the DNA, of the preinitiation complex, which is then able to transcribe from the DNA template at a specific point. Several of the GTFs have now been cloned, and current studies concentrate on determining their role in the preinitiation complex.

The GTFs defined as being absolutely required for basal transcription of a typical pol II gene include transcription factors IIA (TFIIA), TFIIB, TFIID, TFIIE, TFIIF, TFIIH, and TFIIJ. Proteins that bind to the INR, present in only some genes, are not included as general transcription factors as they do not appear to be required for all genes.

The GTFs assemble on the promotor in a sequential order beginning with TFIID, which binds to the TATA sequence present in most protein coding genes (Figure 1). It is interesting to note that TFIID is also the only known DNA-binding protein among the GTFs, and it is believed that the other GTFs are recruited by protein–protein interactions. *In vitro*, at least, TFIIA stabilizes the interaction of TFIID with the TATA element, although it appears to be dispensable for the recruitment of the other GTFs. TFIIB is the next GTF to enter the complex, followed by TFIIF and then RNA polymerase II. TFIIE, TFIIH, and TFIIJ enter the complex last. This assembly at the initiation region of class II genes is now able, in the presence of deoxynucleoside triphosphate (dNTP), to initiate transcription. The formation of a preinitiation complex, however, is ATP independent, although hydrolysis of ATP (or dATP) is required before the synthesis of the first phosphodiester bond (Bunick et al., 1982; Lin and Green, 1991).

To understand the mechanism by which a preinitiation complex forms and how activators may enhance this process has required the purification and cloning of several GTFs. The features and properties of these proteins are now reviewed in the order in which they assemble in the preinitiation complex.

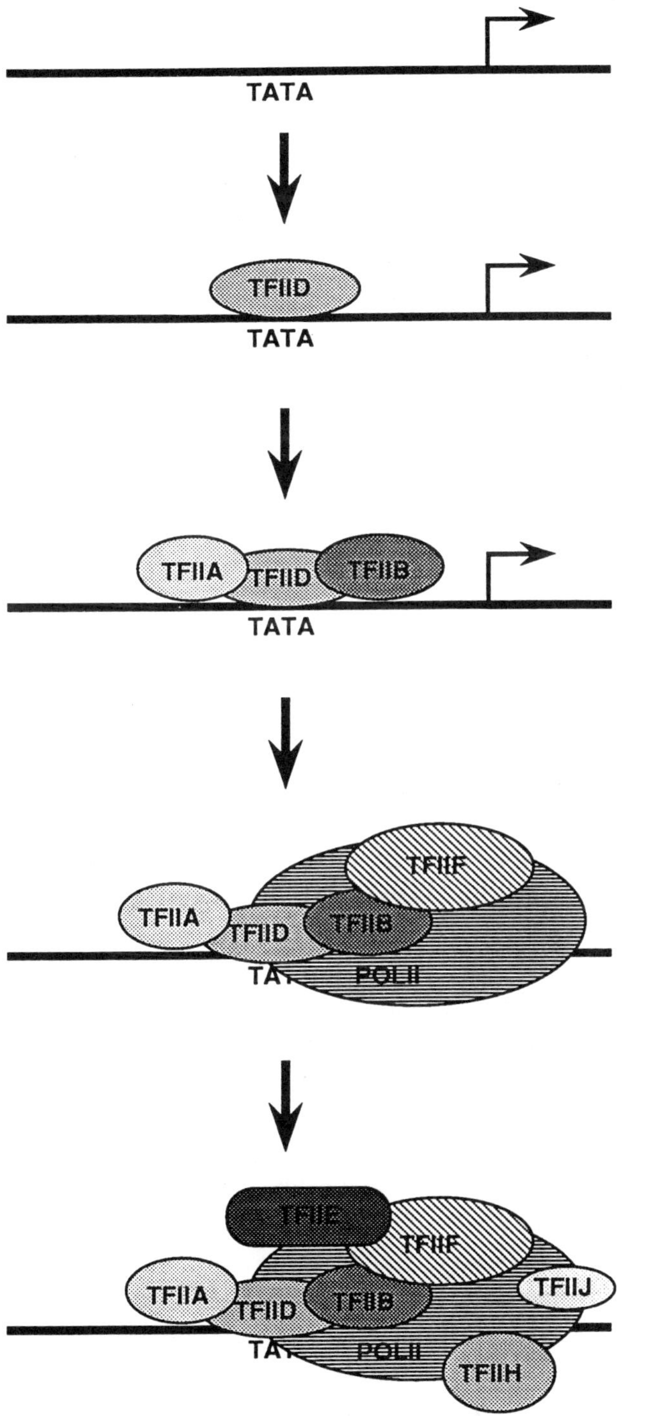

Figure 1   Formation of a preinitiation complex. Order of addition of the GTFs is discussed in the text.

General Transcription Factors

TFIID

In yeast, TFIID exist as a single protein of molecular mass of 27 kilo-daltons (kDa), in higher eukaryotes it appears on chromatography as a multiprotein complex of greater than 100 kDa. Indeed, in HeLa cell nuclear extracts at least two distinct forms of the TFIID multiprotein complex have been described (Timmers and Sharp, 1991). This heterogeneity combined with an apparent instability of the protein rendered problematic the purification of TFIID for sequence analysis from HeLa nuclear extract. Fortunately, yeast TFIID chromatographs as a single polypeptide and could substitute for the human TFIID fraction in basal transcription (Buratowski et al., 1988; Cavallini et al., 1988). Yeast TFIID was purified and the cDNA isolated by reverse genetics (Cavallini et al., 1989; Hahn et al., 1989; Horikoshi et al., 1989; Schmidt et al., 1989). Using degenerate oligonucleotide primers against the TATA-binding core domain of yeast TFIID, the *Drosophila* (Hoey et al., 1990) and human (Hoffman et al., 1990; Kao et al., 1990; Peterson et al., 1990) TFIID clones soon followed. Since that time the cDNAs for TFIID from several other organisms have also been cloned. Interestingly, in all except *Arabidopsis*, in which two genes code for TFIID (Gash et al., 1990), a single gene encodes this fundamental protein. On comparison of these TFIID proteins, the striking observation was that the C-terminal 180 amino acids are highly conserved while the N terminus is highly divergent and increases in length from yeast to human (Figure 2).

The conserved C terminus was shown to be required for TATA binding, with mutational analysis indicating that the integrity of the two direct repeats present in this region are required for binding to DNA (Horikoshi et al., 1990). Nested between the two direct repeats is a region rich in basic residues that will theoretically align on one surface of a hypothetical alpha helix, forming a so-called amphipathic helix. This portion of the protein has been shown to be a contact surface for other proteins (see later discussion). Within the C-terminal direct repeat is a region with high similarity to prokaryotic sigma factors. As will become apparent, sigma factor homologies have been identified in other basal transcription factors, and it is possible that the diverse functions of the prokaryotic sigma factors (including directing bacterial RNA polymerase to the promoter and DNA melting) may have been distributed among several of the eukaryotic GTFs.

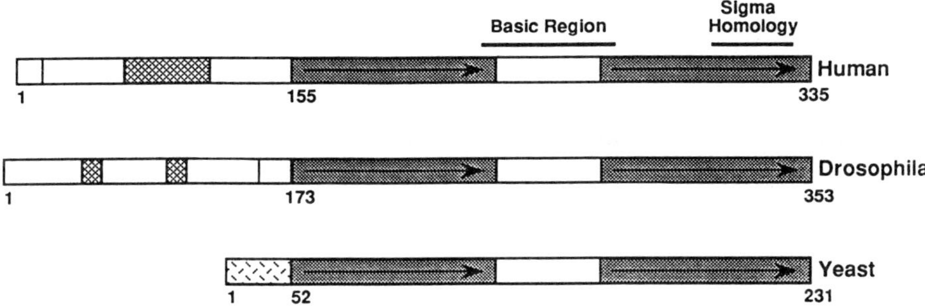

Figure 2   Comparison of the structures of TBP from yeast, *Drosophila*, and human. Adapted with permission from *Nature* 346:387–390, 1990, Macmillan Magazines, Limited.

The N terminus of TFIID is highly divergent across these species for which amino acid sequences have been obtained. Although the nonconserved N terminus of yeast TFIID is only 53 amino acids, the N terminus of human TFIID is 153 amino acids. The human N terminus contains several interesting features, notably a run of 26 glutamine residues flanked by regions rich in serine and threonine residues. To date, no precise function has been attributed to the N terminus of human TFIID although it is absolutely required to support activated transcription.

Using *in vitro* transcription assays, it was found that recombinant TFIID was able to replace the HeLa TFIID fraction in basal transcription but failed to support elevated transcription in the presence of an activator problem (Pugh and Tjian, 1990; reviewed in Lewin, 1990). It was suggested that other factors were required and these were termed coactivators. As discussed earlier, human TFIID exists as a multiprotein complex, and while one of these components, the TATA-binding component, is able to reconstitute basal transcription in conjunction with the other GTFs, additional components of the TFIID complex apparently are required for activation of transcription.

These findings caused a redefinition of TFIID, and these definitions are used for the remainder of this chapter. The cloned protein that binds to the TATA box is now referred to as TATA binding protein (TBP), and the multiprotein complex present in HeLa cell nuclear extract that contains TBP is referred to as TFIID. Thus, the TFIID complex contains factors that are required for activation of transcription, that is, are required to mediate the stimulation of preinitiation complex formation. These proposed "bridges" between activator proteins and TBP have variously been termed adaptors, coactivators, and mediators.

Using band shift assays, TBP binds specifically to the TATA element alone, and is fully capable of providing a platform for the assembly of the other GTFs (Buratowski et al., 1989; Maldonado et al., 1990). Additionally, it has been demonstrated that TBP induces DNA bending at the time of binding, although the significance of this in preinitiation complex formation is unknown (Horikoshi et al., 1992).

TFIIA

TFIIA, an activity present in nuclear protein extracts and shown to associate with TFIID (or TBP) at the TATA box, has been the subject of controversy, with various groups claiming that it was dispensable in *in vitro* transcription assays. This discrepancy has been resolved with the finding that TFIIA activity was present in other GTF fractions, notably the TFIID fraction (Reinberg et al., 1987). In a recombinant system, TFIIA stabilizes the interaction of TBP with the TATA element (Buratowski et al., 1989; Horikoshi et al., 1989). Indeed, using yeast genetics it has been demonstrated that TFIIA interacts with the basic region of TBP (Buratowski and Zhou, 1992). In HeLa cell nuclear extract, TFIIA appears to be composed of several subunits, although this too is subject to controversy. The TFIIA homolog from yeast has been cloned and shown to be composed of two subunits of 32 kDa and 13.5 kDa, which form a heterodimer (Ranish and Hahn, 1991; Ranish et al., 1992). Additionally, yeast TFIIA is capable of substituting for human TFIIA in HeLa nuclear extract transcription assays.

TFIIB

Although TFIID exists as a multiprotein complex in higher eukaryotes, TFIIB activity resides in a single 35-kDa peptide. The cDNA for TFIIB was cloned using reverse genetics and the recombinant protein was shown to possess all the features of native TFIIB, substituting the TFIIB fraction for both basal and activated transcription (Ha et al., 1991; Malik et al., 1991). The yeast homolog of TFIIB has recently been cloned (SUA7; a 38-kDa protein) (Pinto et al., 1992) and exhibits 52% overall similarity to human TFIIB. Both TFIIB and SUA7 contain the same notable features, two direct repeats (a feature in common with TBP) flanking a basic region and a portion homologous to the prokaryotic sigma factors. Indeed, although TBP and TFIIB do not share any significant sequence identity, they are structurally similar molecules in that they both contain two direct repeats separated by a basic region. At the N terminus of both yeast and human TFIIB there is a region with the potential to form a zinc finger, the significance of which is unclear (Pinto et al., 1992). Recombinant TFIIB is able to interact with TBP on a DNA template but does not appear to possess any DNA binding ability itself.

Pinto et al. (1992) found that yeast TFIIB is able to interact with pol II, although the same property has not yet been described for human TFIIB. As the presence of TFIIB in the nucleating preinitiation complex is a prerequisite for recruitment of the remaining GTFs, it is expected that TFIIB should be able to interact with one or both of these recruited factors, TFIIF and pol II.

TFIIF

TFIIF contains two subunits (of 30 kDa and 74 kDa, respectively) and exists in solution as a heterotetramer. The proteins were purified on the basis of their affinity for RNA polymerase II and have acquired the additional name of RAPs (RNA polymerase-associated polypeptides) (Sopta et al., 1985). The two subunits, RAP 30 and RAP 74, have now been cloned and have been shown to replace their purified counterparts in both basal and activated transcription (Aso et al., 1992; Finkelstein et al., 1992; Sopta et al., 1989). Both molecules contain homologies to the prokaryotic sigma factors. Significantly, the sigma homology in RAP 30 is highly conserved across all bacterial species and has been suggested to be required for interaction with bacterial RNA polymerase. Indeed, RAP 30 is able to bind to bacterial RNA polymerase (McCracken and Greenblatt, 1991). Thus, the function of RAP 30 appears to be to guide the polymerase to the promoter, although surprisingly TFIIF is unable to associate with the preinitiation complex in the absence of pol II (Flores et al., 1991). A recent study by Killeen and Greenblatt (1992) demonstrated that RAP 30 prevents pol II from binding nonspecifically to DNA, adding further evidence to the role of RAP 30 in guiding pol II to the promoter.

Both RAP 30 and RAP 74 are phosphorylated *in vivo*, RAP 74 being more extensively phosphorylated. An ATP-dependent DNA helicase activity is associated with HeLa RAP 74 fractions (Sopta et al., 1989), although highly purified or recombinant RAP 74 does not contain such an activity (Flores et al., 1990). Interestingly, RAP 74 is not required to recruit pol II to the promoter (Flores et al., 1991). Moreover, the remainder of the preinitiation complex can be formed in the absence of RAP 74. However, RAP 74 is absolutely required for transcription.

RNA Polymerase II

Mammalian RNA polymerase II is composed of 10 subunits (reviewed by Weinmann, 1992). An interesting feature of the largest subunit of pol II is a repeat of Tyr-Ser-Pro-Thr-Ser-Pro-Ser at the C terminus, the so-called C-terminal domain (CTD; reviewed by Corden, 1990). This portion of the molecule has recently attracted considerable attention. The number of repeats is related to the position of the organism on the evolutionary ladder, with human pol II CTD containing 52 repeats, *Drosophila* 42, and yeast 26.

What then is the function of the CTD? Removal of more than half of the CTD is lethal, indicating that it certainly performs an important function (Allison et al., 1988; Bartolomei et al., 1988; Zehring et al., 1988). It has been noted that the CTD can be highly phosphorylated, and this coincides with the form of pol II present in elongating transcriptional complexes (Laybourn and Dahmus, 1990). Recent results by the groups of Reinberg and Weinmann have found that pol II containing a phosphorylated CTD (pol IIo) enters the preinitiation complex far less efficiently than the dephosphorylated state (pol IIa; Lu et al., 1991). Phosphorylation of the CTD then occurs coincident with the initiation of transcription. Thus, phosphorylation of the CTD in a preinitiation complex may serve as the trigger for elongation. The actual kinase responsible for phosphorylation of the CTD is yet to be determined. As is discussed in the next section, the large subunit of TFIIE contains a consensus kinase domain. However, TFIIE does not phosphorylate the CTD *in vitro* (Lu et al., 1991); and the CTD kinase activity appears to reside within one of the components of the TFIIH fraction (Lu et al., 1992). *In vitro*, the pol II CTD has been shown to be a substrate for several kinases, including the CDC2 kinase, a regulatory protein of the cell cycle (Cisek and Corden, 1989); and the TFIIH fraction (Lu et al., 1992). However, it remains to be determined which of these kinases phosphorylates the CTD *in vivo*.

A recent report that the CTD of pol II could interact with TFIID (Usheva et al., 1992) demonstrated a mechanism of contact with the GTFs. This contact, in addition to that which may be mediated by TFIIF, could form a stable association of pol II with the preinitiation complex. Moreover, it appears that the CTD may contribute to the transcriptional response to acidic activators (Liao et al., 1992).

TFIIE

The TFIIE heterotetramer associates with the preinitiation complex following pol II. The two subunits of 56 kDa (TFIIEα) and 34 kDa (TFIIEβ) have been cloned, and the recombinant proteins substitute the HeLa TFIIE fraction in transcription assays (Ohkuma et al., 1991; Peterson et al., 1991; Sumimoto et al., 1991). TFIIEα contains several interesting features: a protein kinase activity, although substrates have yet to be identified, and a zinc finger, which may be involved in protein–DNA or protein–protein interactions. TFIIEβ contains a consensus nucleotide-binding site. A notable contrast of the two TFIIE subunits is that the C termini of both are highly charged, with TFIIEα containing a highly acidic region and TFIIEβ containing a basic region. It is possible that the two subunits associate via these regions.

The generality of TFIIE as a 'general' transcription factor has recently been challenged. Using an *in vitro* transcription system Parvin et al. (1992) found that antiserum against TFIIE was able to inhibit transcription from the adenovirus major late promoter, while having no effect on transcription of the IgH promoter. Why does one promoter require TFIIE while another does not? The answer is not yet clear, and whether alternative factor(s) are required by the IgH promoter is unknown. What is striking about this observation is that it puts into question the criteria originally used to describe the general transcription factors.

TFIIH and TFIIJ

These two basal transcription factors, which have not been well studied, were purified by further chromatography of the TFIIA and TFIIF fractions and found to be required for transcription *in vitro* (Cortese et al., 1992; Flores et al., 1992). TFIIH has now been purified to an almost homogeneous state, while TFIIJ has remained more elusive. Both TFIIH and TFIIJ enter the preinitiation complex at a late stage, following TFIIE. An exciting finding is the ability of a TFIIH fraction to phosphorylate the CTD (Lu et al., 1992). Interestingly, the phosphorylation of the CTD is strongly stimulated by assembly of the basic preinitiation complex and is further augmented by addition of the TFIIJ fraction. The strong stimulation of this phosphorylation reaction by assembly of the preinitiation complex suggests it may be important for the initiation process. However, pol II whose CTD has been removed still requires TFIIH for basal transcription (Lu et al., 1992).

Transcription of TATA-Less Promoters

TFIID is the only GTF with known DNA-binding activity, and is thought to lay the foundation for assembly of the preinitiation complex. However, several promoters transcribed by RNA polymerase II do not contain a recognizable TATA box, but contain a downstream element now termed the INR (Smale and Baltimore, 1989). TATA-less promoters still require TFIID. Although TFIID is normally the nucleator of the preinitiation complex, in these cases other proteins perform this function (Carcamo et al., 1990; Carcamo et al., 1991; Pugh and Tjian, 1991; Roy et al., 1991; Smale and Baltimore, 1989).

Several groups have reported the finding of proteins that can bind to the INR and recruit TFIID to the transcriptional start point, thus circumventing the requirement for a TATA motif. Roeder's group have identified a factor, TFII-I, that binds to the INR and may nucleate the formation of a preinitiation complex by recruiting TFIID (Roy et al., 1991). Many TATA-less promoters contain upstream recognition sequences for the transcriptional activator protein Sp1. Pugh and Tjian (1991) have reported the presence of a novel factor, which they term a tethering factor, that is able to bind to Sp1 and associate with TFIID. In their hypothesis, promoter-bound Sp1 recruits TFIID to the promoter via this tethering factor.

Transcriptional Activator Proteins

While the TATA and INR elements are required for the accurate initiation of transcription by RNA polymerase II, other sequence elements regulate the rate at which a gene is transcribed. Such gene-specific sequences, on deletion, destroy the responsiveness to activating signals. These sequence elements are the binding sites of proteins that are able to effect an increase in the transcription rate of the gene with which they are associated. In the past decade several such activator proteins have been identified. Transcriptional activator proteins are modular in structure, the DNA-binding and activating regions lying within distinct portions of the protein (reviewed by Frankel and Kim, 1991). These two domains are indeed separable, and interchangeable, a property that has been exploited in their study. The diversity of DNA sequences to which these activator proteins bind correlates with the diversity of the structures of the DNA-binding domains, although some underlying trends are present. Indeed,

DNA-binding domains can be categorized into several well-studied motifs, which include the zinc finger motif (Brown et al., 1985; Miller et al., 1985), helix-turn-helix (reviewed in Pabo and Sauer, 1984), helix-loop-helix (Murre et al., 1989) and basic-leucine zipper (Landsculz et al., 1988; Vinson et al., 1989).

Transcriptional activation domains can also be categorized on the basis of their amino acid composition, although this classification is less well defined than that of DNA-binding domains. These regions were identified by deletion analysis of cloned activators expressed in cell culture. However, there are yet no structural data for any transcriptional activation domain and therefore it is not clear whether classification according to amino acid composition has any structural meaning. Roughly, activation domains have been classified as acidic (Cress and Triezenberg, 1991; Gill and Ptashne, 1987; Hope et al., 1988), glutamine rich (Courey and Tjian, 1988), proline rich (Santoro et al., 1988), and serine/threonine rich (Theill et al., 1989). Acidic activators are characterized by an abundance of acidic residues that lie on one face of an alpha helix, forming an amphipathic alpha helix. Examples of such activators include the herpes simplex VP16 activator and yeast GAL4. The glutamine-rich domain is typified by human Sp1, which contains two regions rich in glutamine residues. CTF/NF-1 is the contender for the proline-rich activation domain and contains a density of proline residues at the C terminus of the protein, while the activation domain of GHF-1 is rich in serine and threonine residues. These differences in the structure of activation domains suggests that distinct classes of activator proteins may operate by different mechanisms.

How Do Activators Increase the Rate of Transcription?

While the DNA binding domain brings the activator protein to the promoter, within the proximity of the transcription initiation site, the activating region presumably enhances the rate of transcription of the associated gene by interacting with a protein that forms a part of the preinitiation complex (Figure 3A). The search for the targets of transcriptional activator proteins has attracted a large amount of interest over the past couple of years.

Preinitiation complex assembly occurs in the absence of activator proteins and presumably at low levels because of a rate-limiting step(s). Activator proteins enhance preinitiation complex formation by overcom-

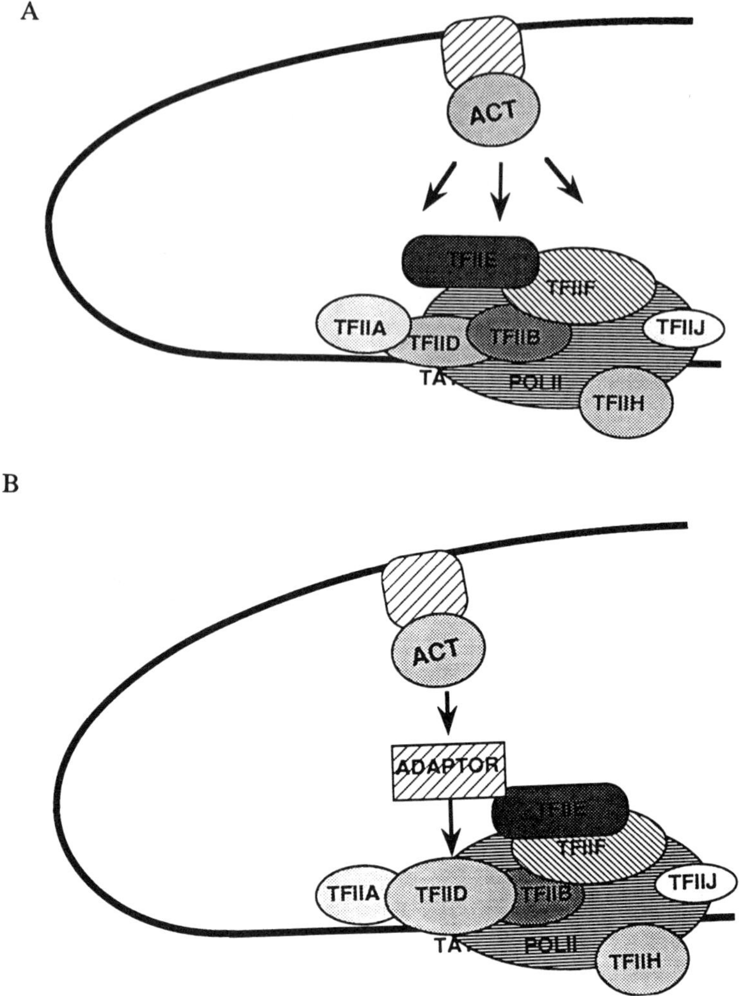

Figure 3   Possible mechanisms by which activators increase the rate of preinitiation complex formation. A: Direct action model. B: Molecular adaptor model.

ing this rate-limiting step. In a study by Lin and Green (1991), assembly of the GTFs in the preinitiation complex was studied in the presence and absence of the acidic activator GAL4-AH. They found that assembly stalled at the TFIIB binding stage, thus indicating that acidic transcriptional activator proteins may enhance transcription by recruiting TFIIB to the promoter. Using protein affinity chromatography, they found that a column containing the acidic activating region of VP16 removed the TFIIB activity from HeLa nuclear extract. Indeed, once TFIIB was cloned

(Ha et al., 1991), it was shown that the acidic activation domain of VP16 interacted directly with recombinant TFIIB (Lin et al., 1991). Moreover, a single point mutation in the activation domain of VP16, which disrupts its activation properties *in vivo* (Berger et al., 1990; Cress and Triezenberg, 1991), also abolishes the VP16–TFIIB interaction (Lin and Green, 1991; Lin et al., 1991).

TFIID is also retained on affinity columns containing VP16, and the same mutation that abolishes the interaction of VP16 with TFIIB also disrupts its interaction with TFIID (Ingles et al., 1991; Stringer et al., 1990). The finding that an activating region can bind to more than one GTF may explain the synergistic properties of transcriptional activation, which is discussed later.

A recent study by Sundseth and Hansen (1992) has shown that the kinetics of preinitiation complex formation point to the TFIIB recruitment as a rate-limiting step and that activators enhance this process. However, whether TFIIB recruitment is the only step at which acidic activators act is not known. Indeed, in addition to the GTFs, acidic activators also require at least one auxiliary protein to achieve transcriptional activation (see later discussion).

The 13S Ela protein of adenovirus has been shown to directly contact TFIID, and its activation domain is required for this interaction (Horikoshi et al., 1991; Lee et al., 1991). Thus, it appears that many activator proteins may interact with GTFs, although TFIIB and TFIID are the only targets that have been described to date.

The Coactivator Enigma

Both acidic activators and Ela 13S have been shown to directly contact GTFs, but there is also evidence that their direct targets are not in fact GTFs. This evidence is based on a property of activator proteins termed squelching, originally described by Gill and Ptashne (1987). Using both *in vitro* and *in vivo* systems, they found that if an activator protein is expressed at high levels it will paradoxically inhibit transcription. Indeed, expression of an activation domain in the absence of a DNA-binding domain is also capable of squelching a DNA-bound activator. It is believed that squelching results from titration of the target of the activator, thus rendering the target in limited supply for the DNA-bound activators and consequently preventing activation.

Using squelching experiments, Martin et al. (1990) found that while Ela activity could be squelched by both Ela and VP16, Ela was incapable of squelching VP16. The logical conclusion from this result was that the target of VP16 was required for activation by Ela but the target of Ela was not required by VP16. Essentially, the target of Ela was dispensable for transcription (at least for acidic activators) and was termed an adaptor proposed to bridge the activation domain of Ela to a component of the transcriptional machinery that binds to acidic activators.

Additional evidence for adaptors has come from *in vitro* squelching experiments. Using an *in vitro* yeast transcription system the associates of Kornberg and Guarente have shown that while VP16 can squelch activated transcription by acidic activators, it does not affect basal transcription (Berger et al., 1990; Kelleher et al., 1990). Thus, these adaptors seem to be dispensable for basal transcription but absolutely required for activated transcription.

At the same time as this indirect evidence for adaptors was unfolding, human TBP was cloned and found to be able to substitute for basal but not activated transcription in *in vitro* transcription assays (Pugh and Tjian, 1990). Only the TFIID fraction was able to support activated transcription. It was concluded that adaptor proteins may be part of the TFIID complex and was speculated that they may bridge contacts between activator proteins and TBP (see Figure 3B; reviewed by Lewin, 1990).

In higher eukaryotes TBP is associated with a complex of proteins and is collectively termed the TFIID complex. The accompanying proteins, TBP-associated factors (TAFs), are at present subject to intensive investigation. Indeed, the TAFs may hold the key to activated transcription. In humans TBP is strongly associated with at least 10 TAFs and in *Drosophila* with 6 (Dynlacht et al., 1991; Tanese et al., 1991). Tjian's group have purified both the *Drosophila* and human factors and shown that they are absolutely required for transcription (Dynlacht et al., 1991; Tanese et al., 1991). Moreover, acidic activators seem to require a different (but possibly overlapping) set of TAFs than proline-rich or glutamine-rich activators (Tanese et al., 1991).

If activator proteins contact the transcriptional machinery via coactivators, then why do the acidic and Ela activation domains directly contact GTFs? The function of coactivators as molecular adaptors has yet to be demonstrated, although the evidence obtained from squelching experiments is difficult to explain in any other way. Indeed, the TAFs may prove to be additional targets of activator proteins and aid in the formation of a stable preinitiation complex. However, neither VP16 nor Ela 13S proteins bind directly to DNA; instead, they are recruited to the promoter by interacting with other DNA-bound proteins (notably Oct 1 and ATF2, respectively) (Liu and Green, 1990; McKnight et al., 1987; Stern et al., 1989). In essence, both these viral proteins can be labeled as adaptors. One thing is certain; the 'coactivator controversy' may not be reconciled until activated transcription is fully reconstituted *in vitro* with recombinant (or at least purified) proteins.

Transcriptional Synergism

A property of all transcriptional activators is that they act synergistically. Both *in vivo* and *in vitro*, a promoter containing only one activator binding site shows only a weak transcriptional activation in the presence of the activator. As the number of binding sites for the activator is increased, the effect on transcription increases (Carey et al., 1990). Indeed, the majority of natural promoters contain more than one site for transcriptional activators. The mechanism by which synergism occurs is unclear, but it is not cooperative binding of the activators to DNA (Carey et al., 1990). Whether a single GTF binds more than one activation domain and is thus more likely to be recruited, or whether several activation domains allow enhanced recruitment of TFIID, TFIIB, and other GTFs, is unclear. It is clear, however, that while some activator proteins will work in synergy with others, some specifically will not (Lin et al., 1990).

Inhibitors of Preinitiation Complex Formation

Mechanisms for both positive and negative control can be found in all biological systems. Transcription is no exception. While most research has focused on transcriptional activation, repression of transcription has also made some progress in the past few years. It is possible to silence a gene by preventing activator proteins from binding to their target sequences and thus stopping activation. Examples of this are DNA-binding proteins that have no activation domain, and DNA methylation, which prevents access to the DNA by transcriptional activators (reviewed by Bird, 1992; Razin and Riggs, 1980).

Recent studies have revealed proteins that contain both a DNA-binding domain and a repression domain that can reduce the basal level of transcription of the associated gene. Examples of these include the Wilms tumor repressor (Madden et al., 1991), Kruppel (Licht et al., 1990), and YY1 (Shi et al., 1991). Like activator proteins, transcriptional repressors are modular in structure. Fusion of the Wilms tumor repressor domain to GAL4 results in a repressor that can function from a synthetic promoter containing GAL4 DNA-binding sites (Madden et al., 1991). Although it is not clear how these repressor domains function, it can be speculated that they do so by interfering with preinitiation complex formation. Another group of repressors that have been observed in HeLa cell nuclear extract do not exert their effects by binding to upstream promoter elements. These proteins NCI1 and NCI2 prevent the formation of a preinitiation complex by association with TFIID (or TBP) at the promoter (Meisterernst and Roeder, 1991). TFIIA can specifically compete for these NCI proteins, recovering transcription. Interestingly, it appears that these proteins do not inhibit activated transcription.

TFIID Complexes and TBP Promiscuity

Elucidation of the TFIID complex has become one of the major goals of research in basic transcriptional mechanisms. Indeed, it is within this complex that essential components required for transcriptional activation reside.

As discussed earlier, in HeLa cell nuclear extract TFIID exists in at least two distinct forms. Recent results have shown that TBP is not, as previously thought, a GTF specific for the transcription of pol II genes. It was noted several years ago that the promoters of some genes transcribed by RNA pol III contained a TATA-like sequence. It has since been found that both pol I (Comai et al., 1992) and pol III (White et al., 1992) genes require TBP for transcription. Indeed, one of the HeLa TFIID fractions appears to specifically contain the associated proteins required for pol III transcription. Thus, the specificity of TFIID complexes for pol I, II, or III genes may be determined by the TAFs contained within the complex (for review, see Green, 1992).

Evidence that these complexes contain different TAFs has been gained using yeast genetics. The groups of Hahn and Struhl have identified point mutations of yeast TBP that specifically inhibit transcription of group I, II, or III pol genes (Cormack and Struhl, 1992; Schultz et al., 1992). Presumably the point mutant inactive in pol II transcription is unable to interact with a TAF(s) required only for pol II genes. As more evidence is gathered, it is becoming apparent that TBP is capable of interacting with a great many cellular proteins. Whether the associations are in a state of flux and whether specific genes require a specific group of TAFs are unknown. Clearly, TBP is a fundamental protein required for transcription of all genes in all eukaryotic organisms.

Future Perspectives

The ultimate goal of workers in the transcription field is to reconstitute both basal and activated transcription using pure components. While only a few of the GTFs remain to be cloned, studies of the proteins in the TFIID complex have only just begun. As these vital components of the transcriptional machinery await cloning, work is progressing with the kinetics and protein contacts involved in the formation of a preinitiation complex and the mechanisms by which activator proteins enhance this process. The comparatively simple acidic activators have received a great deal of attention, and researchers are now beginning to study the mechanisms by which more complicated activators interact with the preinitiation complex. Such activators may be subject to extensive modification such as phosphorylation, which modulates their activation potential (reviewed by Hunter and Karin, 1992). As is discussed in other chapters, several mammalian activator proteins are highly regulated by phosphorylation.

In conclusion, the rapid progress made in purification and cloning of the general transcription factors during the past few years should lay the foundation for the final elucidation of the mechanisms of transcriptional activation.

Acknowledgments.   We thank Drs. B. Choy, H. Kwon, and X. Li for comments on the manuscript. This work in the laboratory was supported by a grant from the National Institutes of Health.

References

Allison LA, Wong JK, Fitzpatrick D, Moyle M, Ingles JC (1988): The C-terminal domain of the largest subunit of RNA polymerase II of *Saccharomyces cerevisiae, Drosophila melanogaster* and mammals: A conserved structure with an essential function. *Mol Cell Biol* 8:321–329.

Aso T, Vasavada HA, Kawaguchi T, Germino FJ, Ganguly S, Kitajima S, Weissman SM, Yasukochi Y (1992): Characterization of cDNA for the large subunit of the transcription initiation factor TFIIF. *Nature* (Lond) 355:461–464.

Bartolomei MS, Halden NF, Cullen CR, Corden JL (1988): Genetic analysis of the repetitive carboxy-terminal domain of the largest subunit of mouse RNA polymerase II. *Mol Cell Biol* 8:330–339.

Berger SL, Cress WD, Cress A, Triezenberg SJ, Guarente L (1990): Selective inhibition of activated but not basal transcription by the acidic activation domain of VP16: Evidence for transcription adaptors. *Cell* 61:1199–1208.

Bird A (1992): The essentials of DNA methylation. *Cell* 70:5–8.

Brown RS, Sander C, Argos P (1985): The primary structure of transcription factor TFIIA has twelve consecutive repeats. *FEBS Lett* 186:271–274.

Bunick D, Zandomeni R, Ackerman S, Weinmann R (1982): Mechanism of RNA polymerase II-specific initiation of transcription in vitro: ATP requirement and uncapped runoff transcripts. *Cell* 29:877–886.

Buratowski S, Hahn S, Guarente L, Sharp PA (1989): Five intermediate complexes in transcription initiation by RNA polymerase II. *Cell* 56:549–561.

Buratowski S, Hahn S, Sharp PA, Guarente L (1988): Function of a yeast TATA-element binding protein in a mammalian transcription system. *Nature* (Lond) 334:37–42.

Buratowski S, Zhou H (1992): Transcription factor IID mutants defective for interaction with transcription factor IIA. *Science* 255:1130–1132.

Carcamo J, Buckbinder L, Reinberg D (1991): The initiator directs the assembly of a transcription factor IID-dependent transcription complex. *Proc Natl Acad Sci USA* 88:8052–8056.

Carcamo J, Maldonado E, Cortes P, Ahn M-H, Ha I, Kasai Y, Flint J, Reinberg D (1990): A TATA-like sequence located downstream of the transcription initiation site is required for expression of an RNA polymerase II transcribed gene. *Genes & Dev* 4:1611–1622.

Carey MF, Lin Y-S, Green MR, Ptashne M (1990): A mechanism for synergistic activation of a mammalian gene by GAL4 derivatives. *Nature* (Lond) 345:361–364.

Cavallini B, Huet J, Plassat JL, Sentenec A, Egly JM, Chambon P (1988): A yeast activity can substitute for the HeLa cell TATA box factor. *Nature* (Lond) 334:77–86.

Cavallini B, Faus I, Matthes H, Chipoulet JM, Winsor B, Egly JM, Chambon P (1989): Cloning of the gene encoding the yeast protein BTF1Y, which can substitute for human TATA box-binding factor. *Proc Natl Acad Sci USA* 86:9803–9807.

Cisek LJ, Corden JL (1989): Phosphorylation of RNA polymerase by the murine homologue of the cell cycle control protein cdc2. *Nature* (Lond) 339:679–684.

Comai L, Tanese N, Tjian R (1992): The TATA-binding protein and associated factors are integral components of the RNA polymerase I transcription factor, SL1. *Cell* 68:965–976.

Corden JL (1990): Tails of RNA polymerase II. *Trends Biochem Sci* 15:383–387.

Cormack BP, Struhl K (1992): The TATA-binding protein is required for transcription by all three nuclear RNA polymerases in yeast cells. *Cell* 69:685–696.

Cortese P, Flores O, Reinberg D (1992): Factors involved in specific transcription by mammalian RNA polymerase II: Purification and analysis of transcription factor IIA and identification of transcription factor IIJ. *Mol Cell Biol* 12:413–421.

Courey AJ, Tjian R (1988): Analysis of Sp1 in vivo reveals multiple transcriptional domains, including a novel glutamine-rich activation motif. *Cell* 55:887–898.

Cress WD, Triezenberg SJ (1991): Critical structural elements of the VP16 transcriptional activation domain. *Science* 251:87–90.

Dynlacht BD, Hoey T, Tjian R (1991): Isolation of coactivators associated with the TATA-binding protein that mediate transcriptional activation. *Cell* 66:563–576.

Finkelstein A, Kostrub CF, Li J, Chavez DP, Wang BQ, Fang SM, Greenblatt J, Burton ZF (1992): A cDNA encoding RAP74, a general initiation factor for transcription by RNA polymerase II. *Nature* (Lond) 355:464–467.

Flores O, Ha I, Reinberg D (1990): Factors involved in specific transcription by mammalian RNA polymerase II: Purification and subunit composition of transcription factor IIF. *J Biol Chem* 265:5629–5634.

Flores O, Lu H, Killeen M, Greenblatt J, Burton ZF, Reinberg D (1991): The small subunit of transcription factor IIF recruits RNA polymerase II to the preinitiation complex. *Proc Natl Acad Sci USA* 88:9999–10003.

Flores O, Lu H, Reinberg D (1992): Factors involved in specific transcription by mammalian RNA polymerase II: Identification and characterization of transcription factor IIH. *J Biol Chem* 267:2786–2793.

Frankel AD, Kim PS (1991): Modular structure of transcription factors: Implications for gene regulation. *Cell* 65:717–719.

Gash A, Hoffman A, Horikoshi M, Roeder RG, Chua N-H (1990): *Arabidopsis thaliana* contains two genes for TFIID. *Nature* (Lond) 346:390–394.

Gill G, Ptashne M (1987): Mutants of GAL4 protein altered in an activation formation. *Cell* 51:121–126.

Green MR (1992): Transcriptional transgressions. *Nature* (Lond) 357:364–365.

Ha I, Lane WS, Reinberg D (1991): Cloning of a human gene encoding the general transcription initiation factor IIB. *Nature* (Lond) 352:689–695.

Hahn S, Buratowski S, Sharp PA, Guarente L (1989): Isolation of the gene encoding the yeast TATA binding protein TFIID. A gene identical to the SPT15 suppressor of Ty element insertions. *Cell* 58:1173–1181.

Hoey T, Dynlacht BD, Peterson MG, Pugh BF, Tjian R (1990): Isolation and characterization of the drosophila gene encoding the TATA box binding protein, TFIID. *Cell* 61:1179–1186.

Hoffman A, Sinn E, Yamamoto T, Wang J, Roy A, Horikoshi M, Roeder RG (1990): Conserved core domain and unique N-terminus with presumptive regulatory motifs in a human TATA factor (TFIID). *Nature* (Lond) 346:387–390.

Hope IA, Mahadevan S, Struhl K (1988): Structural and functional characterization of the short acidic transcriptional activation region of yeast GCN4 protein. *Nature* (Lond) 333:635–640.

Horikoshi M, Bertuccioli C, Takada R, Wang J, Yamamoto T, Roeder RG (1992): Transcription factor TFIID induces DNA bending upon binding to the TATA element. *Proc Natl Acad Sci USA* 89:1060–1064.

Horikoshi M, Wang CK, Fujii H, Cromlish JA, Weil PA, Roeder RG (1989): Cloning and structure of a yeast gene encoding a general transcription factor TFIID that binds to the TATA box. *Nature* (Lond) 341:299–303.

Horikoshi M, Yamamoto T, Ohkuma Y, Weil PA, Roeder RG (1990): Analysis and structure-function relationships of yeast TATA box binding factor TFIID. *Cell* 61:1171–1178.

Horikoshi N, Maguire K, Kralli A, Maldonado E, Reinberg D, Weinmann R (1991): Direct interaction between adenovirus E1A protein and the TATA-box binding transcription factor IID. *Proc Natl Acad Sci USA* 88:5124–5128.

Hunter T, Karin M (1992): The regulation of transcription by phosphorylation. *Cell* 70:375–387.

Ingles CJ, Shales M, Cress WD, Triezenberg SJ, Greenblatt J (1991): Reduced binding of TFIID to transcriptionally compromised mutants of VP16. *Nature* (Lond) 351:588–590.

Kao CC, Lieberman PM, Schmidt MC, Zhou Q, Pei R, Berk AJ (1990): Cloning of transcriptionally active human TATA binding factor. *Science* 248:1646–1650.

Kelleher RJ, Flanagan PM, Kornberg RD (1990): A novel mediator between activator proteins and the RNA polymerase II transcription apparatus. *Cell* 61:1209–1215.

Killeen MT, Greenblatt JF (1992): The general transcription factor RAP30 binds to RNA polymerase II and prevents it from binding nonspecifically to DNA. *Mol Cell Biol* 12:30–37.

Landschulz WH, Johnson PF, McKnight SL (1988): The leucine zipper: A hypothetical structure common to a new class of DNA binding proteins. *Science* 240:1759–1764.

Laybourn PJ, Damus ME (1990): Phosphorylation of RNA polymerase IIA occurs subsequent to interaction with the promoter and before the initiation of transcription. *J Biol Chem* 265:13165–13173.

Lee WS, Kao CC, Bryant GO, Liu X, Berk AJ (1991): Adenovirus E1A activation domain binds the basic repeat in the TATA box transcription factor. *Cell* 67:365–376.

Lewin B (1990): Commitment and activation at pol II promoters: A tail of protein-protein interactions. *Cell* 61:1161–1164.

Liao S-M, Taylor ICA, Kingston RE, Young R (199?): RNA polymerase II carboxy-terminal domain contributes to the response to multiple acidic activators. *Genes & Dev* 5:2431–2440.

Licht JD, Grossel MJ, Figge J, Hansen UM (1990): *Drosophila kruppel* protein is a transcriptional repressor. *Nature* (Lond) 346:76–79.

Lin Y-S, Carey MF, Ptashne M, Green MR (1990): How different eukaryotic transcriptional activators can cooperate promiscuously. *Nature* (Lond) 345:359–361.

Lin Y-S, Green MR (1991): Mechanism of action of an acidic transcriptional activator in vitro. *Cell* 64:971–981.

Lin Y-S, Ha I, Maldonado E, Reinberg D, Green MR (1991): Binding of general transcription factor TFIIB to an acidic activating region. *Nature* (Lond) 353:569–571.

Liu F, Green MR (1990): A specific member of the ATF transcription factor family can mediate transcription activation by the adenovirus Ela protein. *Cell* 61:1217–1224.

Lu H, Flores O, Weinmann R, Reinberg D (1991): The nonphosphorylated form of RNA polymerase II preferentially associates with the preinitiation complex. *Proc Natl Acad Sci USA* 88:10004–10008.

Lu H, Zawel L, Fisher L, Egly J-M, Reinberg D (1992): Human general transcription factor IIH phosphorylates the C-terminal domain of RNA polymerase II. *Nature* (Lond) 358:641–645.

Madden SL, Cook DM, Morris JF, Gashler A, Sukhatme VP, Rauscher FJ III (1991): Transcriptional repression mediated by the WT1 Wilms tumor gene product. *Science* 253:1550–1553.

Maldonado E, Ha I, Cortes P, Weis L, Reinberg D (1990): Factors involved in specific transcription by RNA polymerase II: Role of transcription factors IIA, IID and IIB during formation of a transcription-competent complex. *Mol Cell Biol* 10:6335–6347.

Malik S, Hisatake K, Sumimoto H, Roeder RG (1991): Sequence of general transcription factor TFIIB and relationships to other initiation factors. *Proc Natl Acad Sci USA* 88:9553–9557.

Manley JL, Fire A, Cano A, Sharp PA, Gefter ML (1980): DNA-dependent transcription of adenovirus genes in a soluble whole-cell extract. *Proc Natl Acad Sci USA* 77:3855–3859.

Martin KJ, Lillie JW, Green MR (1990): Evidence for interaction of different eukaryotic transcriptional activators with distinct cellular targets. *Nature* (Lond) 346:147–152.

McCracken S, Greenblatt JF (1991): Related RNA polymerase-binding regions in human RAP30/74 and *Escherichia coli* σ 70. *Science* 253:900–902.

McKnight JLC, Kristie TM, Roizman B (1987): The binding of virion protein mediating α gene induction in herpes simplex virus 1 infected cells to its *cis* site requires cellular proteins. *Proc Natl Acad Sci USA* 84:7061–7065.

Meisterernst M, Roeder RG (1991): Family of proteins that interact with TFIID and regulate promoter activity. *Cell* 67:557–567.

Miller J, McLachlan AD, Klug A (1985): Repetitive zinc-binding domains in the protein transcription factor IIIA from *Xenopus* oocytes. *EMBO J* 4:1609–1614.

Murre C, McCaw PS, Vaessin H, Caudy M, Jan LY, Jan YN, Cabrera CV, Buskin JN, Hauschka SD, Lassar AB, Weintraub H, Baltimore D (1989): Interactions between heterologous helix-loop-helix proteins generate complexes that bind specifically to a common DNA sequence. *Cell* 58:537–544.

Ohkuma Y, Sumimoto H, Hoffmann A, Shimasaki S, Horikoshi M, Roeder RG (1991): Structural motifs and potential sigma homologies in the large subunit of human general transcription factor TFIIE. *Nature* (Lond) 354:398–401.

Pabo CO, Sauer RT (1984): Protein-DNA recognition. *Annu Rev Biochem* 53:293–321.

Parvin JP, Timmers MH, Sharp PA (1992): Promoter specificity of basal transcription factors. *Cell* 68:1135–1144.

Peterson MG, Inostroza J, Maxon ME, Flores O, Admon A, Reinberg D, Tjian R (1991): Structure and functional properties of human general transcription factor IIE. *Nature* (Lond) 354:369–373.

Peterson MG, Tanese N, Pugh BF, Tjian R (1990): Functional domains and upstream activation properties of cloned human TATA binding protein. *Science* 248:1625–1630.

Pinto I, Ware DE, Hampsey M (1992): The yeast SUA7 gene encodes a homolog of human transcription factor TFIIB and is required for normal start site selection in vivo. *Cell* 68:977–988.

Pugh BF, Tjian R (1990): Mechanism of transcriptional activation by Sp1: Evidence for coactivators. *Cell* 61:1187–1197.

Pugh BF, Tjian R (1991): Transcription from a TATA-less promoter requires a multisubunit TFIID complex. *Genes & Dev* 5:1935–1945.

Ranish JA, Hahn S (1991): The yeast general transcription factor TFIIA is composed of two polypeptide subunits. *J Biol Chem* 266:19320–19327.

Ranish JA, Lane WS, Hahn S (1992): Isolation of two genes that encode subunits of the yeast transcription factor IIA. *Science* 255:1127–1130.

Razin A, Riggs AD (1980): DNA methylation and gene function. *Science* 210:604–616.

Reinberg D, Horikoshi M, Roeder RG (1987): Factors involved in specific transcription in mammalian RNA polymerase II: Functional analysis of initiation factors IIA and IID and identification of a new factor operating at sequences downstream of the initiation site. *J Biol Chem* 262:3322–3330.

Roy AL, Miesterernst M, Pognonec P, Roeder RG (1991): Cooperative interaction of an initiator-binding transcription initiation factor and the helix-loop-helix activator USF. *Nature (Lond)* 354:245–248.

Santoro C, Mermod N, Andrews PC, Tjian R (1988): A family of human CCAAT-box-binding proteins active in transcription and DNA replication: Cloning and expression of multiple cDNAs. *Nature* (Lond) 334:218–224.

Schmidt M, Kao CC, Rei R, Berk AJ (1989): Yeast TATA-box transcription factor gene. *Proc Natl Acad Sci USA* 86:7785–7789.

Schultz MC, Reeder RH, Hahn S (1992): Variants of the TATA-binding protein can distinguish subsets of RNA polymerase I, II and III promoters. *Cell* 69:697–702.

Shi Y, Seto E, Chang L-S, Shenk T (1991): Transcriptional repression by YY1, a human GLI-kruppel-related protein, and relief of repression by adenovirus E1A protein. *Cell* 67:377–388.

Smale ST, Baltimore D (1989): The 'initiator' as a transcription control element. *Cell* 57:103–113.

Sopta M, Burton ZF, Greenblatt J (1989): Structure and associated DNA-helicase activity of a general transcription initiation factor that binds to RNA polymerase II. *Nature (Lond)* 341:410–414.

Sopta M, Carthew RW, Greenblatt J (1985): Isolation of three proteins that bind to mammalian RNA polymerase II. *J Biol Chem* 260:10353–10360.

Stern S, Tanaka M, Herr W (1989): The Oct-1 homeodomain directs formation of a multiprotein-DNA complex with the HSV transactivator VP16. *Nature (Lond)* 341:624–630.

Stringer KF, Ingles CJ, Greenblatt J (1990): Direct and selective binding of an acidic transcriptional activator domain to the TATA-box factor TFIID. *Nature (Lond)* 345:783–786.

Sumimoto H, Ohkuma Y, Sinn E, Jato H, Shimasaki S, Horikoshi M, Roeder RG (1991): Conserved sequence motifs in the small subunit of human general transcription factor TFIIE. *Nature (Lond)* 354:401–404.

Sundseth R, Hansen U (1992): Activation of RNA polymerase II transcription by the specific DNA-binding protein LSF. *J Biol Chem* 267:7845–7855.

Tanese N, Pugh BF, Tjian R (1991): Coactivators for a proline-rich activator purified from the multisubunit human TFIID complex. *Genes Dev* 5:2212–2224.

Theill LE, Castrillo J-L, Wu D, Karin M (1989): Dissection of functional domains of the pituitary-specific transcription factor GHF-1. *Nature* (Lond) 342:945–948.

Timmers MH, Sharp PA (1991): The mammalian TFIID protein is present in two functionally distinct complexes. *Genes & Dev* 5:1946–1956.

Usheva A, Maldonado E, Goldring A, Lu H, Houbavi C, Reinberg D, Aloni Y (1992): Specific interaction between the non-phosphorylated form of RNA polymerase II and the TATA binding protein. *Cell* 69:871–881.

Vinson CR, Sigler PB, McKnight SL (1989): Scissors-grip model for DNA recognition by a family of leucine zipper proteins. *Science* 246:911–916.

Weinmann R (1992): The basic RNA polymerase II transcriptional machinery. *Gene Express* 2:81–91.

Weil PA, Luse DS, Segall J, Roeder RG (1979): Selective and accurate initiation of transcription at the Ad2 major late promoter in a soluble system dependent on purified RNA polymerase II and DNA. *Cell* 18:469–484.

White RJ, Jackson SP, Rigby PW (1992): A role for the TATA-box-binding protein component of the transcription factor IID complex as a general RNA polymerase III transcription factor. *Proc Natl Acad Sci USA* 89:1949–1953.

Zehring WA, Lee JM, Weeks JR, Jokerst RS, Greenleaf AL (1988): The C-terminal repeat domain of RNA polymerase II largest subunit is essential in vivo but is not required for accurate transcription initiation in vitro. *Proc Natl Acad Sci USA* 85:3698–3702.

Chapter 2

Plasticity of the Differentiated State

Helen M. Blau

As the embryo develops, cells specialize for function in tissues. From the time the sperm fertilizes the egg, all cells of the body contain the same DNA, yet they only express a subset of their genetic material. Two types of signals that play a critical role in instructing a cell to assume a specific differentiated state are positional information and induction. Gradients of morphogenetic molecules provide positional cues from a distance, the morphogens diffusing from a source and establishing a continuous range of concentrations. The composite of signals received by cells in different locations differs, leading to variations in gene expression. Inductive signals are provided by immediate cell neighbors. The nature and mode of action of such distal and proximal signals in establishing the differentiated state has been extensively discussed (Gurdon, 1992; St. Johnston and Nusslein-Volhard, 1992; Wolpert, 1969, 1989). But the question of how, once established, the differentiated state is maintained has received less attention. By what molecular mechanism is the differentiated state of a cell regulated? That question is the focus of this chapter.

Overview

Evidence from a number of species supports the hypothesis that differentiation requires continuous regulation. This evidence suggests that gene expression in the differentiated cells of eukaryotes, as in prokaryotes (Jacob and Monod, 1961), is dynamic (Blau, 1992; Blau and Baltimore, 1991). Accordingly, an active regulatory decision is made continuously for each differentiation-specific gene, a decision determined by the protein composition of the cell at any given time. In principle, such decisions

GENE EXPRESSION: GENERAL AND CELL-TYPE-SPECIFIC
Michael Karin, Editor
© 1993 Birkhäuser Boston

should be readily reversible. In view of the stability of the differentiated state *in vivo*, this mechanism appears to allow for unacceptable plasticity. In addition, this mechanism seems unduly cumbersome. Because only a small fraction of a cell's total genes are expressed at any given time, the investment in negative regulators required to maintain the majority of genes in a silent state appears disproportionately large. How is the transcription and expression of the bulk of genetic material prevented? Is it necessary to fix genes stably in an inactive state? I propose that it is not. Indeed, as I discuss in this chapter, although possibly counterintuitive, a mechanism for controlling differentiation that requires continuous regulation affords plasticity and yet provides essential stability, is readily heritable, and can be achieved by a limited number of regulators.

Evidence That Differentiation Requires Continuous Regulation

Nuclear Transplantation Experiments

Gurdon's classic nuclear transplantation experiments (Gurdon, 1962) showed that silent genes in differentiated cells could be expressed in *Xenopus*. When the nuclei of differentiated amphibian intestinal cells were transplanted into enucleated eggs, entire feeding tadpoles developed. These experiments demonstrated that the genes in intestinal cells are reversibly inactivated. When passed back through egg cytoplasm, diverse genes such as globin genes typical of highly specialized red blood cells are expressed by the progeny of the reconstituted egg destined for erythropoiesis.

However, the passage of the differentiated nuclei through the egg could have stripped the DNA of all regulatory influences and allowed reprogramming. Indeed, DiBerardino et al. (1986) showed that the frequency of obtaining feedback tadpoles was increased from approximately 2% to 75% if the nuclei were initially injected into maturing oocytes, conditioned by oocyte cytoplasm, and allowed to progress to the blastula stage before transplantation into enucleated eggs. This finding suggested that an incubation step was necessary to alter chromatin structure or to remove methyl groups, allowing access of transcription factors. Thus, although genes that were normally shut off were rendered accessible, the unusual environment provided by the maturing oocyte could have played a critical role.

Transdetermination Experiments

Experiments in *Drosophila* have shown that silent genes could be activated and expressed in specialized cells without exposure to oocyte components. Following serial transplantation of imaginal disk cells, "transdetermination," or a change in determined state, often occurs. Cells originally destined for genital structures first give rise to leg or head structures and if transplanted again, give rise to wings (Hadorn, 1963). All these changes in differentiated state occur in the environment of the adult abdomen. The nature of the structure produced is highly predictable and depends on the prior history of the transplanted imaginal disk. These experiments provide evidence for plasticity in gene expression. However, the ordered progression of structures that are generated suggests a sequence of regulatory steps is involved. A similarly predictable developmental transformation of one body part into another occurs when *Drosophila* homeotic genes are mutated (Lewis, 1978).

Transdifferentiation Experiments

At sites of wounds, changes in differentiated state can occur. This transdifferentiation, by contrast with nuclear transplantation and transdetermination, occurs *in situ* and does not require relocation of cells to ectopic sites. Following excision of tissue in jellyfish, a single cell type, striated muscle, generates six different cell types. The types of cells produced transcend lineage barriers, mesodermal muscle giving rise to ectodermal neural sensory cells (Schmid and Alder, 1984). Transdifferentiation, however, is not a phenomenon restricted to simple organisms. When the iris of the eye of amphibian, chicken, and human species is damaged, a process of repair is induced. To regenerate the lens, melanocytes dedifferentiate, divide, and redifferentiate. This process entails a dramatic change in differentiated state during which the cellular machinery required for pigment synthesis is shut off and that necessary for crystallin synthesis is induced (Eguchi and Okada, 1973). As in jellyfish, in response to novel signals the cells of the iris change their differentiated state in their normal position within the body.

Somatic Cell Hybrid Experiments

Cell fusion experiments allowed an analysis of changes in gene expression independent of an *in vivo* environment. In somatic cell hybrids, the influence of one cell type on the function of another can be examined.

In general, tissue-specific genes were repressed in hybrids formed by fusing two disparate cell types (Carlsson et al., 1974; Konieczny et al., 1983; McCormick et al., 1988). However, when the number of genomes contributed by a differentiated cell type such as melanocyte or hepatoma cell was increased, silent genes encoding enzymes involved in melanin synthesis or albumin were activated in fibroblasts (Davidson, 1972; Peterson and Weiss, 1972). Such experiments provided early evidence that gene expression could be altered by diffusible transacting regulators. In addition, they predicted that transformation was recessive to the normal differentiated state (Harris and Klein, 1969), a prediction borne out in recent molecular studies of retinoblastoma. However, at times such experiments were difficult to interpret because of the extensive chromosome loss and rearrangement that accompanied genetic selection.

Heterokaryon Experiments

Heterokaryon experiments have demonstrated that a change in differentiated state can occur simply by altering the protein composition of the cytoplasm. In all the experiments described previously, differentiation was altered following extensive cell division. Thus, the results were compatible with the models of Brown (1984) and Weintraub (1985), which suggested that changes in chromatin associated with DNA replication were necessary to activate silent genes. Heterokaryon experiments showed that this was not the case.

Heterokaryons are nondividing cell hybrids in which the nuclei of the fused cells remain separate and distinct (Harris et al., 1969; Ringertz et al., 1971). DNA replication either does not occur or can be efficiently inhibited by addition of cytosine arabinoside. After fusion of muscle cells with primary human cell types representing all three embryonic lineages (endoderm, ectoderm, and mesoderm), silent genes that encoded a diversity of membrane and contractile products were activated (Blau et al., 1983, 1985). Gene dosage, or the balance of regulators contributed by the two fused cell types, was critical. For example, when the number of muscle cell nuclei exceeded the number of liver cell nuclei, extinction of liver genes and activation of muscle genes were observed; conversely, when the number of liver cell nuclei exceeded the number of muscle cell nuclei, muscle gene expression was repressed. By using a similar heterokaryon approach, these results were corroborated for muscle (Wright, 1984a, 1984b) and extended to a variety of tissue-specific genes including hematopoietic (Baron and Maniatis, 1986), hepatic (Spear and

Tilghman, 1990), and pancreatic (Wu et al., 1991) genes, which were activated in fibroblasts after fusion with cell types derived from blood, liver, and pancreas, respectively.

Heterokaryon experiments provide strong support for a mechanism of differentiation based on continuous regulation for the following reasons. (1) They show that silent genes are accessible and can be activated in numerous cell types merely by altering the protein composition of the cytoplasm without recourse to DNA replication. (2) The capacity to activate previously silent tissue-specific genes is a characteristic not only of cells initiating differentiation, but also of cells in which differentiation is well under way. From this finding it appears that the activity of transacting regulators is not required transiently at the onset of differentiation, but is required continuously to maintain it. (3) Nuclear ratio, or gene dosage, contributed by the two fused cell types determines the outcome. Thus, the balance of regulators is critical in establishing which genes are repressed or activated. (4) Differences among cell types are observed in the frequency and kinetics of gene activation or repression in heterokaryons. This finding is expected because the combination and relative ratio of proteins that interact in each type of heterokaryon differs.

Experiments with Single Regulators

Heterokaryon experiments showed that intracellular protein composition determines gene expression. Other experiments have shown that, in some cases, a single protein when present at a relatively high concentration is capable of gaining access to and activating the expression of silent genes that the cells would normally never express. When constitutively expressed, MyoD induces myosin heavy chain expression in a range of nonmuscle cell types (Davis et al., 1987). Myosin activation is observed not only in fibroblasts, which like muscle are mesodermal in origin, but also in less closely related cells such as melanocytes and neuroblastoma cells (Weintraub et al., 1989). This property can now be ascribed to a group of four regulators, all members of a helix-loop-helix family of transcription factors (Murre et al., 1989), that bind to the Ebox consensus sequence found in many muscle-specific genes (Olson, 1990).

To date, this property has not been shared by tissue-specific transcription factors other than the MyoD family, for reasons that are currently unclear. The lack of detectable gene activation may result from the choice of cell type in which the factors were expressed. Even in the case of MyoD, although myosin expression could be forced by high

MyoD concentrations, the program of myogenesis was not induced in most cell types; a heritable change in the majority of myogenic genes expressed was only observed with fibroblasts (Schaefer et al., 1990). The partial response or lack of response observed with regulators such as pituitary PIT-1 (GHF-1) (Bodner et al., 1988; Ingraham et al., 1988), liver HNF-1 (Baumhueter et al., 1990), and neural MASH-1 (Johnson et al., 1990) may occur because regulator concentrations are insufficient, essential endogenous cell proteins are lacking, or proteins that interfere with regulator activity are present. Such inhibitory proteins may complex with, modify, or compete directly with the transactivator. Thus, the effects of a single regulator on the differentiated state are generally buffered by the protein composition of the cell. Nonetheless, the finding that forced high-level expression of individual myogenic regulators such as MyoD activates silent genes showed that those genes are accessible in nonmuscle cells and that their expression state can be altered by a change in the concentration of one protein.

Temperature-Sensitive Mutants and Somatic Mosaics

Perhaps the most compelling evidence that differentiation is continuously regulated derives from experiments in the normal developing intact organism. The most stringent test for a requirement for regulator function is to eliminate the regulator. This can be achieved in temperature-sensitive mutants of *Drosophila* or *Caenorhabditis elegans* simply by a shift in temperature. In *Drosophila*, regulator production can also be aborted by x-ray-induced mitotic recombination or chromosome loss that leads to the production of individuals which are somatic mosaics. These individuals contain patches of homozygous mutant cells amid heterozygous (normal) cells. Both of these experimental manipulations allow investigation of the requirement of a gene product during a precise temporal window.

Experiments of this type have shown that the expression of a gene such as *ultrabithorax* that controls segment identity is required throughout development (Lewis, 1964; Vogt, 1946). Other genes involved in pattern formation, such as those in the large polycomb family that encode negative regulators, are also required continuously; if expression is altered even at late larval stages, the pattern is disrupted (Duncan and Lewis, 1982). Similarly, both positive and negative regulators of sex determination must be expressed continuously or the sexual characteristics of the cells will change, even in adulthood (Belote et al., 1985a; Kimble et al., 1984; Wieschaus and Nothiger, 1982). Ongoing gene expression is also required to maintain the identity of neurosensory cells (Way and Chalfie, 1989). These genetic experiments are corroborated by molecular experiments showing that the transcripts and proteins encoded by sex determination and homeotic selector genes are present continuously throughout development (Belote et al., 1985b; Duncan, 1987).

Perhaps the most dramatic example of plasticity is the change in behavior induced in *Drosophila* by a shift in temperature that disrupts the expression of the *tra-2* gene (Belote and Baker, 1987). If this regulator is not expressed continuously, an adult female will change its courtship behavior to that typical of an adult male. Thus, even in the regulation of sex and reproduction, which is critical to the survival of the species, there is a remarkable degree of plasticity. These findings demonstrate that developmental decisions that are central to the welfare of the organism, such as patterning or sex, are not locked in place by stable heritable mechanisms but require continuous regulation.

Implications of Continuous Regulation

If differentiation requires continuous regulation, how is knowledge of the differentiated state stably propagated to cell progeny? In particular, because only a small fraction of the total genes of a differentiated cell is expressed at any given time, by what mechanisms are most genes at most times prevented from being expressed?

Mechanisms for Controlling the Differentiated State

In theory, the differentiated state of a cell could be fixed by a mechanism that shuts off a cadre of genes by preventing access of transcription factors. A mechanism of this type is attractive in that it could afford necessary stability and should require a finite number of regulators. However, there is little evidence for such a mechanism in differentiation. Histones were suggested to play a role in the stable repression of a multitude of unneeded genes, but histone binding need not be stable. It is now apparent that in the absence of DNA replication, nucleosomes are displaced and DNAse hypersensitive sites are induced. Moreover, these changes are readily reversible and can all be accounted for by a change in the stoichiometry of *trans*-acting factors (Grunstein, 1990). Another candidate mechanism for stably controlling the expression state of tissue-specific genes is heritable methylation, as occurs in the silencing of genes that have been inactivated on the X-chromosome or by imprinting (Mohandas et al., 1981; Surani et al., 1990). However, most evidence suggests that methylation does not play a causal role in gene repression even in these cases. In addition, the methylation status of differentiation-specific genes is not always correlated with expression. Finally, the significance of methylation as a mechanism of gene control has been brought into question by the fact that it is not conserved in evolution; the DNA of *Drosophila* is not methylated (Bird and Taggart, 1980).

Alternatively, the differentiated state could be controlled by a mechanism that requires continuous regulation, as suggested by the experiments cited in the first half of this review. Gene expression patterns would be actively maintained by the composition of proteins present in a cell at any given time. This type of mechanism appears precarious, allowing plasticity and change where it is not needed or wanted. Moreover, the number of regulators required to repress the large number of unnecessary genes in each differentiated cell type would appear to be unwieldy. Nonetheless, as I describe next, a stable and heritable differentiated state can be achieved by continuous regulation of gene expression.

Numbers of Regulators

A series or hierarchy of regulators leads to the establishment of each distinct differentiated state. However, the progression from totipotent to differentiated cell does not, in general, restrict the range of possible genes that a cell can express, as concepts such as "committed stem cell," "terminally differentiated cell," and "fate maps" might suggest. Thus, the number of regulators required to control the expression state of so many genes seems prohibitive. However, feedback loops can limit the number of necessary regulators to achieve the same end by ensuring that the entire hierarchy of regulators need not be continuously expressed and that earlier steps can be bypassed by maintaining threshold concentrations of key regulators. Thus, regulators in the hierarchy that act for short periods of time to establish a differentiated state, like those that act continuously to maintain it, need not lead to permanent changes.

A feedback loop for which there is some evidence is autoregulation of transcription factors. Autoregulation has been well documented for bacteriophage lambda repressor, some of the *Drosophila* homeotic selector gene products, the signal transducer c-*jun* and the helix-loop-helix family of myogenic regulators (Angel et al., 1988; Kuziora and McGinnis, 1988; Ptashne, 1986; Thayer et al., 1989). Once gene expression is activated, autoregulation of transcription serves to maintain these regulators at a critical threshold concentration, providing stability. At cell division, the dose of regulator partitioned to daughter cells is sufficient to maintain the pattern of gene expression typical of the differentiated state of its parent. Autocatalytic calcium-calmodulin-dependent protein kinase provides another feedback mechanism (Lisman and Goldring, 1988). Extracellular matrix components also appear to act back on the cell types that produce them, reinforcing and stabilizing the differentiated state (Greenburg and Hay, 1988; Streuli et al., 1991). In years to come, it seems likely that many more feedback mechanisms will be identified that actively promote and stabilize the differentiated state. Such feedback controls could ensure that positive factors are produced at a critical concentration or that negative factors are titrated, allowing differentiation to proceed. These factors could either act directly at the level of transcription or indirectly on growth control pathways.

"Global repression binding sites" could also serve to limit the number of requisite regulators. Although, to date, relatively few such repetitive sequences have been identified, they could, in theory, constitute a highly effective mechanism by which a number of related tissue-specific genes could be silenced by a few negative regulators. For example, it seems likely that related DNA sequences mediate the effects of the Polycomb class of negative regulators, which bind to at least 60 sites within polytene chromosomes at which diverse homeotic or other Polycomb target genes reside (Zink and Paro, 1989). Repetitive elements in the vicinity of several chicken and rat genes have also been reported (Baniahmad et al., 1987; Laimins et al., 1986).

Negative Regulation

A major regulatory dilemma is that most genes at most times must be maintained in an inactive state because differentiated cells express only a small fraction of their DNA. That negative regulation is an active ongoing process is suggested by the extinction of differentiated traits observed in many cell fusion experiments (for review, see Ringertz and Savage, 1976), a property that can be mapped to specific chromosomal loci (Killary and Fournier, 1984). Once these loci are lost, the genes they silenced are reexpressed. Loss of function mutants in *Drosophila* have revealed loci such as *hairy* and *extramacrochaete* that act as negative regulators of the *achaete-scute* complex and are required continuously to repress the differentiation of sensory organs (Moscoso del Prado and Garcia-Bellido, 1984). Similarly, the continuous expression of the homeotic polycomb genes prevents the expression of the *Drosophila* ultrabithorax gene, which is central to establishing segment identity early in development (Lewis, 1978; Struhl and Akam, 1985). If ultrabithorax expression is disrupted in temperature shift mutants, pattern is altered.

Mechanisms for keeping genes shut off are therefore of particular interest. Negative regulators can affect the expression of specific genes by preventing the expression of genes encoding positive regulators (McCormick et al., 1988) or by competing with positive regulators for DNA-binding sites (Jaynes and O'Farrell, 1988). Alternatively, negative regulators can interact directly with positive regulators and prevent DNA binding (Baeuerle and Baltimore, 1988; Benezra et al., 1990) or render activators nonfunctional once bound to DNA (Ma and Ptashne, 1987). A mechanism known as silencing permits negative regulation at a distance and provides a means for shutting off gene expression even in the presence of positive regulators. First described in yeast (Brand et al., 1985), silencers have now been identified in a number of tissue-specific genes in a wide variety of species (Baniahmad et al., 1987; Winoto and Baltimore, 1989). Although it has been postulated that they may inhibit transcription or define chromatin loops, the mechanism by which silencers act in mammalian cells and to what extent that mechanism parallels that described in yeast remains to be determined. In addition, as described, repetitive elements or "global repression binding sites" associated with a number of different genes could provide a means for silencing a battery of genes with relatively few types of negative regulators (Baniahmad et al., 1987; Laimins et al., 1986).

Balance of Positive and Negative Regulators is Critical

If differentiation is continuously regulated, the relative concentration of positive and negative regulators is of central importance. How is protein concentration altered? Recent evidence indicates that many regulatory proteins form complexes, for example, heterodimers via leucine zipper or helix-loop-helix motifs (Landschulz et al., 1989; Murre et al., 1989a). Such interactions can either promote or inhibit the function of a regulator. For instance, the transcription factor MyoD requires the protein E12 to bind DNA efficiently (Murre et al., 1989b), but is prevented from binding DNA when complexed to the protein Id (Benezra et al., 1990). Thus, the effective concentration of a regulator is altered not only when its rate of synthesis or degradation is changed, but also when the concentration of the proteins with which it interacts is altered. Another case in point is the transcription factor NFκB which is inhibited from entering the nucleus and is therefore inactive when it is complexed to IκB in the cytoplasm (Baeuerle and Baltimore, 1988a, 1988b). The complexity of these interactions increases as the number of different partners with which a protein can associate increases, as is the case in intact cells (Peterson et al., 1990; Schaefer et al., 1990). Clearly, in addition to their abundance the relative affinity and cooperative interactions of regulators at DNA-binding sites will have a profound impact on gene expression. Recent evidence that synergism among diverse transcriptional regulators occurs even at concentrations at which their DNA-binding sites are saturated suggests that regulators have cooperative effects, not just as heterodimers but also as multimeric complexes (Carey et al., 1990; Lin et al., 1990). Because proteins act in combinations, small changes in the relative concentrations of single regulators can have large effects on the expression of the differentiated state of the cell, by shifting a critical balance, reaching a threshold, and setting off a cascade of events. These predictions have been borne out *in vivo*. The dosage of genes encoding the helix-loop-helix proteins *daughterless*, *hairy*, and *achaete-scute* determines sex in *Drosophila* (Parkhurst et al., 1990). Gene dosage also determines sex in *C. elegans* (Hodgkin, 1990) and the phenotype of neurosensory cells in *Drosophila* (Botas et al., 1982), and is responsible for several hereditary developmental disorders in humans (Epstein, 1986).

The Need for Plasticity

Plasticity of gene expression in differentiated cells might seem precarious. If the differentiated state is controlled by mechanisms that are dynamic and reversible, it could change. Perhaps plasticity is essential, however. This possibility is suggested by findings that the same regulatory genes appear to be used at different times in development to specify different processes. These genes must therefore be accessible and subject to reactivation after repression. The *engrailed* gene in *Drosophila* is a case in point. *Engrailed* is expressed at two different developmental stages, and its function and the regulatory network that governs its expression at these two stages differ (DiNardo et al., 1988). The need for plasticity is also evident in wound repair. Following injury, even mammalian cells change their phenotype or transdifferentiate; for example, melanin-producing iris cells give rise to crystallin-producing lens cells (Eguchi and Okada, 1973).

Indeed, by using sensitive single cell markers to monitor the fate of cells in normal development or following implantation into novel sites, additional evidence of plasticity has been revealed. Rodent myogenic precursor cells specialized for contraction in a particular muscle fiber type express novel genes if they fuse with a different fiber type during normal muscle development (Hughes and Blau, 1992). These experiments show plasticity of gene expression in heterokaryons that form naturally *in vivo*, corroborating the results of *in vitro* cell fusions. Similarly, quail neural crest cells give rise to a multiplicity of unexpected cell types, including representatives of different embryonic germ layers (Le Douarin, 1986). In addition, neural cells, induced to proliferate by expression of a temperature-sensitive SV40 T-antigen, assume the phenotype of the region of the mouse brain into which they are subsequently transplanted (Rentranz et al., 1990). Application of these types of experimental approaches to other experimental systems may well demonstrate further plasticity of the differentiated state.

Are there genes that are not continuously regulated? Possibly all gene expression, even that of genes on the inactive X-chromosome or those genes subject to imprinting, is actively controlled. That this may be the case is suggested by studies of position-effect variegation in *Drosophila*, which shares properties with X-chromosome inactivation in mammals. Gene inactivity in both cases is associated with heterochromatin, a region of the chromosome that appears to be permanently condensed (Alberts and Sternglanz, 1990; Reuter et al., 1990). Chromosome translocations lead to the inactivation by heterochromatin of genes not normally subject to this type of regulation in species ranging from *Drosophila* to humans (Reuter et al., 1990).

A link between the spreading of heterochromatin in position-effect variegation and the expression of the polycomb family of genes that encodes trans-acting negative regulators has recently been suggested. First, the effects of both types of regulation are dose dependent: the negative regulation of the *bithorax* complex by the polycomb gene products depends on the number of copies of the complex (Duncan and Lewis, 1982). Similarly, the dosage of the *Su(var)* gene determines the size of the domain encompassed in heterochromatin (Reuter et al., 1990). Second, the heterochromatin-associated protein HP1 encoded by a member of the Su(var) family responsible for position-effect variegation shares homology with a protein encoded by a member of the polycomb family (Paro and Hogness, 1991). Thus, even such apparently stable mechanisms for shutting off genes as heterochromatin may require continuous regulation through feedback loops that maintain threshold concentrations of critical regulators.

Summary

As described in this chapter, substantial evidence suggests that the differentiated state is controlled continuously. The plasticity in gene expression that this type of mechanism affords could be advantageous, and may even be essential. Indeed, there are remarkably few changes during differentiation that are completely irreversible, the gene rearrangements leading to immunoglobulin expression being one clear exception. A prediction of this model for differentiation is that any nucleus exposed to the appropriate constellation of proteins at the appropriate concentration should be able to perform functions typical of any given differentiated cell type. Feedback controls presumably prevent cells from spontaneously changing by ensuring that the requisite dosage of regulators is maintained and that a stable pattern of gene expression is propagated to cell progeny. To date, few such feedback mechanisms have been defined; most remain to be elucidated. A knowledge of the molecular nature of these control mechanisms should aid in the design of therapeutic intervention in disease states such as malignancy in which the balance of differentiation and proliferation controls goes awry.

References

Alberts B, Sternglanz R (1990): Gene expression: Chromatin contract to silence. *Nature* (Lond) 344:193–194.

Angel P, Hattori K, Smeal T, Karin M (1988): The *jun* proto-oncogene is positively autoregulated by its product, jun/AP-1. *Cell* 55:875–885.

Baeuerle PA, Baltimore D (1988a): IκB: A specific inhibitor of the NF-κB transcription factor. *Science* 242:540–546.

Baeuerle PA, Baltimore D (1988b): Activation of DNA-binding activity in an apparently cytoplasmic precursor of the NF-κB transcription factor. *Cell* 53:211–217.

Baniahmad A, Muller M, Steiner C, Renkawitz R (1987): Activity of two different silencer elements of the chicken lysozyme gene can be compensated by enhancer elements. *EMBO J* 6:2297–2303.

Baron MH, Maniatis T (1986): Rapid reprogramming of globin gene expression in transient heterokaryons. *Cell* 46:591–602.

Baumhueter S, Mendel DB, Conley PB, Kuo CJ, Turk C, Graves MK, Edwards CA, Courtois G, Crabtree GR (1990): HNF-1 shares three sequence motifs with the POU domain proteins and is identical to LF-B1 and APF. *Genes & Dev* 4:372–379.

Belote JM, Baker BS (1987): Sexual behavior: Its genetic control during development and adulthood in *Drosophila melanogaster. Proc Natl Acad Sci USA* 84:8026–8030.

Belote JM, Handler AM, Wolfner MF, Livak KJ, Baker BS (1985a): Sex-specific regulation of yolk protein gene expression in *Drosophila. Cell* 40:339–348.

Belote JM, McKeown MB, Andrew DJ, Scott TN, Wolfner MF, Baker BS (1985b): Control of sexual differentiation in *Drosophila melanogaster. Cold Spring Harbor Symp Quant Biol* 50:605–614.

Benezra R, Davis RL, Lockshon D, Turner DL, Weintraub H (1990): The protein Id: A negative regulator of helix-loop-helix DNA binding proteins. *Cell* 61:49–59.

Bird AP, Taggart MH (1980): Variable patterns of total DNA and rDNA methylation in animals. *Nucleic Acids Res* 8:1485–1497.

Blau HM, (1992): Differentiation requires continuous active control. *Annu Rev Biochem* 61:1213–1230.

Blau HM, Baltimore D (1991): Differentiation requires continuous regulation. *J Cell Biol* 112:781–783.

Blau HM, Chiu C-P, Webster C (1983): Cytoplasmic activation of human nuclear genes in stable heterocaryons. *Cell* 32:1171–1180.

Blau HM, Pavlath GK, Hardeman EC, Chiu C-P, Silberstein L, Webster SG, Miller SC, Webster C (1985): Plasticity of the differentiated state. *Science* 230:758–766.

Bodner M, Castrillo J-L, Theill LE, Deerinck T, Ellisman M, Karin M (1988): The pituitary-specific transcription factor GHF-1 is a homeobox-containing protein. *Cell* 55:505–518.

Botas J, Moscoso del Prado J, Garcia-Bellido A (1982): Gene-dose titration analysis in the search of trans-regulatory genes in *Drosophila. EMBO J* 1:307–311.

Brand AH, Breeden L, Abraham J, Sternglanz R, Nasmyth K (1985): Characterization of a "silencer" in yeast: A DNA sequence with properties opposite to those of a transcriptional enhancer. *Cell* 41:41–48.

Brown DD (1984): The role of stable complexes that repress and activate eucaryotic genes. *Cell* 37:359–365.

Carey M, Lin Y-S, Green MR, Ptashne M (1990): A mechanism for synergistic activation of a mammalian gene by GAL4 derivatives. *Nature* (Lond) 345:361–364.

Carlsson S-A, Luger O, Ringertz NR, Savage RE (1974): Phenotypic expression in chick erythrocyte × rat myoblast hybrids and in chick myoblast × rat myoblast hybrids. *Exp Cell Res* 84:47–55.

Davidson RL (1972): Regulation of melanin synthesis in mammalian cells: Effect of gene dosage on the expression of differentiation. *Proc Natl Acad Sci USA* 69:951–955.

Davis RL, Weintraub H, Lassar AB (1987): Expression of a single transfected cDNA converts fibroblasts to myoblasts. *Cell* 51:987–1000.

DiBerardino MA, Orr NH, McKinnell RG (1986): Feeding tadpoles cloned from *Rana* erythrocyte nuclei. *Proc Natl Acad Sci USA* 83:8231–8234.

DiNardo S, Sher E, Heemskerk-Jongens J, Kassis JA, O'Farrell PH (1988): Two-tiered regulation of spatially patterned engrailed gene expression during *Drosophila* embryogenesis. *Nature* (Lond) 332:604–609.

Duncan I (1987): The bithorax complex. *Annu Rev Genet* 21:285–319.

Duncan I, Lewis EB (1982): Genetic control of body segment differentiation in *Drosophila*. In: *Developmental Order: Its Origin and Regulation*, pp. 533–554, New York: Alan Liss.

Eguchi G, Okada TS (1973): Differentiation of lens tissue from the progeny of chick retinal pigment cells cultured *in vitro*: A demonstration of a switch of cell types in clonal cell culture. *Proc Natl Acad Sci USA* 70:1495–1499.

Epstein CJ (1986): The theoretical mechanisms and issues: The primary and secondary effects of aneuploidy. In: *The Consequences of Chromosome Imbalance: Principles, Mechanisms, and Models*, pp. 65–79, London: Cambridge University Press.

Greenburg G, Hay ED (1988): Cytoskeleton and thyroglobulin expression change during transformation of thyroid epithelium to mesenchyme-like cells. *Development* 102:605–622.

Grunstein M (1990): Histones and nucleosomes: Repressors of transcription initiation. *Annu Rev Cell Biol* 6:643–678.

Gurdon JB (1962): The developmental capacity of nuclei taken from intestinal epithelium cells of feeding tadpoles. *J Embryol Exp Morphol* 10:622–640.

Gurdon JB (1992): The generation of diversity and pattern in animal development. *Cell* 68:185–199.

Hadorn E (1963): Differenzierungsleistungen wiederholt fragmentierter Teilstücke männlicher Genitalscheiben von *Drosophila melanogaster* nach Kultur in vivo. *Dev Biol* 7:617–629.

Harris H, Klein G (1969): Malignancy of somatic cell hybrids. *Nature* (Lond) 224:1314–1316.

Harris H, Sidebottom E, Grace DM, Bramwell ME (1969): The expression of genetic information: A study with hybrid animal cells. *J Cell Sci* 4:499–525.

Hodgkin J (1990): Sex determination compared in *Drosophila* and *Caenorhabditis*. *Nature (Lond)* 344:721–728.

Hughes SM, Blau HM (1992): Muscle fiber pattern is independent of cell lineage in postnatal rodent development. *Cell* 68:659–671.

Ingraham HA, Chen RP, Mangalam HJ, Elsholtz HP, Flynn SE, Lin CR, Simmons DM, Swanson L, Rosenfeld MG (1988): A tissue-specific transcription factor containing a homeodomain specifies a pituitary phenotype. *Cell* 55:519–529.

Jacob F, Monod J (1961): Genetic regulatory mechanisms in the synthesis of proteins. *J Mol Biol* 3:318–356.

Jaynes JB, O'Farrell PH (1988): Activation and repression of transcription by homoeodomain-containing proteins that bind a common site. *Nature* (Lond) 336:744–749.

Johnson JE, Birren SJ, Anderson DJ (1990): Two rat homologues of *Drosophila achaete-scute* specifically expressed in neuronal precursors. *Nature* (Lond) 346:858–861.

Killary AM, Fournier REK (1984): A genetic analysis of extinction: Trans-dominant loci regulate expression of liver-specific traits in hepatoma hybrid cells. *Cell* 38:523–534.

Kimble J, Edgar L, Hirsh D (1984): Specification of male development in *Caenorhabditis elegans*: The *fem* genes. *Dev Biol* 105:234–239.

Konieczny SF, Lawrence JB, Coleman JR (1983): Analysis of muscle protein expression in polyethylene glycol-induced chicken: Rat myoblast heterokaryons. *J Cell Biol* 97:1348–1355.

Kuziora MA, McGinnis W (1988): Autoregulation of a *Drosophila* homeotic selector gene. *Cell* 55:477–485.

Laimins L, Holmgren-Konig M, Khoury G (1986): Transcriptional "silencer" elements in rat repetitive sequences associated with the rat insulin 1 gene locus. *Proc Natl Acad Sci USA* 83:3151–3155.

Landschulz WH, Johnson PF, McKnight SL (1989): The DNA binding domain of the rat liver nuclear protein C/EBP is bipartite. *Science* 243:1681–1688.

Le Douarin NM (1986): Cell line segregation during peripheral nervous system ontogeny. *Science* 231:1515–1522.

Lewis EB (1964): Genetic control and regulation of developmental pathways. In: *The Role of Chromosomes in Development*, Locke M, ed. New York: Academic Press.

Lewis EB (1978): A gene complex controlling segmentation in *Drosophila*. *Nature* (Lond) 276:565–570.

Lin Y-S, Carey M, Ptashne M, Green MR (1990): How different eukaryotic transcriptional activators can cooperate promiscuously. *Nature* (Lond) 345:359–361.

Lisman JE, Goldring MA (1988): Feasibility of long-term storage of graded information by the Ca^{2+}/calmodulin-dependent protein kinase molecules of the postsynaptic density. *Proc Natl Acad Sci USA* 85:5320–5324.

Ma J, Ptashne M (1987): The carboxy-terminal 30 amino acids of GAL4 are recognized by GAL80. *Cell* 50:137–142.

McCormick A, Wu D, Castrillo J-L, Dana S, Strobl J, Thompson EB, Karin M (1988): Extinction of growth hormone expression in somatic cell hybrids involves repression of the specific trans-activator GHF-1. *Cell* 55:379–389.

Mohandas T, Sparkes RS, Shapiro LJ (1981): Reactivation of an inactive human X chromosome: Evidence for X inactivation by DNA methylation. *Science* 211:393–396.

Moscoso del Prado J, Garcia-Bellido A (1984): Genetic regulation of the achaete-scute complex of *Drosophila melanogaster*. *Roux's Arch Dev Biol* 193:242–245.

Murre C, McCaw PS, Baltimore D (1989a): A new DNA binding and dimerization motif in immunoglobulin enhancer binding, daughterless, MyoD, and myc proteins. *Cell* 56:777–783.

Murre C, McCaw PS, Vaessin H, Caudy M, Jan LY, Jan YN, Cabrera CV, Buskin JN, Hauschka SD, Lassar AB, Weintraub H, Baltimore D (1989b): Interactions between heterologous helix-loop-helix proteins generate complexes that bind specifically to a common DNA sequence. *Cell* 58:537–544.

Olson EN (1990): The MYOD family: A paradigm for development? *Genes & Dev* 4:1454–1461.

Parkhurst SM, Bopp D, Ish-Horowicz D (1990): X: A ratio, the primary sex-determining signal in *Drosophila*, is transduced by helix-loop-helix proteins. *Cell* 63:1179–1191.

Paro R, Hogness DS (1991): The Polycomb protein shares a homologous domain with a heterochromatin-associated protein of Drosophila. *Proc Natl Acad Sci USA* 88:263–267.

Peterson CA, Gordon H, Hall ZW, Paterson BM, Blau HM (1990): Negative control of helix-loop-helix family of myogenic regulators in NFB mutant. *Cell* 62:493–502.

Peterson JA, Weiss MC (1972): Expression of differentiated functions in hepatoma cell hybrids: Induction of mouse albumin production in rat hepatoma—mouse fibroblast hybrids. *Proc Natl Acad Sci USA* 69:571–575.

Ptashne M (1986): *A Genetic Switch: Gene Control and Phage* λ. Oxford: Cell Press & Blackwell Scientific.

Rentranz PJ, Cunningham MG, McKay RDG (1990): Region-specific differentiation of the hippocampal stem cell line HiB5 upon implantation into the developing mammalian brain. *Cell* 66:713–729.

Reuter G, Giarre M, Farah J, Gausz J, Spierer A, Spierer P (1990): Dependence of position-effect variegation in *Drosophila* on dose of a gene encoding an unusual zinc-finger protein. *Nature* (Lond) 344:219–223.

Ringertz NR, Carlsson S-A, Ege T, Bolund L (1971): Detection of human and chick nuclear antigens in nuclei of chick erythrocytes during reactivation in heterokaryons with HeLa cells. *Proc Natl Acad Sci USA* 68:3228–3232.

Ringertz NR, Savage RE (1976): *Cell Hybrids*. New York: Academic Press.

Schaefer BW, Blakely BT, Darlington G, Blau HM (1990): Effect of cell history on response to helix-loop-helix family of myogenic regulators. *Nature* (Lond) 344:454–458.

Schmid V, Alder H (1984): Isolated mononucleated, striated muscle can undergo pluripotent transdifferentiation and form a complex regenerate. *Cell* 38:801–809.

Spear BT, Tilghman SM (1990): The role of the α-fetoprotein regulatory elements in transcriptional activation in transient heterokaryons. *Mol Cell Biol* 10:5047–5054.

St. Johnston D, Nusslein-Volhard C (1992): The origin of pattern in the *Drosophila* embryo. *Cell* 68:201–219.

Streuli CH, Bailey N, Bissell MJ (1991): Control of mammary epithelial differentiation: Basement membrane induced tissue-specific gene expression in the absence of cell-cell interaction and morphological polarity. *J Cell Biol* 115:1383–1395.

Struhl G, Akam M (1985): Altered distributions of ultrabithorax transcripts in extra sex combs mutant embryos of *Drosophila*. *EMBO J* 4:3259–3264.

Surani MA, Allen ND, Barton SC, Fundele R, Howlett SK, Norris ML, Reik W (1990): Developmental consequences of imprinting of parental chromosomes by DNA methylation. *Philos Trans R Soc Lond B* 326:313–327.

Thayer MJ, Tapscott SJ, Davis RL, Wright WE, Lassar AB, Weintraub H (1989): Positive autoregulation of the myogenic determination gene myoD1. *Cell* 58:241–248.

Vogt VM (1946): Zur labilen determination der imaginalscheiben von Drosophila: VI. die umwandlung präsumptiven rüsselgewebes in bein-oder fühlergewebe. *Z Naturforsch* 1:469–475.

Way JC, Chalfie M (1989): The mec-3 gene of *Caenorhabditis elegans* requires its own product for maintained expression and is expressed in three neuronal cell types. *Genes Dev* 3:1823–1833.

Weintraub H (1985): Assembly and propagation of repressed and depressed chromosomal states. *Cell* 42:705–711.

Weintraub H, Tapscott SJ, Davis RL, Thayer MJ, Adam MA, Lassar AB, Miller AD (1989): Activation of muscle specific genes in pigment, nerve, fat, liver and fibroblast cell lines by forced expression of MyoD. *Proc Natl Acad Sci USA* 86:5434–5438.

Wieschaus E, Nöthiger R (1982): The role of the transformer genes in the development of genitalia and analia of *Drosophila melanogaster*. *Dev Biol* 90:320–334.

Winoto A, Baltimore D (1989): $\alpha\beta$ Lineage-specific expression of the α T cell receptor gene by nearby silencers. *Cell* 59:649–655.

Wolpert L (1969): Positional information and the spatial pattern of cellular differentiation. *J Theor Biol* 25:1–47.

Wolpert L (1989): Positional information revisited. *Development* 107:Suppl 3–12.

Wright WE (1984a): Induction of muscle genes in neural cells. *J Cell Biol* 98:427–435.

Wright WE (1984b): Expression of differentiated functions in heterokaryons between skeletal myocytes, adrenal cells, fibroblasts and glial cells. *Exp Cell Res* 151:55–69.

Wu KJ, Samuelson LC, Howard G, Meisler MH, Darlington GJ (1991): Transactivation of pancreas-specific gene sequences in somatic cell hybrids. *Mol Cell Biol* 11:4423–4430.

Zink B, Paro R (1989): In vivo binding pattern of a trans-regulator of homeotic genes in *Drosophila melanogaster*. *Nature* (Lond) 337:468–471.

Chapter 3

Gene Regulation by Steroid Hormones

Miguel Beato

The ability of cells to specify the fraction of their genetic information that they express in a particular spatiotemporal context is essential for adaptation to the changing conditions of their surroundings. In higher organisms, cells must respond to stimuli from the outer world and to signals from other cells directed to coordinate their state of activity for the proper development and functioning of the whole animal. Many of these signals impinge on receptors located at the cell membrane, where they elicit changes in the intracellular concentration of key molecules, second messengers, that ultimately modulate the expression of genetic programs. The signal transduction mechanism is simpler in the case of molecules acting through nuclear receptors able to recognize the signal and to interact directly with the nuclear genome. To this class belong the steroid hormones that influence the expression of a great variety of genes in many different cells.

Gene regulation by steroid hormones is accomplished by a variety of different mechanisms leading to induction or repression of particular genes. These mechanisms are all mediated by a single class of intracellular hormone receptors that, in the unliganded state, are maintained in an inactive form by poorly understood mechanisms probably involving their association with other cellular proteins. In this chapter, I review the structural and functional organization of the nuclear receptors and how they discriminate the DNA sequences to which they bind. Using the hormone induction of mouse mammary tumor virus (MMTV) as an example, I discuss the synergistic interactions among receptor molecules on a complex hormone-responsive element (HRE) as well as the interaction between receptors and other transcription factors, in particular nuclear

GENE EXPRESSION: GENERAL AND CELL-TYPE-SPECIFIC
Michael Karin, Editor
© 1993 Birkhäuser Boston

factor I (NFI) and the octamer-binding OTF-1. Particular attention is devoted to the use of cell-free transcription assays to dissect the different aspects of regulated transcription and to the role of nucleosome positioning in determining the accessibility of the MMTV promoter elements to transcription factors.

Induction of several genes by steroid hormones can be prevented by inhibitors of protein synthesis, and this has been taken as suggesting an indirect effect of the hormones on gene expression requiring the induction of an intermediate. I show here that in some systems this type of effect is mediated by a direct binding of the receptor to the corresponding promoter, although interaction with an auxiliary labile factor is required for induction.

Steroid hormones, and in particular glucocorticoids, not only induce but also inhibit gene expression at the transcriptional level. I also discuss the various mechanisms used by nuclear receptors in mediating transcriptional repression, including competition with other transcription factors for binding to DNA, direct protein–protein interactions, and the phenomenon of "squelching."

Structure of Steroid Hormone Receptors

The molecular cloning of steroid hormone receptors has clearly indicated that they belong to a large superfamily of nuclear receptors which includes the receptors for retinoic acid, thyroid hormones, and several genes for which a physiological ligand is not yet known (Beato, 1989). This latter class of "orphan" receptors includes genes of known function, such as the *knirps* gene of *Drosophila*, and several genes of unknown function. (For a recent review of orphan receptor, see O'Malley, 1990.) All members of the superfamily are organized according to a modular pattern and include at least three structural and functional domains able to act relatively independently of one another (Evans, 1988).

Members of the nuclear receptor superfamily are identified by the presence of a short DNA-binding domain composed of some 70 amino acid residues containing 10 conserved cysteines. Eight of these cysteines can be organized into two so-called zinc fingers, each encompassing four cysteine residues tetrahedrally coordinating a zinc ion. There has been some confusion as to the actual cysteine residues involved in coordinating the carboxy terminal zinc ion, but x-ray analysis has clarified this issue (Luisi et al., 1991) and shown that the zinc ion coordination deduced from mutation analysis is correct (Severne et al., 1988). The zinc finger motif was originally proposed for the transcription factor TFIIIA from *Xenopus laevis*, in which a pair of cysteines and a pair of histidines, instead of four cysteines, serve to coordinate each of the nine zinc ions that build up the basic repeated structure of the protein (Miller et al., 1985). Subsequently, many other genes for regulatory proteins, as well as enzymes, have been found to contain zinc fingers. At present, the family of zinc finger proteins is the largest class of DNA-binding proteins. In steroid hormone receptors, the two fingers are encoded by two different exons and appear to have different function. Contrary to the zinc fingers of TFIIIA that can bind individually to a short stretch of DNA, the two zinc fingers of the steroid hormone receptors form a single DNA-binding structure (see following), although the first zinc finger has been reported to bind DNA as an isolated peptide (Archer et al., 1990).

A comparison of the amino acid sequence in the DNA-binding domain of the different nuclear receptor genes allows their classification into two subfamilies (Figure 1). The glucocorticoid receptor (GR) is the prototype of the smaller subfamily that includes the progesterone receptor (PR), the androgen receptor, and the mineralocorticoid receptor. The larger subfamily is more heterogeneous, and its prototype is the estrogen receptor (ER). This group includes the receptors for vitamin D_3, ecdysone, thyroid hormone, retinoic acid, and the peroxisomal proliferator activators (Dreyer et al., in press; Issemann and Green, 1990), as well as many of the so-called orphan receptors for which no ligand has been identified (O'Malley, 1990). The main differences between the members of the two subgroups of nuclear receptors reside in the knuckles of the two zinc fingers, in regions that have been shown to be important for DNA sequence recognition and receptor dimerization (Figure 2a).

Most mutations in the conserved amino acids of the DNA-binding domain lead to reduced transactivation activity that correlates with a decreased affinity for DNA (Hollenberg and Evans, 1988; Schena et al., 1989). An exception is the first lysine following the N-terminal zinc finger. Mutation of this amino acid to glycine interferes with transactivation without significantly reducing DNA-binding activity (Hollenberg and Evans, 1988; Oro et al., 1988). This mutation seems to influence the ability of the DNA binding domain to functionally synergize with the transactivation function located in the N-terminal half of the receptor (S. Rusconi, personal communication). Interestingly, several mutations in

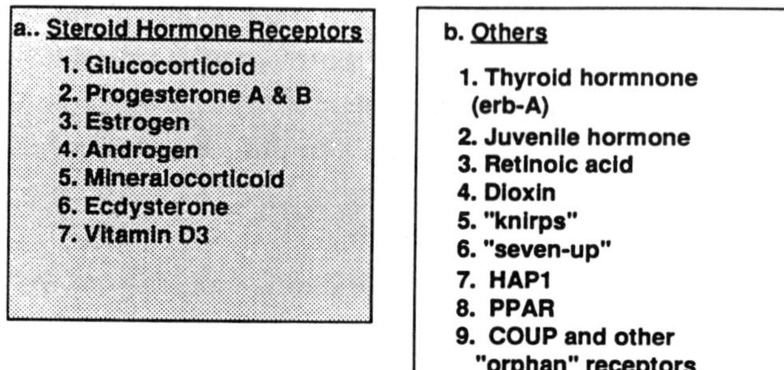

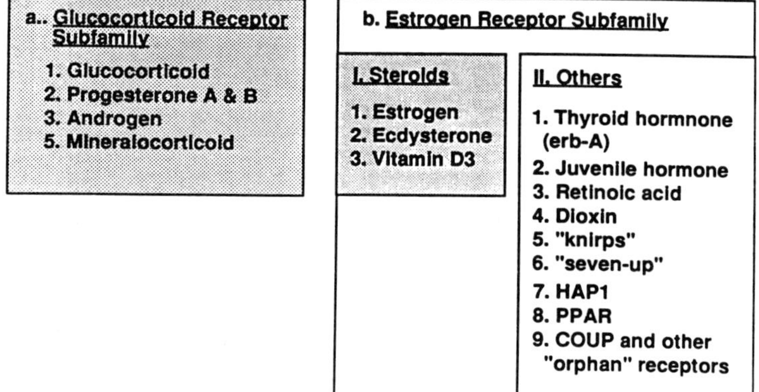

Figure 1   Classification of nuclear receptors according to ligand binding and receptor structure.

the second zinc finger selectively affect transactivation without reducing transrepression (Schena et al., 1989).

The small DNA-binding domain seems to contain a dimerization function (Forman and Samuels, 1990) and a nuclear localization signal that, in the case of the glucocorticoid receptor, acts independently of ligand binding (Picard and Yamamoto, 1987). This domain has also been postulated to be transcriptionally active in the absence of hormone, although to a much lesser extent than the intact receptor (Miesfeld et al., 1987), and is able to mediate DNA-binding cooperativity to adjacent hormone-responsive elements (HREs) (Baniahmad et al., 1991).

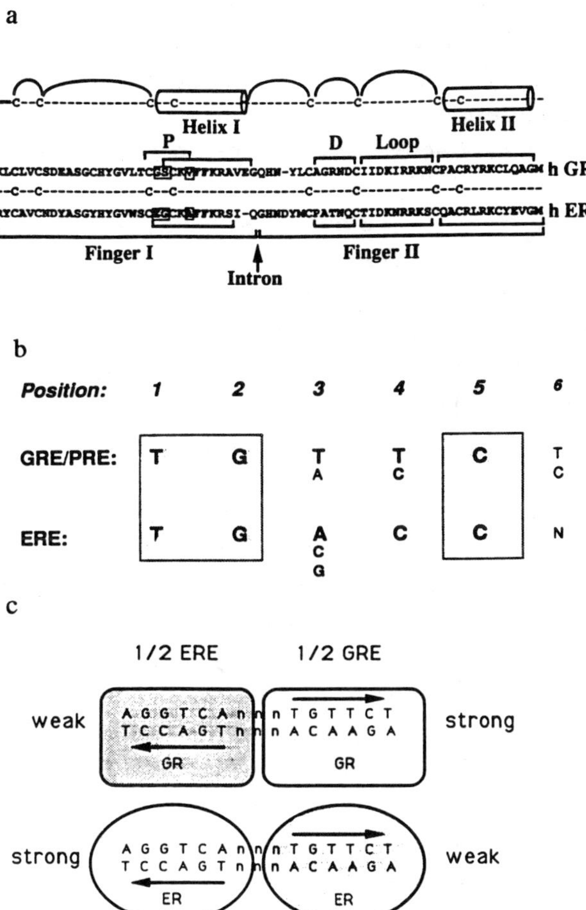

Figure 2   Structure of the DNA-binding domains of steroid hormone receptors and their target DNA sequences. **a.** The amino acid sequences of the human glucocorticoid receptor (GR) and estrogen receptor (ER) are shown using the one-letter amino acid code. The positions of the cysteine residues involved in zinc coordination are indicated. The P and D boxes, and helices I and II, are also indicated. The amino acid residues responsible for sequence discrimination between GRE/PRE (glucocorticoid progesterone responsive element) and ERE (estrogen responsive element) are indicated by shadowed boxes. **b.** Structure of the half palindrome of GR/PR and ER. **c.** Binding of GR or ER to hybrid regulatory elements.

Another relatively well-conserved region of steroid hormone receptors is the large carboxy-terminal domain responsible for binding of the hormone ligand, referred to as the hormone-binding domain. The region required for specific binding of the hormone appears to be relatively large. In the case of the glucocorticoid and progesterone receptors, some of the amino acid residues that are contacted by the hormone ligand have been identified in photo-cross-linking experiments (Carlstedt-Duke et al., 1988; Simons et al., 1987). Genetic analysis of the ER has also been used to map the region of the hormone-binding domain responsible for direct binding of estrogens (Fawell et al., 1990). In these latter studies, it was possible to separate the region implicated in hormone binding from that responsible for ligand-induced receptor dimerization. The region of the estrogen receptor required for homodimerization has been identified as a 22-amino-acid peptide, conserved in other steroid receptors as well as in the homodimeric steroid-binding protein uteroglobin (Beato et al., 1983; Lees et al., 1990).

The hormone-binding domain also contains a so-called transcriptional activation function that is dependent for activity on binding of an agonistic ligand (Hollenberg and Evans, 1988; Tora et al., 1989). A refined analysis of this transcription activation function is complicated by the existence, in the same region, of a nuclear translocation signal that is also hormone dependent (Picard and Yamamoto, 1987). Moreover, this region contains amino acid sequences that are important for the interaction between the receptors and the heat shock protein hsp 90 (Chambraud et al., 1990; Housley et al., 1990; Howard et al., 1990; Pratt et al., 1988).

The amino-terminal half of the nuclear receptors is their most variable region and the function of this region is less well defined. Some receptors, such as the progesterone receptor, exist in multiple forms with N-terminal regions of different lengths (Conneely et al., 1986; Kastner et al., 1990; Loosfelt et al., 1986). In GR, ER, and PR a transactivation function, which seems to be independent of ligand binding, has been assigned to this region (Bocquel et al., 1989; Hollenberg and Evans, 1988; Tora et al., 1989). Part of the agonistic activity of some antisteroids appears to be mediated by the transactivation function in this region of the receptor protein (Kastner et al., 1990). This region plays an essential role in the synergistic interactions among receptor dimers bound to DNA (Baniahmad et al., 1991), and this effect is mediated by cooperative DNA binding (Wright and Gustafsson, 1991). In several cases the requirement for the transactivation function at the N-terminal region appears to be promoter- and cell type specific (Bocquel et al., 1989; Tora et al., 1988). This offers a possible way to explain the complexity and specificity of the hormonal response in different cells.

Hormone-Responsive Elements

The nucleotide sequences recognized by the steroid hormone receptors and responsible for their effects on gene activity, the so-called hormone-responsive or hormone-regulatory elements (HREs), have been studied in great detail (Beato, 1989). In most cases, these sequences exhibit a partial palindromic structure with two unequally well conserved halves separated by three nonconserved base pairs. This palindromic structure and the symmetry of the contacts between receptors and the N-7 positions of guanines already suggested that a dimer of the receptor interacts with each HRE (Scheidereit and Beato, 1984; Scheidereit et al., 1986). This idea has been confirmed (Kumar and Chambon, 1988; Tsai et al., 1988), although the GR can also interact as monomer with half-palindromic sites, albeit with reduced affinity (Chalepakis et al., 1990). Moreover, efficient binding of the GR and the PR to their specific sites on DNA requires an unspecific interaction with the flanking base pairs, as demonstrated by hydroxyl radical protection and interference experiments (Chalepakis et al., 1988a). In those cases where several HREs are clustered in a hormone-regulatory region, one finds a functional interaction between receptor dimers bound to the individual HREs that requires a stereospecific alignment of the sites and is, at least in part, mediated by DNA-binding cooperativity (Schmid et al., 1989; Schüle et al., 1988; Tsai et al., 1989). Cooperative DNA binding to adjacent palindromes appears to be mediated by the DNA-binding domain of the receptor, although functional synergism can also occur with receptors molecules linked to the DNA-binding region of the yeast transregulator GAL4 (Baniahmad et al., 1991).

In compliance with the division of the nuclear receptors into two subgroups according to the structure of their DNA-binding domain, the HREs can also be divided into two sugroups: the GRE/PRE subgroup and the ERE subgroup. The half palindrome of the GRE/PRE has the general structure TGTYCT, whereas the prototype ERE half is TGACC (see Figure 2b). We have recently shown that the last base of the half GRE/PRE, although highly conserved, is not essential for receptor binding or for functional activity (Truss et al., 1990). Therefore, the main differences between the two types of HREs reside in the third and fourth positions. The fourth position of the GRE/PRE is a C in 40% of the cases, and in transfection experiments this position tolerates any of the four bases without influencing glucocorticoid response (Nordeen et al.,

1990). Therefore, the most important distinction is the T in the third position of GRE/PRE as opposed to the A of the ERE (Klock et al., 1987). Moreover, the ERE of the rabbit uteroglobin promoter contains a C in the third position (Slater et al., 1990), suggesting that ER does not contact directly this position of the ERE. We know that GR and PR contact the 5'-methyl group of the T in the third position, but this base can be replaced by an A, provided that a T is located in the fourth position. It seems that the sequence context of the individual positions allows some flexibility in the way the DNA-binding domain of the receptors contacts the HRE. This flexibility is further documented by the observation that hybrid HREs composed of a half GRE/PRE linked to a half ERE are recognized by GR and PR as well as by ER, and respond to glucocorticoids, progestins, and estrogens, respectively, in transfection experiments (Truss et al., 1990). In this case we assume that a monomer of the receptor binds tightly to the "correct" half of the palindrome, the one corresponding to its recognition helix, whereas the other monomer binds weakly to the "incorrect" half and is stabilized by protein–protein interactions in the homodimer (see Figure 2c) (Dahlman-Wright et al., 1990).

The amino acid side chains responsible for the distinction between GRE/PRE and ERE are apparently located at the knuckle of the first zinc finger, between cysteines 3 and 4, and immediately downstream (see Figure 2a, helix 1) (Danielson et al., 1989; Green et al., 1988; Mader et al., 1989; Umesono and Evans, 1989). We have confirmed the significance of these amino acids in DNA-binding experiments. The ER point mutant HE84 [Glu to Gly in position 203 of the ER (Mader et al., 1989)] is able to recognize both a GRE and an ERE as well as a hybrid GRE/ERE, and in transfections experiments can transactivate promoters carrying any of these three HREs (Truss et al., 1990).

In addition to the amino acids at the knuckle of the first zinc finger, a few amino acid residues located between the cysteins 1 and 2 of the second finger (see Figure 2a, region D) are also important in specifying HRE target recognition, probably by dictating the precise orientation of receptor dimers (Danielsen et al., 1989). The three-dimensional structure of the zinc finger region of GR and of the ER have been elucidated in nuclear magnate resonance (NMR) studies (Härd et al., 1990; Schwabe et al., 1990), and the crystal structure of complexes of the GR bound to a GRE has been solved (Luisi et al., 1991). From these data it is clear that the amino acids at the knuckle of the first zinc finger that were identified as relevant for DNA sequence recognition in mutation experiments are oriented toward the major groove of the DNA double helix. In particular, the valine at position 442 of the rat GR (position 642 of the human GR) contacts the methyl group of the T at the third position of the half palindrome that is essential for HRE discrimination (Luisi et al., 1991). In addition to the identification of the amino acid residues that contact DNA, the crystal structural data confirm the location of the D loop at the interface between monomers in the DNA-bound homodimer (Dahlman-Wright et al., 1991; Luisi et al., 1991).

Modulation of the Receptor Function: Ligand Binding hsp90, and Phosphorylation

According to the classical two-step model of hormone action, ligand binding induces a conformational change in the receptor that enables the complex to translocate to the cell nucleus and to bind to chromatin (Jensen et al., 1968). A logical prediction of this model is that the ligand will be required for the receptor to bind to DNA. Contrary to this prediction, we found that the GR in crude liver cytosol was able to bind specifically to the HRE of MMTV in the absence of ligand or even in the presence of the antagonist RU486 (Willmann and Beato, 1986). This finding was confirmed by several other groups working with PR, ER, and thyroid hormone receptor. In kinetic experiments we found that binding of the agonist accelerates both the on-rate and the off-rate of the binding reaction between receptor and DNA (Schauer et al., 1989). This kinetic effect was observed both with nonspecific DNA and with DNA fragments carrying an HRE, and suggested that one of the roles of the hormone ligand could be to facilitate the scanning of genomic DNA in search of the correct HREs (Schauer et al., 1989).

These *in vitro* results were in apparent contradiction with a published report on DNA binding of the receptor *in vivo*. Using the technique of genomic footprinting with dimethyl sulfate (DMS), the HRE of the rat tyrosine amino transferase gene is found to be occupied in hepatocytes only after hormone administration (Becker et al., 1986). Although the identity of the factor causing the footprint over the HRE following induction has been questioned (Rigaud et al., 1991), these results suggest that *in vivo* the HRE is not occupied in the absence of hormone. There are two explanations for this finding that are not mutually exclusive. First, in agreement with our *in vitro* findings, the DNA-binding kinetics of the hormone-free receptor could be too slow to allow the protein to identify the HRE in the context of genomic DNA. Alternatively, the hormone-free receptor could be maintained in a conformation unable to translocate to the cell nucleus and to bind DNA *in vivo* because of its interaction with some other factor. Circumstantial evidence supporting the latter mechanism is derived from work on the heat shock protein hsp90.

The identification of the heat shock protein hsp90 as a component of the nonactivated steroid hormone receptor complex has been reported by different groups (for review, see Godowski and Picard, 1988). In agreement with a role of hsp90 in modulating the DNA-binding activity of the free receptors *in vivo*, the thyroid hormone receptor, which does not interact with hsp90 (Dalman et al., 1990), is bound to DNA in the absence of hormone. According to the prevalent model, binding of the hormone agonist will induce a conformational change in the receptor, leading to dissociation of the complex with hsp90 and other proteins, and to DNA binding (see Godowski and Picard, 1988). In addition, it seems that the association with hsp90 may even facilitate the binding of ligand to the receptor. This is suggested by the observation that bacterially expressed GR, free of hsp90, has a much lower affinity for the hormone ligand than the same protein expressed in the reticulocyte lysate that contains abundant hsp90 (Ohara-Nemoto et al., 1990). This observation may explain the fact that yeast strains carrying the GR but deficient in the yeast homolog of hsp90 show a lower response to hormone (Picard et al., 1990). However, in these yeast strains, no constitutive expression of reporter genes containing a GRE is observed, suggesting that in addition to dissociation from hsp90 the hormone ligand is actively needed for other steps in transactivation by the GR (Picard et al., 1990).

Preliminary attempts to map the region responsible for the interaction with hsp90 suggest a location in the carboxy-terminal half of the receptors, between the DNA-binding domain and the hormone-binding domain (Chambraud et al., 1990; Housley et al., 1990; Howard et al., 1990; Pratt et al., 1988). It seems that the antagonist RU486 is much less efficient in inducing dissociation of the receptor from hsp90, and this would explain in part its antihormonal activity. However, additional mechanisms must account for the antagonistic action of RU486, because the receptor–antagonist complex is unable to activate transcription *in vivo* or *in vitro* even after binding to the HRE (El-Ashry et al., 1989; Guiochon-Mantel et al., 1988; Kalff et al., 1990).

Another mechanism that influences receptor activity is phosphorylation. There are various reports claiming that phosphorylation of serines or threonines is needed for hormone binding (see Auricchio, 1989). The phosphorylation sites for the progesterone receptor have been mapped, and hormone binding not only increases phosphorylation at sites already phosphorylated before hormone administration but also induces a new phosphorylation site (Denner et al., 1990a). The estrogen receptor appears to require phosphorylation on tyrosine for optimal estrogen binding (Migliaccio et al., 1989, 1991), but there is no indication for similar requirements for any other of the steroid hormone receptors. The role of phosphorylation on dimerization, interaction with hsp90 (that is itself a phosphoprotein), nuclear translocation, DNA binding, or transactivation is unknown. In the case of the progesterone receptor there are indications that a cAMP-dependent protein kinase may enhance the ability of the hormone-free receptor to activate transcription (Denner et al., 1990b; Power et al., 1991). In addition, phosphorylation could play an important role in the process of receptor recycling after accomplishing its function in the cell nucleus.

The intracellular location of the unoccupied receptors and the mechanism of their nuclear translocation after hormone treatment have been the subject of intensive debate. Contrary to the original proposal (Jensen et al., 1968), the receptors for estrogen and progesterone appear to be already nuclearly located before hormone binding, whereas the glucocorticoid receptors show a mixed distribution between cytoplasm and nucleus in the absence of hormone and become predominantly nuclear after hormone treatment (see King, 1987). A shuttle mechanism between cytoplasm and nucleus has been described for the progesterone receptor that requires ATP and involves a continuous and active relocation of the hormone–receptor complex (Guiochon-Mantel et al., 1991), which could be regulated by protein kinases and phosphatases (DeFranco et al., 1991).

Mechanism of Hormonal Induction of MMTV

The initial identification of DNA elements able to mediate hormone induction in gene transfer experiments was accomplished with the MMTV system (for original references, see Beato, 1991). This system has also provided the first demonstration for specific binding of a steroid hormone receptor to DNA (Payvar et al., 1983; Scheidereit et al., 1983). Consequently, the interaction of GR and PR with the HRE of the MMTV was studied in great detail (Chalepakis et al., 1988b). In the GR mouse strain, there are four copies of the hexanucleotide motif TGTTCT between positions −190 and −75 of the MMTV promoter, and each of these motifs is contacted by GR or PR (Scheidereit and Beato, 1984). Although the upstream motif is part of an imperfect palindrome between −190 and −160, the other three motifs do not have a corresponding half with the correct spacing. Thus, a dimer of the receptor binds to the promoter-distal palindrome, but the exact stoichiometry of receptor binding to the promoter proximal sites is unclear (Chalepakis et al., 1988b; Perlmann et al., 1990; Wrange et al., 1986).

DNA Topology and Receptor Binding

The results of insertions between the distal and the proximal sites suggest that there is a strong functional cooperativity between receptor molecules bound to these two regions (Chalepakis et al., 1988b). This cooperativity is strongly dependent on the topology of the transfected plasmid (Piña et al., 1990c) and is not accompanied by corresponding changes in the affinity of the receptors for linear DNA fragments (Chalepakis et al., 1988b). We assume, therefore, that a functional interaction among receptor molecules requires bending or other deformations of the MMTV DNA that are favored by negative supercoiling (Piña et al., 1990c). A similar argument applies for the interaction between the hormone receptors and other transcription factors mediating induction of the adjacent promoter (Piña et al., 1990c).

A role for DNA topology on receptor binding to the MMTV-HRE has been suggested by binding experiments with plasmids of various topologies (Piña et al., 1990c) and confirmed using minicircles. We found that the affinity of the receptor is low for relaxed minicircles and increases with the degree of negative supercoiling, with an abrupt increase between topoisomers −2 and −3 (M. Truss and M. Beato, unpublished data). Interestingly, a similar topology-dependent transition is observed with respect to the sensitivity of the minicircles to the nuclease BAL 31. This suggests that a conformational transition is induced by supercoiling, which favors binding of the hormone receptor to the HRE. The reverse effect should therefore be predicted and has been observed, namely, that binding of the receptor favors a topological transition of plasmids carrying the MMTV-HRE (Carballo and Beato, 1990). Whether this alteration of DNA topology is directly related to the mechanism of transcriptional activation remains unknown, but it certainly offers an attractive possibility.

Involvement of Transcription Factors NFI and OTF-1

Do the receptors directly activate the basic transcriptional machinery or are other factors required? (Schatt et al., 1990; Strähle et al., 1988). Several reports had been published describing the participation of the transcription factor NFI in hormonal induction of the MMTV promoter (Brüggemeier et al., 1990, and references therein). In complementation experiments with NFI-deficient choriocarcinoma cell lines, we have demonstrated that NFI indeed acts as a transcription factor *in vivo* (Brüggemeier et al., 1990). However, *in vitro* binding of NFI to its cognate sequence between -75 and -63 on the MMTV promoter is not favored by binding of the hormone receptors. On the contrary, both proteins compete for DNA binding, in agreement with the overlap of their respective binding sites (Brüggemeier et al., 1990). Therefore, a simple mechanism involving DNA-binding cooperatively between hormone receptors and NFI is not in agreement with our findings.

In cell-free transcription experiments with nuclear extracts from HeLa cells, we have shown that addition of the PR to templates driven by the MMTV promoter enhances their transcriptional efficiency about 10 fold (Kalff et al., 1990). In this sytem, deletion of the NFI binding site, or addition of an excess of oligonucleotide carrying the NFI consensus sequence, reduces the basic expression of the MMTV promoter dramatically, but does not influence the stimulatory effect of preincubation with the PR (Kalff et al., 1990). These *in vitro* data confirmed the *in vivo* results and in addition showed that NFI acts as a basal transcription factor on the MMTV promoter, without any indication for synergism or cooperativity with the hormone receptors. We concluded that factors other than NFI must mediate the transactivation of the MMTV promoter by hormone receptors *in vitro*.

A search for other possible factors involved in transcriptional activation of the MMTV promoter led to the identification of two octamer motifs between the NFI binding sites and the TATA box. Mutations at these sites resulted in a significant reduction of the hormonal induction of the MMTV promoter in gene transfer experiments (Brüggemeier et al., 1991). More interestingly, these mutated promoters were completely unresponsive to the addition of PR *in vitro*, suggesting that the effect of PR on cell-free transcription of the MMTV promoter is actually mediated by an octamer binding factor. As OTF-1 (Oct1) is the main octamer binding factor in HeLa cells, and purified OTF-1 binds to the two octamer motifs of the MMTV promoter, we assume that the effect of PR on transcription of MMTV-DNA is mediated by OTF-1 (Brüggemeier et al., 1991). In DNase I footprinting experiments with purified proteins, we showed that binding of the glucocorticoid or the progesterone receptor to the HRE of MMTV facilitates the otherwise weak binding of OTF-1 to the two octamer motifs of the MMTV promoter (Brüggemeier et al., 1991). It seems therefore that the effect of added PR on the cell-free transcription of MMTV promoter is mediated by a direct protein–protein interaction between the hormone receptor and OTF-1.

Role of Chromatin Structure in Constitutive Repression: Promoter Occlusion and Its Relief

The lack of DNA binding cooperativity between hormone receptors and NFI was unexpected, because *in vivo* hormone treatment does induce NFI binding to the MMTV promoter (Cordingley et al., 1987) although there is the same amount of NFI in nuclear extracts from induced and control cells (Cordingley and Hager, 1988). On the other hand, the MMTV-LTR region is precisely organized into nucleosomes, and following hormone adminstration, a DNaseI-hypersensitive site appears over the HRE (Richard-Foy and Hager, 1987). On the basis of these findings one could postulate that the organization of the MMTV sequences in chromatin prevents its transcription. This idea is supported by the results of microinjection experiments in *Xenopus oocytes*, as inhibition of chromatin assembly by excess carrier DNA selectively enhances transcription from the coinjected MMTV promoter (Perlmann and Wrange, 1991).

It seemed possible, therefore, that changes in chromatin structure could mediate the effect of hormone treatment on NFI binding. To directly test this hypothesis we have performed nucleosome reconstitution experiments with the relevant region of the MMTV promoter. We confirmed the previous *in vivo* (Richard-Foy and Hager, 1987) and *in vitro* (Perlmann and Wrange, 1988) observations that a nucleosome core is precisely positioned over the region containing the HRE (Piña et al., 1990b). In addition, we found that the binding site for NFI is included in this nucleosome that extends from -198 to -45. The orientation of the NFI binding sequences is such that their major groove is pointing inward toward the histone octamer, making the site unaccessible for the protein (Figure 3). This prediction was confirmed experimentally with partially purified or recombinant NFI that binds very efficiently to naked

MMTV-DNA, but is unable to recognize the MMTV promoter organized in nucleosomes (Archer et al., 1991; Piña et al., 1990b). On the contrary, the hormone receptors do bind efficiently to reconstituted nucleosomes (Perlmann and Wrange, 1988; Piña et al., 1990b), with an affinity only four- to fivefold lower than their affinity for naked MMTV-DNA. This is in agreement with the prediction based on the structure of the reconstituted nucleosome, because in only two of the four TGTTCT motifs are the major grooves exposed to the exterior and accessible for receptor binding (Figure 3). In DNase I footprinting experiments, only the two receptor binding sites with major grooves oriented toward the exterior of the nucleosomes were protected by purified hormone receptors (Piña et al., 1990b). As seen in Figure 4, receptor molecules bound to these two sites (GRE1 and GRE4) should be in close proximity on the nucleosome enabling a direct protein–protein interaction.

Thus, it seems that the precise positioning of the DNA double helix on the surface of the histone octamer could account for the lack of binding of NFI before hormone treatment. In the absence of hormone, the promoter is inaccessible to NFI and silent *in vivo*. If the model is correct, and promoter occlusion results from nucleosome positioning, binding of receptor to the nucleosomally organized MMTV promoter should alter its structure and expose the recognition sequence of NFI. We have preliminary evidence supporting this concept. On binding of receptor, the nucleosome is not disassembled (Perlmann and Wrange, 1988), but the region containing the NFI binding site becomes more accessible to digestion by exonuclease III (Piña et al., 1990b). This model would also explain why transcription of the naked MMTV promoter *in vitro* is rather efficient even in the absence of hormone receptor (Kalff et al., 1990). In the *in vitro* transcription system the DNA is not organized into nucleosomes, and the repression from chromatin structure is not observed. Transcription of efficiently reconstituted chromatin will be needed to reproduce the derepression mechanism *in vitro*.

In further reconstitution experiments with different linear and circular DNA fragments of the MMTV promoter, we have tried to determine the factors responsible for the precise positioning of the nucleosome core on the DNA double helix. The results of DNaseI digestion of minicircles and the analysis of the degree of bending of linear DNA molecules suggest that it is the nucleotide sequence of the MMTV promoter itself that dictates the position of the DNA helix on the surface of the histone octamer (Piña et al., 1990a). Based on the differential compressibility of

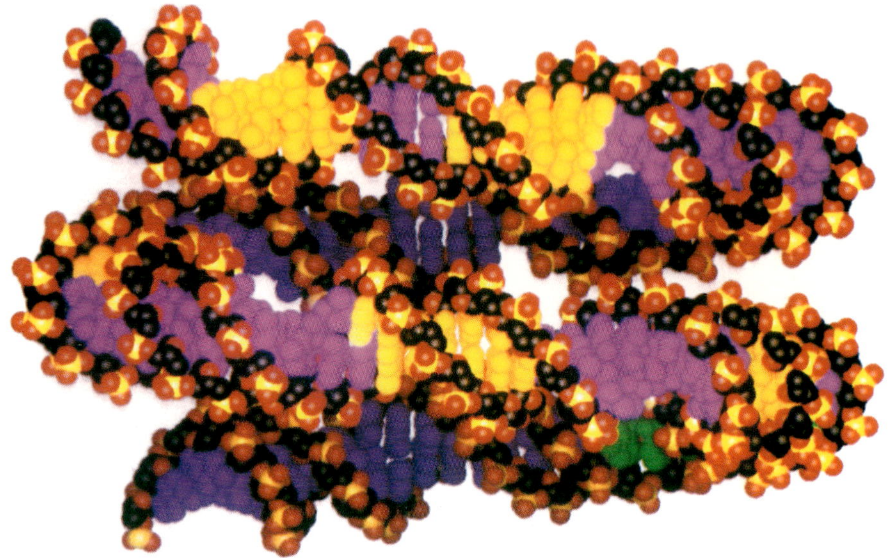

Figure 3

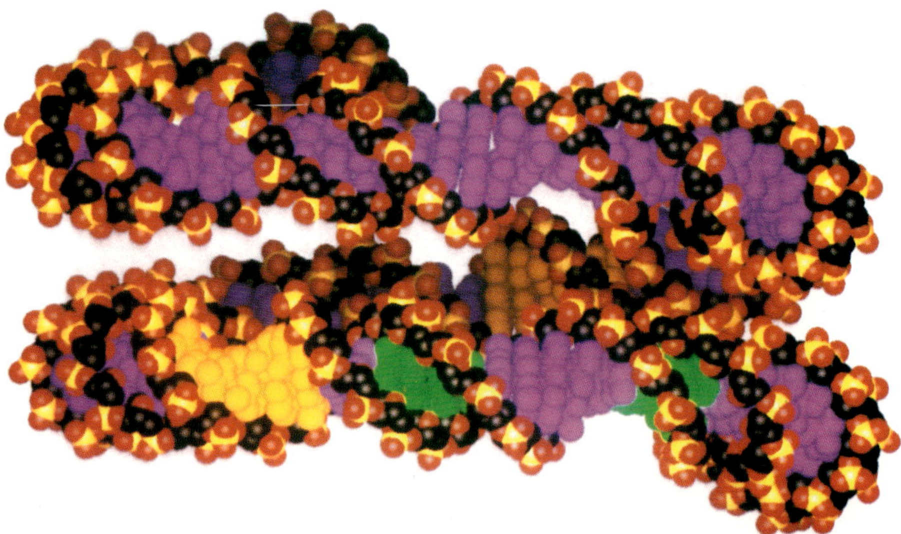

Figure 3   Path of the DNA double helix on the histone octamer. Computer graphic representation of the two DNA superhelical turns of the MMTV nucleosome B. The base pairs recognized by GR (GRE 1 to 4) are shown in yellow and those recognized by NFI are shown in green. The upper panel shows a view of the nucleosome with the exposed upstream receptor binding site; the lower panel has been turned to show the region of the accessible promoter proximal receptor binding site and the masked NFI binding site.

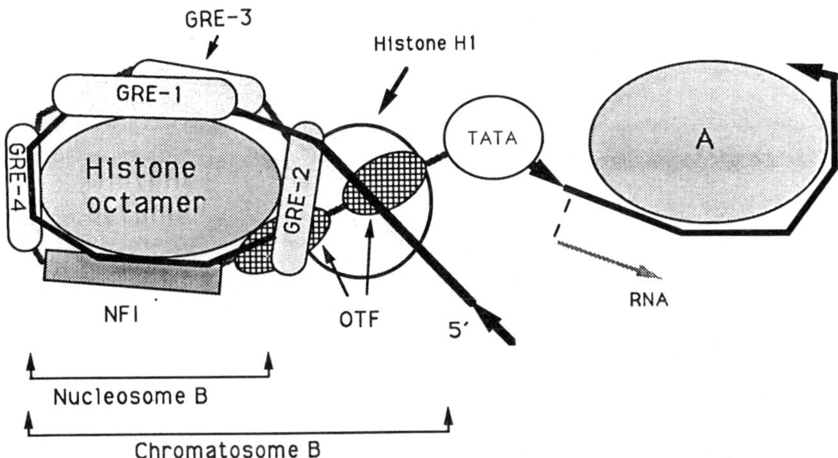

Figure 4   Relative orientation of binding sites for hormone receptors and transcription factors on the nucleosomally organized MMTV promoter. The two superhelical turns of the DNA double helix on nucleosome B and the beginning of nucleosome A are shown in a top view to indicate the relative orientation of the binding sites for hormone receptors (GRE 1 to 4), NFI, octamer transcription factors (OTF), and the TATA box. The position occupied by histone H1 is indicated.

the minor groove in different dinucleotides, it is possible to formulate an algorithm that predicts the preferred path of the DNA helix (Drew and Calladine, 1987; Piña et al., 1990d; Satchwell et al., 1986). The three-dimensional structure adopted by the DNA in chromatin seems to be an intrinsic property of the nucleotide sequence that should be amenable to genetic analysis.

The concept that nucleosome structure determines the accessibility of target sites for transcription and regulatory factors is appealing because it offers a simple explanation for the inheritance of a phenotype during cell division. How general could this mechanism be? There are several other systems in which a detailed analysis of regulatory regions has provided evidence for their precise organization into nucleosomes and for specific alterations of chromatin structure following induction. These include the PHO5 gene of yeast, the heat shock genes of *Drosophila*, and the tyrosine amino transferase gene of rat liver. The 5'-regulatory region of the PHO5 gene contains five positioned nucleosomes, two of them flanking a DNaseI-hypersensitive region. In the linker between these two nucleosomes there is a UAS that is recognized by a protein factor on low phosphate induction. This leads to destabilization of two nucleosomes on each side, exposing other UASs and leading to the formation of a stable transcription complex (Almer and Hörz, 1986; Almer et al., 1986). In the promoter region of the hsp26 gene of *Drosophila*, a precisely positioned nucleosome brings two binding sites for the heat shock activation factor into the correct orientation to enable efficient binding and interaction with the TATA box binding factor (Elgin, 1988). In the tyrosine amino transferase gene, the region around -2.4 kb contains an array of HREs that have been shown to be important for induction by glucocorticoids. This region is organized into precisely positioned nuclesomes and becomes hypersensitive to DNaseI after hormone treatment (Becker et al., 1984; Carr and Richard-Foy, 1990). Maintainance of the DNaseI-hypersensitive site requires the continuous presence of the hormone and correlates with transcriptional activity of the promoter (Reik et al., 1991). A similar mechanism may operate in the rabbit uteroglobin gene; here again is an hormone-dependent DNaseI-hypersensitive region at around -2.4 kb that overlaps a cluster of binding sites for the hormone receptors (Jantzen et al., 1987). On induction, not only the region overlapping the HRE, but also a region in the promoter where no binding sites for PR are found, become DNaseI hypersensitive. Thus, the possibility exists that binding of transcription factors in the proximity of the upstream HRE induces a long-distance effect on the promoter, an effect that could be mediated by protein–protein interaction on chromatin.

The general validity of this gene regulation mechanism is suggested from experiments with yeast strains carrying mutated histone genes or an altered stoichiometry of core histones leading to altered nucleosome structure (Han and Grunstein, 1988). Although the mutated cells are still viable, they show alterations in several regulatory pathways including GAL4, mating type control, and PHO5. In strains with low levels of histone H4 that exhibit nucleosome depletion, the chromatin structure of the PHO5 regulatory region is distorted, and this is accompanied by expression of the PHO5 gene, even under conditions of repression at high phosphate (Han et al., 1988). In this and other regulated genes, transcriptional stimulation by nucleosome depletion is observed even in the absence of UAS, suggesting that one of the functions of the UAS complex is to remove repression caused by chromatin structure (Han and Grunstein, 1988). Mutant yeast strains carrying deletions in the conserved N-terminal region of histone H4 exhibit alterations of both repression (Kayne et al., 1988) and induction of gene activity (Durrin et al., 1991). A similar phenotype has been detected in independently selected mutants of yeast (Natsoulis et al., 1991; Vidal and Gaber, 1991; Vidal et al., 1991), and could reflect the existence of general factors that are involved in mediating the effect of chromatin structure on gene expression (Kornberg and Lorch, 1991).

In the foregoing discussion, no mention has been made of histone H1 and its possible role in regulating access to DNA sequences. However, the region recognized by the transcription factor OTF-1 lies in the linker DNA between nucleosomes B and A, the same region that will interact with histone H1 (see Figure 4). There is evidence that histone H1 may also influence the precise positioning of the DNA double helix on the surface of the nucleosome (Meersseman et al., 1991), and that transcription of free DNA and chromatin templates is inhibited by addition of histone H1 (Croston et al., 1991; Laybourn and Kadonaga, 1991; Shimamura et al., 1989; Wolffe, 1989). Moreover, histone H1 will also contact the dyad axis of the nucleosome B around -113, in a region that is important for binding and function of the hormone receptors (see Figure 4). It is therefore possible that histone H1 plays an important role in masking accessibility of the promoter sequences for various factors, and removal of histone H1 may be a prerequisite for further interactions with the nucleosome core.

Impairment of Hormonal Induction by Inhibitors of Protein Synthesis

In many systems where steroid hormones induce the expression of a particular gene, this effect can be prevented by simultaneous treatment with inhibitors of protein synthesis. This observation has been accepted as an indication for an indirect effect on gene expression requiring the synthesis of an intermediary labile factor. Alternatively, it has been interpreted as suggestive of posttranscriptional effect of hormones. The induction of α-1-acidic glycoprotein (AGP) by glucocorticoids in rat liver is a typical example (Reinke and Feigelson, 1985). Contrary to the induction of MMTV expression the effect of dexamethasone on transcription of the α-1-acidic glycoportein gene can be inhibited by cycloheximide treatment and is not detectable in run-on transcription assays (Vannice et al., 1984). However, analysis of the α-1-acidic glycoportein promoter shows that it contains sequences to which the glucocorticoid receptor binds selectively (Baumann and Maquat, 1986) and that mutation of these sequences eliminates induction (Klein et al., 1988). In addition to this GRE, flanking sequences are important for hormonal induction (Klein et al., 1988), and these sequences bind the AGP/EBP transcription factor

(Williams et al., 1991). Thus, the inhibition of hormonal induction by cycloheximide could reflect the lability of a second protein factor required for promoter expression rather than an indirect induction mechanism. If this explanation applies, there will be, in principle, no difference in the mechanism of induction of AGP and MMTV, as in both cases induction is mediated by factors binding near the HRE. The only difference that determines sensitivity to inhibitors of protein synthesis is the half-life of the auxiliary factor. The general validity of this mechanism remains to be established, but binding sites for hormone receptors have been described in other genes exhibiting the phenomenon of delayed induction (Chan et al., 1991; Hess et al., 1990).

Mechanism of Transcriptional Repression

Steroid hormones are also able to repress the expression of specific genes. This is particularly important for glucocorticoids that are used clinically because of their suppressive effect on inflammation and immune response. Cloning of the receptors demonstrated that for most steroid hormones a single class of receptor molecules mediates all biological effects. The question arose whether the same type of DNA regulatory elements that mediates gene induction is also responsible for repression.

Among the first glucocorticoid-repressed genes to be studied was the gene for the α-subunit of the glycoprotein hormones. The expression of this gene in pituitary and placental cells is stimulated by cAMP and is inhibited by glucocorticoids at the transcriptional level (Akerblom et al., 1988, and references therein). Induction by cAMP is mediated by two copies of the cAMP regulatory element that are recognized by a specific protein called CREB (Delegeane et al., 1987; Deutsch et al., 1987; Jameson et al., 1987; Silver et al., 1987). In transfection experiments the negative effect of glucocorticoid is only observed with cells in which the cAMP regulatory element is active, whereas in cells not responding to cAMP glucocorticoids have a stimulatory effect on the α-subunit promoter (Akerblom et al., 1988). These findings suggest that the effect of glucocorticoids is to counteract the induction by cAMP. In agreement with this concept, DNA binding experiments with purified GR identified three binding sites on the α-subunit promoter, two of which overlap with the cAMP responsive elements (Akerblom et al., 1988). Moreover, DNA binding seems to be the only required function of GR; the amino-terminal and carboxy-terminal halves can be removed or replaced by β-galactosidase without impairing repression (Oro et al., 1988). Thus, in this particular case, glucocorticoid repression could result from competition by GR and positive factors for overlapping target sites on the promoter. A similar mechanism seems to apply for glucocorticoid inhibition of the cAMP-inducible proopiomelanocortin gene (Drouin et al., 1987). How binding competition takes place is not known, but one should keep in mind that efficient binding of GR requires sequences flanking the HRE (Chalepakis et al., 1990). Occupancy of these sequences would generate steric hindrance, a mechanism that could also apply to the competition between GR and NFI on the MMTV promoter (see the foregoing discussion). There are other negatively regulated promoters in which binding sites for the hormone receptors have been identified, but the mechanism of transrepression remains obscure (Sakai et al., 1988).

An apparently different mechanism of repression is operative in the glucocorticoid inhibition of AP1-mediated induction (for review, see Miner and Yamamoto, 1991; Schüle and Evans, 1991). Although there is some controversy as to the requirement of the DNA-binding domain of GR, it seems clear that binding of GR to DNA is not essential for transrepression. Rather, GR seems to inhibit the activity of the Fos–Jun complex by an interaction with the AP1 complex (Jonat et al., 1990). This idea is supported by the reverse effect, namely the inhibition of GR-mediated induction of a GRE by Fos, in the absence of Fos binding to the GRE (Lucibello et al., 1990; Schüle et al., 1990; Yang-Yen et al., 1990). One possible mechanism to explain this type of repression is a postulated heterodimerization between a receptor monomer and other components of the AP1 complex. Curiously enough the outcome of the interaction can be a reciprocal inhibition or a potentiation, depending on the cell type and the promoter used (Diamond et al., 1990; Shemshedini et al., 1991; Yoshinaga and Yamamoto, 1991). This type of cross-talk between nuclear receptors and nuclear factors could also be involved in other signal transduction pathways, and probably exemplifies a common behavior of transregulatory proteins.

Another mechanism that has been postulated to explain repression by hormone receptors is the phenomenon of "squelching" (Gill and Ptashne, 1988). This interesting idea, introduced by Marc Ptashne, holds that transactivators acting through some common factor can compete for limiting amounts of the factor in solution. Therefore, an excess of a particular transactivator would inhibit, or "squelch," its own transactivation as well as transactivation by other regulators that use the same intermediate. While this mechanism can be readily demonstrated in transient transfection experiments, its role in physiological regulation remains to be established. There are indications that it could also participate in the repression of the alpha subunit gene mentioned (Chatterjee et al., 1991).

Conclusions

We have seen that the precise positioning of the MMTV DNA double helix on the surface of the histone octamer is an intrinsic property of the nucleotide sequence. As this positioning has profound consequences on the accessibility and function of the promoter, we conclude that, like proteins, DNA has a tertiary structure that determines part of its function. This implies that in addition to coding and regulatory information there is conformational information in DNA that modulates the accessibility of regulatory information and is therefore critical for the realization of the genetic program. Appropriate genetic manipulations should help to evaluate the significance of this conformational information.

The nucleosome structure described here suggests the existence of two types of DNA-binding regulatory proteins: those that are able to interact with nucleosomally organized DNA, and those that are not. Provided the DNA remains in the B form, if a protein needs to contact more than five consecutive base pairs through the major groove it will be unable to "see" its target sequences organized in a nucleosome, independently of their precise rotational positioning; no matter the phase of the double helix, a part of the recognized major groove will be masked. The same will apply if the protein contacts two sets of short sequences that are not on the same face of the double helix. On the contrary, a protein that contacts short stretches of fewer than five base pairs located on one side of the double helix will be able to recognize its cognate sequence on nucleosomes, provided the major grooves are properly oriented.

The available data on the regulated transcription of the MMTV promoter suggest that there are at least two different, and possibly alternative, mechanisms by which steroid hormones induce promoter activation. One follows a pathway similar to that reported for prokaryotic transactivators (Ptashne and Gann, 1990). According to this model, on binding to the HRE the hormone receptors facilitate the interaction of OTF-1 with the two octamer motifs of the MMTV promoter. Whether OTF-1 acts directly on the TATA box binding complex or alternatively requires an adaptor with an acidic transactivation domain is not completely clear. The second mechanism involves removal of promoter repression caused by nucleosome positioning, and is mediated mainly by NFI. In this mechanism there is no direct interaction between the hormone receptors and the responsible transcription factor, and the only function of the receptor may be to remodel chromatin. The two pathways may overlap, as it is not clear to what extent binding of OTF-1 to the MMTV promoter is modulated by its organization in nucleosomes. The stronger distal octanucleotide motif is located between −50 and −43, exactly at the 3′-border of the nucleosome covering the HRE (see Figure 4). According to our prediction, it should be unaccessible to OTF-1. It is therefore conceivable that receptor binding may be required *in vivo* to allow acess not only to NFI but also to OTF-1.

A comprehensive description of the induction process requires knowledge of the chromatin organization of the MMTV promoter at the nucleotide level during the different phases of hormonal induction *in vivo*. The sequential formation of protein–DNA complexes could be eventually followed after appropriate cross-linking by short pulses of UV light. Ultimately, a precise understanding of the transactivation mechanism will only be possible when correctly reconstituted chromatin templates are successfully transcribed *in vitro* using purified receptors, NFI, OTF-1, TFIID and accessory factors, and RNA polymerase II. Recent progress in chromatin reconstitution techniques and the availability of most of the factors in cloned form have brought us closer to this goal.

*Acknowledgments.*I thank Emily P. Slater for carefully reading the manuscript. The experimental work summarized in this review was supported by grants from the Deutsche Forschungsgemeinschaft and the Fonds der Chemischen Industrie.

References

Akerblom IW, Slater EP, Beato M, Baxter JD, Mellon PL (1988): Negative regulation by glucocorticoids through interference with a cAMP responsive enhancer. *Science* 241:350–353.

Almer A, Hörz W (1986): Nuclease hypersensitive regions with adjacent positioned nucleosomes mark the gene boundaries of the PHO5/PHO3 locus in yeast. *EMBO J* 5:2681–2687.

Almer A, Rudolph H, Hinnen A, Hörz W (1986): Removal of positioned nucleosomes from the yeast PHO5 promoter upon induction releases additional

66   M. Beato

upstream activating DNA elements. *EMBO J* 5:2689–2696.

Archer TK, Cordingley MG, Wolford RG, Hager GL (1991): Transcription factor access is mediated by accurately positioned nucleosomes on the mouse mammary tumor virus promoter. *Mol Cell Biol* 11:688–698.

Archer TK, Hager GL, Omichinski JG (1990): Sequence-specific DNA binding by glucocorticoid receptor "zinc finger peptides." *Proc Natl Acad Sci USA* 87:7560–7564.

Auricchio F (1989): Phosphorylation of steroid receptors. *J Steroid Biochem* 32:613–622.

Baniahmad C, Muller M, Altschmied J, Renkawitz R (1991): Co-operative binding of the glucocorticoid receptor DNA binding domain is one of at least two mechanisms for synergism. *J Mol Biol* 222:155–165.

Baumann H, Maquat LE (1986): Localization of DNA sequences involved in dexamethasone-dependent expression of the alpha$_1$-acid gycoprotein gene. *Mol Cell Biol* 7:2551–2561.

Beato M (1989): Gene regulation by steroid hormones. *Cell* 56:335–344.

Beato M (1991): Transcriptional regulation of mouse mammary tumor virus by steroid hormones. *Crit Rev Oncogen* 2:195–210.

Beato M, Arnemann J, Menne C, MAller H, Suske G, Wenz M (1983): Regulation of the expression of the uteroglobin gene by ovarian hormones. In: *Regulation of Gene Expression by Hormones*, McKerns KW, ed. pp. 151–175, New York: Plenum.

Becker P, Renkawitz R, Schütz G (1984): Tissue-specific DNaseI hypersensitive sites in the 5'-flanking sequences of the tryptophan oxygenase and tyrosine aminotransferase genes. *EMBO J* 3:2015–2020.

Becker PB, Gloss B, Schmid W, Strähle U, Schütz G (1986): *In vivo* protein-DNA interactions in a glucocorticoid response element require the presence of the hormone. *Nature* (Lond) 324:686–688.

Bocquel MT, Kumar V, Stricker C, Chambon P, Gronemeyer H (1989): The contribution of the N- and C-terminal regions of steroid receptors to activation of transcription is both receptor and cell-specific. *Nucleic Acids Res* 17:2581–2595.

Brüggemeier U, Kalff M, Franke S, Scheidereit C, Beato M (1991): Ubiquitous transcription factor OTF-1 mediates induction of the mouse mammary tumor virus promoter through synergistic interaction with hormone receptors. *Cell* 64:565–572.

Brüggemeier U, Rogge L, Winnacker EL, Beato M (1990): Nuclear factor I acts as a transcription factor on the MMTV promoter but competes with steroid hormone receptors for DNA binding. *EMBO J* 9:2233–2239.

Carballo M, Beato M (1990): Binding of the glucocorticoid receptor induces a topological changes in plasmids containing the hormone-responsive element of mouse mammary tumor virus. *DNA Cell Biol* 9:519–521.

Carlstedt-Duke J, Strömstedt Persson PEB, Cederlund E, Gustafsson JA, Jörnvall H (1988): Identification of hormone-interacting amino acid residues within the steroid-binding domain of the blucocorticoid receptor in relation to other steroid hormone receptors. *J Biol Chem* 263:6842–6846.

Carr KD, Richard-Foy H (1990): Glucocorticoids locally disrupt an array of positioned nucleosomes on the rat amino transferase promoter in hepatoma cells. *Proc Natl Acad Sci USA* 87:9300–9304.

Chalepakis G, Postma JPM, Beato M (1988a): A model for hormone receptor binding to the mouse mammary tumor virus regulatory element based on hydroxyl radical footprinting. *Nucleic Acids Res* 16:10237–10247.

Chalepakis G, Arnemann J, Slater EP, Brüller HJ, Gross B, Beato M (1988b): Differential gene activation by glucocorticoids and progestins through the hormone regulatory element of mouse mammary tumor virus.   *Cell* 53: 371–382.

Chalepakis G, Schauer M, Cao X, Beato M (1990): Efficient binding of glucocorticoid receptor to its responsive element requires a dimer and DNA flanking sequences. *DNA Cell Biol* 9:355–368.

Chambraud B, Berry M, Redeuilh G, Chambon P, Baulieu EE (1990): Several regions of human estrogen receptor are involved in the formation of receptor-heat shock protein 90 complexes. *J Biol Chem* 265:20686–20691.

Chan GCK, Hess P, Meenakshi T, Carlstedtduke J, Gustafsson JA, Payvar F (1991): Delayed secondary glucocorticoid response elements—unusual nucleotide motifs specify glucocorticoid receptor binding to transcribed regions of α-2u-globulin DNA. *J Biol Chem* 266:22634–22644.

Chatterjee VKK, Madison LD, Mayo S, Jameson JL (1991): Repression of the human glycoprotein hormone α-subunit gene by glucocorticoids: Evidence for receptor interaction with limiting transcriptional activators. *Mol Endocrinol* 5:100–110.

Conneely O, Sullivan WP, Toft DO, Birnbaumer M, Cook RG, Maxwell BL, Zarucki-Schulz T, Greene GL, Schrader WT, O'Malley BW (1986): Molecular cloning of the chicken progesterone receptor. *Science* 233:767–770.

Cordingley MG, Hager GL (1988): Binding of multiple factors to the MMTV promoter in crude and fractionated nuclear extracts. *Nucleic Acid Res* 16:609–630.

Cordingley MG, Riegel AT, Hager GL (1987): Steroid-dependent interaction of transcription factors with the inducible promoter of mouse mammary tumor virus *in vivo*. *Cell* 48:261–270.

Croston GE, Kerrigan LA, Lira LM, Marshak DR, Kadonaga JT (1991): Sequence-specific antirepression of histone H1-mediated inhibition of basal RNA polymerase II transcription. *Science* 251:643–649.

Dahlman-Wright K, Siltala-Roos H, Carlstedt-Duke J, Gustafsson JA (1990): Protein-protein interactions facilitate DNA binding by the glucocorticoid receptor DNA-binding domain. *J Biol Chem* 265:14030–14035.

Dahlman-Wright K, Wright A, Gustafsson, J-A, Carlstedt-Duke J (1991): Interaction of the glucocorticoid receptor DNA-binding domain with DNA as a dimer is mediated by a short segment of five amino acids. *J Biol Chem* 266:3107–3112.

Dalman FC, Koenig RJ, Perdew GH, Massa E, Pratt WB (1990): In contrast to the glucocorticoid receptor, the thyroid hormone receptor is translated in the DNA binding state and is not associated with hsp90. *J Biol Chem* 265:2615–2618.

Danielsen M, Hinck L, Ringold GM (1989): Two amino acids within the knuckle of the first zinc finger specify DNA response element activation by the glucocorticoid receptor. *Cell* 57:1131–1138.

DeFranco DB, Qi M, Borror KC, Garabedian MJ, Brautigan DL (1991): Protein phosphatase types 1 and/or 2A regulated nucleocytoplasmic shuttling of glucocorticoid receptors. *Mol Endocrinol* 5:1215–1228.

Delegeane AM, Ferland LH, Mellon PL (1987): Tissue-specific enhancer of the human glycoprotein hormone α-subunit gene: Dependence on cyclic AMP-inducible elements. *Mol Cell Biol* 7:3994–4002.

Denner LA, Scharder WT, O'Malley BW, Weigel NL (1990a): Hormonal regulation and identification of chicken progesterone receptor phosphorylation sites. *J Biol Chem* 265:16548–16555.

Denner LA, Weigel NL, Maxwell BL, Schrader WT, O'Malley BW (1990b): Regulation of progesterone receptor-mediated transcription by phosphorylation. *Science* 250:1740–1743.

Deutsch PJ, Jameson JL, Habener JF (1987): Cyclic AMP responsiveness of human gonadotropin-α gene transcription is directed by a repeated 18-base pairs enhancer. *J Biol Chem* 262:12169–12174.

Diamond MI, Miner JN, Yoshinaga SK, Yamamoto KR (1990): Transcription factor interactions: Selectors of positive or negative regulation from a single DNA element. *Science* 249:1266–1272.

Drew HR, Calladine CR (1987): Sequence-specific positioning of core histones on an 860 base-pair DNA. *J Mol Biol* 195:143–173.

Dreyer C, Krey G, Keller H, Givel F, Helftenbein G, Wahli W: Control of the peroxisomal β-oxidation of a novel family of nuclear hormone receptors. *Cell* 68:879–887.

Drouin J, Charron J, Gagner JP, Jeannotte L, Nemer M, Plante RK, Wrange (1987): The pro-opiomelanocortin gene: A model for negative regulation of transcription by glucocorticoids. *J Cell Biochem* 35:293–304.

Durrin LK, Mann RK, Kayne PS, Grunstein M (1991): Yeast histone H4 N-terminal sequence is required for promoter activation in vivo. *Cell* 65:1023–1031.

El-Ashry D, Oñate SA, Nordeen SK, Edwards DP (1989): Human progesterone receptor complexed with antagonist RU486 binds to hormone response elements in a structurally altered form. *Mol Endocrinol* 3:1545–1558.

Elgin SCR (1988): The formation and function of DNaseI hypersensitive sites in the process of gene activation. *J Biol Chem* 263:19259–19262.

Evans RM (1988): The steroid and thyroid hormone receptor superfamily. *Science* 240:889–895.

Fawell SE, Lees JA, White R, Parker MG (1990): Characterization and colocalization of steroid binding and dimerization activities in the mouse estrogen receptor. *Cell* 60:953–962.

Forman BM, Samuels HH (1990): Dimerization among nuclear hormone receptors. *New Biol* 2:587–594.

Gill G, Ptashne M (1988): Negative effect of the transcriptional activator GAL4. *Nature* (Lond) 334:721–724.

Godowski PJ, Picard D (1989): Steroid receptors. How to be both a receptor and a transcription factor. *Biochem Pharmacol* 38:3135–3143.

Green S, Kumar V, Theulaz I, Wahli W, Chambon P (1988): The N-terminal DNA-binding "zinc finger" of the oestrogen and glucocorticoid receptors determines target gene specificity. *EMBO J* 7:3037–3044.

Guiochon-Mantel A, Lescop P, Christinmaitre S, Loosfelt H, Perrotapplanat M, Milgrom E (1991): Nucleocytoplasmic shuttling of the progesterone receptor. *EMBO J* 10:3851–3859.

Guiochon-Mantel A, Loosfelt H, Ragot T, Bailly A, Atger M, Misrahi M, Perricaudet M, Milgrom E (1988): Receptors bound to antiprogestin form abortive complexes with hormone responsive elements. *Nature* (Lond) 336:695–698.

Han K, Grunstein M (1988): Nucleosome loss activates yeast downstream promoters *in vivo*. *Cell* 55:1137–1145.

Han K, Kim UJ, Kayne P, Grunstein M (1988): Depletion of histone H4 and nucleosomes activates the PHO5 gene in *Saccharomyces cerevisiae*. *EMBO J* 7:2221–2228.

Härd T, Kellenbach E, Boelens R, Maler BA, Dahlman K, Freedman LP, Carlsted-Duke J, Yamamoto KR, Gustafsson J, Kaptein R (1990): Solution structure of the glucocorticoid receptor DNA-binding domain. *Science* 249:157–160.

Hess P, Meenakshi T, Chan GCK, Carlstedt-Duke J, Gustafsson JA, Payvar F (1990): Purified glucocorticoid receptors bind selectively in vitro to a cloned DNA fragment that mediates a delayed secondary response to glucocorticoids in vivo. *Proc Natl Acad Sci USA* 87:2564–2568.

Hollenberg SM, Evans RM (1988): Multiple and cooperative trans-activation domains of the human glucocorticoid receptor. *Cell* 55:899–906.

Housley P, Sanchez E, Danielsen M, Ringold G, Pratt W (1990): Evidence that the conserved region in the steroid binding domain of the glucocorticoid receptor is required for both optimal binding of hsp90 and protection from proteolytic cleavage. *J Biol Chem* 265:12778–12781.

Howard KJ, Holley SJ, Yamamoto KR, Disselhorst CW (1990): Mapping the hsp90 binding region of the glucocorticoid receptor. *J Biol Chem* 265:11928–11935.

Issemann I, Green S (1990): Activation of a member of the steroid hormone receptor superfamily by peroxisome proliferators. *Nature* (Lond) 347:645–650.

Jameson JL, Deutsch PJ, Gallagher GD, Jaffe RC, Habener JF (1987): Trans-acting factors interact with a cAMP response element to modulate expression of the human gonadotropin a gene. *Mol Cell Biol* 7:3032–3040.

Jantzen C, Fritton HP, Igo-Kemenes T, Espel E, Janich S, Cato ACB, Mugele K, Beato M (1987): Partial overlapping of binding sequences for steroid hormone receptors and DNaseI hypersensitive sites in the rabbit uteroglobin gene region. *Nucleic Acid Res* 15:4535–4552.

Jensen EV, Suzuki T, Kawashima T, Stumpf WE, Jungblut PW, DeSombre ER (1968): A two-step mechanism for the interaction of estradiol with rat uterus. *Proc Natl Acad Sci USA* 59:632–636.

Jonat C, Rahmsdorf H, Park K, Cato A, Gebel S, Ponta H, Herrlich P (1990): Antitumor promotion and antiinflamation: Down-regulation of AP-1 (Fos/Jun) activity by glucocorticoid hormone. *Cell* 62:1189–1204.

Kalff M, Gross B, Beato M (1990): Progesterone receptor stimulates transcription of mouse mammary tumor virus in a cell-free system. *Nature* (Lond) 344:360–362.

Kastner P, Krust A, Turcotte B, Stropp U, Tora L, Gronemeyer H, Chambon P (1990): Two distinct estrogen-regulated promoters generate transcripts encoding the two functionally different human progesterone receptor forms A and B. *EMBO J* 9:1603–1624.

Kayne PS, Kim UJ, Han M, Mullen JR, Yoshizaki F, Grunstein M (1988): Extremely conserved histone H4 N terminus is dispensable for growth but essential for repressing the silent mating loci of yeast. *Cell* 55:27–39.

King RJB (1987): Structure and function of steroid receptors. *J Endocrinol* 114:341–349.

Klein ES, DiLorenzo D, Posseckert G, Beato M, Ringold GM (1988): Sequences downstream of the glucocorticoid regulatory element mediate cycloheximide inhibition of steroid induced expression from the rat α_1-acid glycoprotein promoter: Evidence for a labile transcription fact. *Mol Endocrinol* 2:1343–1351.

Klock G, Strähle U, Schütz G (1987): Oestrogen and glucocorticoid responsive elements are closely related but distinct. *Nature* (Lond) 329:734–736.

Kornberg RD, Lorch Y (1991): Irresistible force meets immovable object: Transcription and the nucleosome. *Cell* 67:833–836.

Kumar V, Chambon P (1988): The estrogen receptor binds tightly to its responsive element as a ligand-induced homodimer. *Cell* 55:145–156.

Laybourn PJ, Kadonaga JT (1991): Role of nucleosomal cores and histone-H1 in regulation of transcription by RNA polymerase-II. *Science* 254:238–245.

Lees J, Fawell S, White R, Parker M (1990): A 22-amino-acid peptide restores DNA-binding activity to dimerization-defective mutants of the estrogen receptor. *Mol Cell Biol* 10:5529–5531.

Loosfelt H, Atger M, Misrahi M, Guiochon-Mantel A, Meriel C, Logeat F, Benarous R, Milgrom E (1986): Cloning and sequence analysis of rabbit progesterone receptor complementary DNA. *Proc Natl Acad Sci USA* 83:9045–9049.

Lucibello FC, Slater EP, Jooss KU, Beato M, Müller R (1990): Mutual transrepression of Fos and the glucocorticoid receptor: Involvement of a functional domain in Fos which is absent in FosB. *EMBO J* 9:2827–2834.

Luisi BF, Xu WX, Otwinowski Z, Freedman LP. Yamamoto KR, Sigler PB (1991): Crystallographic analysis of the interaction of the glucocorticoid receptor with DNA. *Nature* (Lond) 352:497–505.

Mader S, Kumar V, Verneuil H, Chambon P (1989): Three amino acids of the oestrogen receptor are essential to its ability to distinguish an oestrogen from a glucocorticoid-responsive element. *Nature* (Lond) 338:271–274.

Meersseman G, Pennings S, Bradbury M (1991): Chromatosome positioning on assembled long chromatin. Linker histones affect nucleosome placement on 5 S rDNA. *J Mol Biol* 220:89–100.

Miesfeld R, Godowski PJ, Maler BA, Yamamoto KR (1987): Glucocorticoid receptor mutants that define a small region sufficient for enhancer activation. *Science* 236:423–427.

Migliaccio A, Castoria G, Falco AD, DiDomenica M, Galdiero M, Nola E, Chambon P, Auricchio F (1991): In vitro phosphorylation and hormone binding activation of the synthetic wild type human estradiol receptor. *J Steroid Biochem Mol Biol* 38:407–413.

Migliaccio A, DiDomenico M, Green S, Falco A, Kajtaniak EL, Blasi F, Chambon P, Auricchio F (1989): Phosphorylation on tyrosine of in vitro synthesized human estrogen receptor activates its hormone binding. *Mol Endocrinol* 3:1061–1069.

Miller J, McLachlan AD, Klug A (1985): Repetitive zinc-binding domains in the protein transcription factor IIIA from *Xenopus* oocytes. *EMBO J* 4:1609–1614.

Miner JN, Yamamoto KR (1991): Regulatory crosstalk at composite response elements. *Trends Biochem Sci* 16:423–426.

Natsoulis G, Dollard C, Winston F, Boeke JD (1991): The products of the *SPT10* and *SPT21* genes of *Saccharomyces cerevisiae* increase the amplitude of transcriptional regulation at a large number of unlinked loci. *New Biol* 3:1249–1259.

Nordeen SK, Suh BJ, Kühnel B, Hutchison C (1990): Structural determinants of a glucocorticoid receptor recognition element. *Mol Endocrinol* 4:1866–1873.

Ohara-Nemoto Y, Strömstedt P-E, Dahlman-Wright K, Nemoto T, Gustafsson JA, Carlstedt-Duke J (1990): The steroid-binding properties of recombinant glucocorticoid receptor: A putative role for heat shock protein hsp90. *J Steroid Biochem* 37:481–490.

O'Malley BW (1990): The steroid receptor superfamily: More excitement predicted for the future. *Mol Endocrinol* 4:363–369.

Oro AE, Hollenberg SM, Evans RM (1988): Transcriptional inhibition by a glucocorticoid receptor-β-galactosidase fusion protein. *Cell* 55:1109–1114.

Payvar F, DeFranco D, Firestone GL, Edgar B, Wrange Okret S, Gustafsson JA, Yamamoto KR (1983): Sequence-specific binding of glucocorticoid receptor to MMTV DNA at sites within and upstream of the transcribed region. *Cell* 35:381–392.

Perlmann T, Erikson P, Wrange Ö (1990): Quantitative analysis of the glucocorticoid receptor-DNA interaction at the mouse mammary tumor virus glucocorticoid response element. *J Biol Chem* 265:17222–17229.

Perlmann T, Wrange Ö (1988): Specific glucocorticoid receptor binding to DNA reconstituted in a nucleosome. *EMBO J* 7:3073–3079.

Perlmann T, Wrange Ö (1991): Inhibition of chromatin assembly in *Xenopus* oocytes correlates with derepression of the mouse mammary tumor virus promoter. *Mol Cell Biol* 11:5259–5265.

Picard D, Khursheed B, Garabedian MJ, Fortin MG, Lindquist S, Yamamoto KR (1990): Reduced levels of hsp90 compromise steroid receptor action in vivo. *Nature* (Lond) 348:166–168.

Picard D, Yamamoto KR (1987): Two signals mediate hormone-dependent nuclear localization of the glucocorticoid receptor. *EMBO J* 3333–3340.

Piña B, Barettino D, Truss M, Beato M (1990a): Structural features of a regulatory nucleosome. *J Mol Biol* 216:975–990.

Piña B, Brüggeimeier U, Beato M (1990b): Nucleosome positioning modulates accessibility of regulatory proteins to the mouse mammary tumor virus promoter. *Cell* 60:719–731.

Piña B, Haché RJG, Arnemann J, Chalepakis G, Slater EP, Beato M (1990c): Hormonal induction of transfected genes depends on DNA topology. *Mol Cell Biol* 10:625–633.

Piña B, Truss M, Ohlenbusch H, Postma J, Beato M (1990d): DNA rotational positioning in a regulatory nucleosome is determined by base sequence. An algorithm to model the preferred superhelix. *Nucleic Acids Res* 18:6981–6987.

Power RF, Lydon JP, Conneely OM, O'Malley BW (1991): Dopamin activation of an orphan of the steroid receptor superfamily. *Science* 252:1546–1548.

Pratt WB, Jolly DJ, Pratt DV, Hollenberg SM, Giguerre V, Cadepond FM, Schweizer-Groyer G, Catelli MG, Evans RM, Baulieu EE (1988): A region in the steroid binding domain determines formation of the non-DNA binding, 9S glucocorticoid receptor complex. *J Biol Chem* 263:267–273.

Ptashne M, Gann AAF (1990): Activators and targets. *Nature* (Lond) 346:329–331.

Reik A, Schütz G, Stewart AF (1991): Glucocorticoids are required for establishment and maintenance of an alteration in chromatin structure: Induction leads to a reversible disruption of nucleosomes over an enhancer. *EMBO J* 10:2569–2576.

Reinke R, Feigelson P (1985): Rat alpha-1-acid glytoprotein. Gene sequence and regulation by glucocorticoids in transfected L-cells. *J Biol Chem* 260:4397–4403.

Richard-Foy H, Hager GL (1987): Sequence-specific positioning of nucleosomes over the steroid-inducible MMTV promoter. *EMBO J* 6:2321–2328.

Rigaud G, Roux J, Pictet R, Grange T (1991): In vivo footprinting of rat TAT gene: Dynamic interplay between the glucocorticoid receptor and a liver-specific factor. *Cell* 67:977–986.

Sakai DD, Helms S, Carlstedt-Duke J, Gustafsson JA, Rottman FM, Yamamoto KR (1988): Hormone-mediated repression of transcription: A negative glucocorticoid response element from the bovine prolactin gene. *Genes & Dev* 2:1144–1154.

Satchwell SC, Drew HR, Travers AA (1986): Sequence periodicities in chicken nucleosome core DNA. *J Mol Biol* 191:659–675.

Schatt MD, Rusconi S, Schaffner W (1990): A single DNA-binding transcription factor is sufficient for activation from a distant enhancer and/or from a promoter position. *EMBO J* 9:481–487.

Schauer M, Chalepakis G, Willmann T, Beato M (1989): Binding of hormone accelerates the kinetics of glucocorticoid and progesterone receptor binding to DNA. *Proc Natl Acad Sci USA* 86:1123–1127.

Scheidereit C, Beato M (1984): Contacts between receptor and DNA double helix within a glucocorticoid regulatory element of mouse mammary tumor. *Proc Natl Acad Sci USA* 81:3029–3033.

Scheidereit C, Geisse S, Westphal HM, Beato M (1983): The glucocorticoid receptor binds to defined nucleotide sequences near the promoter of mouse mammary tumor. *Nature* (Lond) 304:749–752.

Scheidereit C, Westphal HM, Carlson C, Bosshard H, Beato M (1986): Molecular model of the interaction between the glucocorticoid receptor and the regulatory elements of inducible genes. *DNA* 5:383–391.

Schena M, Freedman LP, Yamamoto KR (1989): Mutations in the glucocorticoid receptor zinc finger that distinguish interdigitated DNA binding and transcriptional enhancement activities. *Genes & Dev* 3:1590–1601.

Schmid W, Strähle U, Schütz G, Schmitt J, Stunnenberg H (1989): Glucocorticoid receptor binds cooperatively to adjacent recognition sites. *EMBO J* 8:2257–2263.

Schüle R, Evans RM (1991): Cross-coupling of signal transduction pathways—zinc finger meets leucine zipper. *Trends Genet* 7:377–381.

Schüle R, Muller M, Otsuka-Murakami H, Renkawitz R (1988): Cooperativity of the glucocorticoid receptor and the CACCC-box binding factor. *Nature* (Lond) 332:87–90.

Schüle R, Rangarajan P, Kliewer S, Ransone LJ, Bolado J, Yang N, Verma IM, Evans RM (1990): Functional antagonism between oncoprotein c-*jun* and the glucocorticoid receptor. *Cell* 62:1217–1226.

Schwabe J, Neuhaus D, Rhodes D (1990): Solution structure of the DNA-binding domain of the estrogen receptor. *Nature* (Lond) 348:458–461. Severne Y, Wieland S, Schaffner W, Rusconi S (1988): Metal binding "finger" structures in the glucocorticoid receptor defined by site-directed mutagenesis. *EMBO J* 7:2503–2508.

Shemshedini L, Knauthe R, Sassonecorsi P, Pornon A, Gronemeyer H (1991): Cell-specific inhibitory and stimulatory effects of Fos and Jun on transcription activation by nuclear receptors. *EMBO J* 10:3839–3849.

Shimamura A, Sapp M, Rodriguez-Campos A, Worcel A (1989): Histone H1 represses transcription from minichromosomes assembled *in vitro*. *Mol Cell Biol* 9:5573–5584.

Silver BJ, Bokar JA, Virgin JB, Vallen EA, Milsted A, Nilson JH (1987): Cyclic AMP regulation of the human glycoprotein hormone α-subunit gene is mediated by an 18-bp element. *Proc Natl Acad Sci USA* 84:2198–2202.

Simons SS, Pumphrey J, Rudikoff S, Eisen HJ (1987): Identification of cyteine 656 as the amino acid of hepatoma tissue culture cell glucocorticoid receptors that is covalently labeled by dexamethasone 21-mesylate. *J Biol Chem* 262:9676–9680.

Slater EP, Redeuilh G, Theis K, Suske G, Beato M (1990): The uteroglobin promoter contains a noncanonical estrogen responsive element. *Mol Endocrinol* 4:604–610.

Strähle U, Schmid W, Schütz G (1988): Synergistic action of the glucocorticoid receptor with transcription factors. *EMBO J* 7:3389–3395.

Tora L, Gronemeyer H, Turcotte B, Gaub MP, Chambon P (1988): The N-terminal region of the chicken progesterone receptor specifies target gene activation. *Nature* (Lond) 333:185–188.

Tora L, White J, Brou C, Tasser D, Webster N, Scheer E, Chambon P (1989): The human estrogen receptor has two independent nonacidic transcriptional activation functions. *Cell* 59:477–487.

Truss M, Chalepakis G, Beato M (1990): Contacts between steroid hormone receptors and thymines in DNA: An interference method. *Proc Natl Acad Sci USA* 87:7180–7184.

Tsai SY, Carlsted-Duke J, Weigel NL, Dahlman K, Gustafsson JA, Tsai MJ, O'Malley BW (1988): Molecular interactions of steroid hormone receptor with its enhancer element: Evidence for receptor dimer formation. *Cell* 55:361–369.

Tsai SY, Tsai MJ, O'Malley BW (1989): Cooperative binding of steroid hormone receptors contributes to transcriptional synergism at target enhancer elements. *Cell* 57:443–448.

Umesono K, Evans RM (1989): Determinants of target gene specificity for steroid/thyroid hormone receptors. *Cell* 57:1139–1146.

Vannice JL, Taylor JM, Ringold GM (1984): Glucocorticoid-mediated induction of alpha-1-acid glycoprotein: Evidence for hormone-regulated RNA processing. *Proc Natl Acad Sci USA* 81:4241–4245.

Vidal M, Gaber RF (1991): *RPD3* encodes a second factor required to achieved maximum positive and negative transcriptional states in *Saccharomyces cerevisiae*. *Mol Cell Biol* 11:6317–6327.

Vidal M, Strich R, Esposito RE, Gaber RF (1991): *RPD1* (*SIN3/UME4*) is required for maximal activation and repression of diverse yeast genes. *Mol Cell Biol* 11:6306–6316.

Williams PM, Ratajczak T, Lee SC, Ringold GM (1991): AGP/EBP(LAP) expressed in rat hepatoma cells interacts with multiple promoter sites and is necessary for maximal glucocorticoid induction of the rat alpha-1 acid glycoprotein gene. *Mol Cell Biol* 11:4959–4965.

Willmann T, Beato M (1986): Steroid-free glucocorticoid receptor binds specifically to mouse mammary tumour virus DNA. *Nature* (Lond) 324:688–691.

Wolffe AP (1989): Dominant and specific repression of Xenopus oocyte 5S RNA genes and satellite I DNA by histone H1. *EMBO J* 8:527–537.

Wrange Ö, Carlstedt-Duke J, Gustafsson JA (1986): Stoichiometric analysis of the specific interaction of the glucocorticoid receptor with DNA. *J Biol Chem* 261:11770–11778.

Wright APH, Gustafsson JA (1991): Mechanism of synergistic-transcriptional transactivation by the human glucocorticoid receptor. *Proc Natl Acad Sci USA* 88:8283–8287.

Yang-Yen H-F, Chambard J-C, Sun Y-L, Smeal T, Schmidt TJ, Drouin J, Karin M (1990): Transcriptional intereference between c-Jun and the glutocorticoid receptor: Mutual inhibition of DNA binding due to direct protein-protein interaction. *Cell* 62:1205–1215.

Yoshinaga SK, Yamamoto KR (1991): Signaling and regulation by a mammalian glucocorticoid receptor in *Drosophila* cells. *Mol Endocrinol* 5:844–853.

Chapter 4

Control of Transcription and Cellular Proliferation by cAMP

Marc R. Montminy

Cyclic AMP (cAMP) regulates a striking number of physiological processes, including intermediary metabolism, cellular proliferation, and
neuronal signaling, by altering basic patterns of gene expression. In
the liver, for example, cAMP stimulates glucose production in part by
inducing the transcription of the gene for phosphoenol-pyruvate carboxykinase (PEPCK), a rate-limiting gluconeogenic enzyme, more than 15
fold (Lamers et al., 1982; Sasaki et al., 1984). The transcriptional induction by cAMP is rapid, peaking at 30 min and declining gradually over
24 h. This burst in transcription does not depend on new protein synthesis and suggests, therefore, that transcriptional modulation by cAMP
involves the covalent modification of a preexisting nuclear factor. Because all the known cellular effects of cAMP occur via cAMP-dependent
protein kinase, it appears that phosphorylation is the most likely mechanism by which cAMP would regulate gene expression.

In the process of characterizing *cis*-acting sequences underlying
cAMP responsiveness of the somatostatin gene, we (Montminy et al.,
1986) and others (Comb et al., 1986; Short et al., 1986) observed a
short palindromic core motif, 5'-TGACGTCA-3', that was highly conserved among several cAMP-inducible promoters, including those for
c-*fos*, somatostatin, tyrosine hydroxylase, vasoactive intestinal peptide,
proenkephalin, and PEPCK (Figure 1). The cAMP response element
(CRE) displayed properties of a classical enhancer sequence, stimulating
transcription in a distance- and orientation-independent manner. The CRE
also conferred cAMP inducibility when placed upstream of a nonrespon-
sive gene, indicating that CRE binding factors could interact with com-

GENE EXPRESSION: GENERAL AND CELL-TYPE-SPECIFIC
Michael Karin, Editor
© 1993 Birkhäuser Boston

Gene	Sequence
Somatostatin*	CTGGGGGCGCCTCCTTGGC**TGACGTCA**GAGAGAGAG (-32)
PEPCK*	TGATCCAAAGGCCGGCCCC**TTACGTCA**GAGGCGAGC (-74)
VIP*	TCCCATGGCCGTCATACTG**TGACGTC**TTTCAGAGCA (-60)
Parathyroid hormone	GGGAG**TGACGTCA**TCT (-65)
Proenkephalin*	GGGCC**TGCGTCA**GC (-87)
α-Chorionic gonadotropin	AAAATT**GACGTCA**TGG (-113)
c-*fos**	CCGCCCAG**TGACGTA**GGA (-57)
Cytomegalovirus enhancer	CCACCCCAT**TGACGTCA**ATGGGAGTT (-124)
BLV LTR	ACCAGACAGA**GACGTCA**GCTGCCAGA (-144)
HTLV-II LTR	CCACGGCCC**TGACGTC**CCTCCCCCCC (-162)
Intracisternal A particle	CCTCTCCCG**TGACGTCA**TCTGGGG (-86)

Figure 1   Homology in 5′-flanking regions of genes containing somatostatin palindrome; a short sequence 5′-TGACGTCA-3′ is conserved among cAMP-responsive genes. *Asterisks* indicate genes known to be transcriptionally regulated by cAMP; **boldface** type indicates palindromic sequences; and position of 3′-most nucleotide is indicated in parentheses next to each sequence. PEPCK, phosphoenolpyruvate carboxykinase; VIP, vasoactive intestinal peptide; BLV, bovine leukemia virus; LTR, long terminal repeat.

ponents of the transcriptional machinery (e.g., TATA factors) that were not specific to cAMP-inducible promoters. As expected, transcriptional regulation by cAMP appears to be mediated by cAMP-dependent protein kinase (kinase-A). Kinase A-deficient cell lines, for example, are unable to support somatostatin transcription in response to forskolin (Grove et al., 1987; Montminy et al., 1986). Microinjection of catalytic subunit into cells, furthermore, can directly activate CRE-dependent transcription without simultaneous addition of cAMP (Riabowol et al., 1988).

The observation that CRE sequences are highly conserved prompted us to ask whether these elements would recognize a common nuclear factor. Using a DNase I protection assay, we were able to detect CRE-binding (CREB) activity in nuclear extracts of PC12 cells (Montminy and Bilezikjian, 1987). By employing DNA sequence-specific affinity chromatography we purified a 43-kDa protein to apparent homogeneity from PC12 cells and brain tissue (Montminy and Bilezikjian, 1987; Yamamoto et al., 1988) (Figure 2). When added to nuclear extracts, purified CREB specifically stimulated transcription of CRE-containing genes such as somatostatin, suggesting that CREB was a transcription factor as well as a DNA-binding protein. Further biochemical experiments have revealed

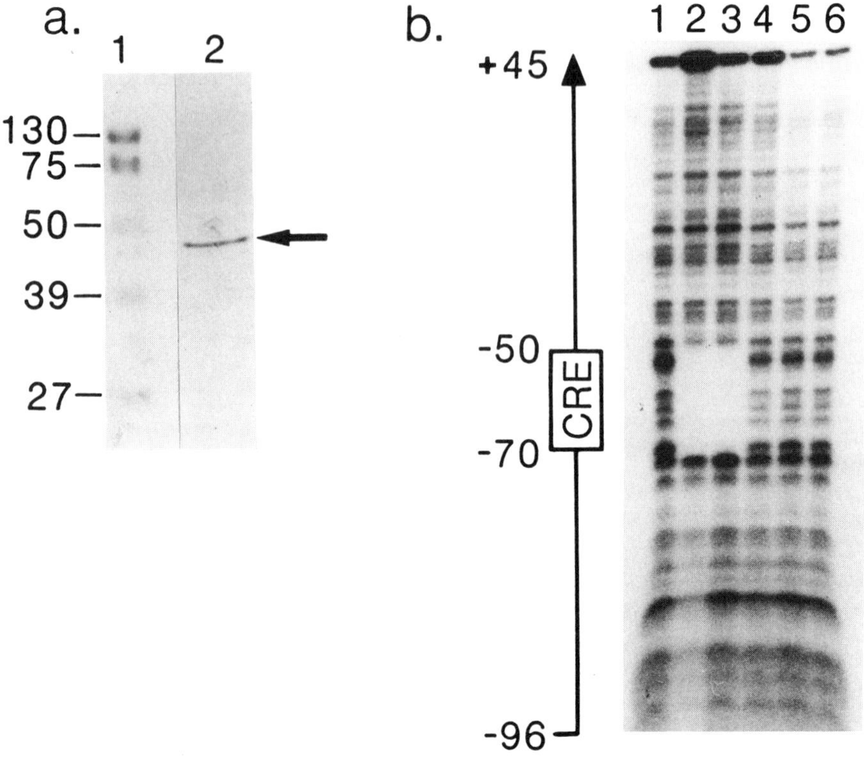

Figure 2   Purification of 43-kDa nuclear protein with CRE-binding activity. **Panel a:** SDS-polyacrylamide gel shows CREB from rat brain after purification over CRE oligonucleotide affinity column (lane 2). *Arrow* points to purified CREB. Lane 1, molecular weight markers (in kilodaltons). **Panel b:** DNase I protection assay using ^{32}P-labeled c-*fos* probe extending from −96 to +45. Position of CRE sequence shown diagramatically. After incubating DNA probe with protein fractions, samples were digested with DNase I. Products of digestion were resolved on urea-polyacrylamide gel. Lane 1, no protein; Lane 2, purified CREB. Lanes 3–6, protein fractions eluted and renatured from SDS-polyacrylamide gel shown in panel A. Only lane 3, which contains the 43-kDa protein, shows CRE-binding activity.

that CREB stimulates transcription of these genes by binding to DNA as a dimer. Moreover, the transcriptional efficacy of CREB is regulated by protein kinase-A phosphorylation (Yamamoto et al., 1988).

Experiments showing that the dimerization and transcriptional efficacy of CREB are each stimulated by phosphorylation at distinct sites suggest that CREB is regulated by multiple kinases *in vivo*. Nuclear extracts prepared from cAMP-treated PC12 cells appear identical to untreated cells in both DNA-binding and transcriptional activity; however, the absence of detectable changes in these parameters may arise from (1) the liberation of kinases during extract preparation, which causes extensive phosphorylation of CREB and subsequently "deregulates" the extract, or (2) the loss of other nuclear factors that are required for transcriptional regulation by cAMP. These experiments demonstrate the inherent limitations of transcription studies without purified components that have been thoroughly characterized.

Creb Structure

We have isolated a cDNA clone for CREB using amino acid sequence information from purified CREB protein (Gonzalez et al., 1989) (Figure 3). When expressed in *E. coli*, the protein encoded by this cDNA possesses CRE-binding activity that is identical to that of purified CREB. Sequence analysis of the CREB cDNA predicts a cluster of protein kinase-A, protein kinase-C, and casein kinase II consensus recognition sites near the N terminus of the protein. The closeness of these sites to one another suggests that they may interact to regulate CREB activity. A stretch of regularly spaced leucine residues, characteristic of a potential "leucine-zipper" dimerization domain (Landschultz et al., 1988), is located near the C terminus of the CREB molecule. CREB shares substantial homology in this region with other "leucine-zipper" proteins including c-Fos, c-Myc, and c-Jun. The prevalence of such dimerization domains among nuclear factors illustrates the potential for combinatorial regulation through the formation of heterodimers with factors such as CREB. Another group has also characterized a CREB cDNA clone by screening a lambda-gt11 cDNA expression library from human placenta with a CRE oligonucleotide (Hoeffler et al., 1988).

The isolation of a cDNA encoding CREB, the production of a bacterial CREB fusion protein, and the characterization of antisera that immunoprecipitate this factor have allowed us to examine the interaction of CREB with other proteins which are necessary to form a cAMP-responsive transcription complex. The presence of two domains, phosphorylation and dimerization, in the CREB molecule offers two different, although not mutually exclusive, conceptual models for regulation by cAMP.

In model I, cAMP stimulates phosphorylation of CREB, thereby causing it to undergo a conformational change that exposes the transcrip-

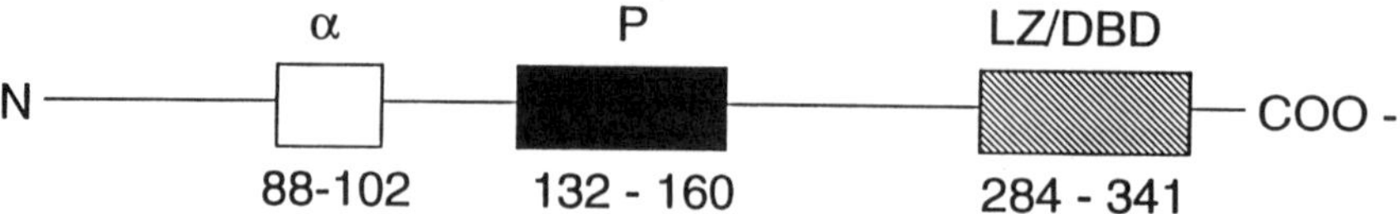

Figure 3   Diagram of CREB protein as predicted from cDNA clone obtained from rat PC-12 mRNA. Open box (with α above it) indicates alternately spliced transactivating region (amino acids 88–102); solid box (with P above it) (amino acid 132–160) indicates position of phosphorylation sites for kinase-C, kinase-A, and casein kinase II; shaded box (with DBD/LZ above) (amino acid 284–341) indicates position of DNA binding and leucine-zipper dimerization domains.

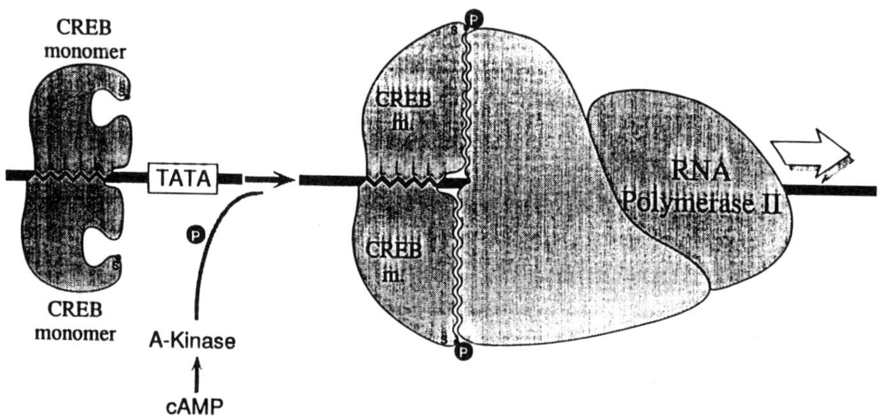

Figure 4   Model for induction of eukaryotic genes by cAMP. CREB is shown bound to DNA as dimer. The Ls represent leucine residues that form leucine-zipper dimerization domain; S represents a single serine residue that is phosphorylated by C subunit *in vitro*. After hormonal stimulation, cAMP levels increase, leading to release of active C subunit from inactive kinase-A holoenzyme. C-subunit is directly transported to the nucleus where it phosphorylates CREB. Phosphorylation (P) cause changes in conformation of CREB molecule, allowing putative CREB transactivating domain to interact with its target protein in RNA polymerase II complex. As a result of this interaction, the pol II complex can initiate transcription.

tion-activating domain of the molecule to its target protein (Figure 4). Current evidence supports the concept that such activating domains are usually acidic in character and have been termed "negative noodles" (Ma and Ptashne, 1987; Ptashne, 1988). The six glutamate residues C terminal to the A-kinase domain would constitute such a potential structure. On the basis of the ability of CREB to transactivate promoters not normally regulated by cAMP, the target protein for this activating domain would presumably be a nonspecific TATA factor like TFIID or an RNA polymerase II-associated protein. In fact, CREB, also referred to as ATF (Lee and Green, 1987), has been shown to interact with TFIID, forming a preinitiation complex, and to increase the affinity of TFIID for its recognition site (Horikoshi et al., 1988).

Model II suggests that, by analogy with transcription factor AP-1, CREB regulates transcription primarily through interactions (possibly at its dimerization domain) with other nuclear factors. This model may be particularly applicable to the tissue-specific regulation of genes such as chorionic gonadotropin and somatostatin. The chorionic gonadotropin gene promoter, for example, contains a tissue-specific element flanked by two consensus CRE motifs (Delegeane et al., 1987; Jameson et al., 1986). Deletion of these CREs completely inactivates tissue-specific expression. As with TFIID, the tissue-specific factor can only bind to its cognate sequence when CREB is present. Studies showing that the tissue-specific expression of the somatostatin gene is also dependent on the CRE sequence (Andrisani et al., 1987) suggest that CREB may perform a more general function, serving perhaps as an anchor for tissue-specific factors.

To support the hypothesis that CREB is regulated by kinase-A, we have examined the phosphorylation of CREB in PC12 cells in response to forskolin treatment (Gonzalez and Montminy, 1989). PC12 cells were incubated in inorganic ^{32}P for 5 h and then treated with forskolin (10 mM) or ethanol vehicle for 30 min. Cells were then harvested in SDS-lysis buffer and immunoprecipitates prepared using CREB antiserum 219. After SDS-PAGE, the 43-kDa, ^{32}P-labeled bands representing CREB protein were identified and excised. Two-dimensional tryptic maps (Figure 5A) revealed a single spot that migrated at the same position as the A-kinase-phosphorylated CREB peptide. Phosphorylation of this site was enhanced approximately sixfold in samples treated with forskolin. PC12 cells labeled with [^{35}]methionine showed no increase in immunoprecipitable CREB during the same time period, indicating that the increase in phosphorylation which we observed was independent of new protein synthesis (Figure 5B). In contrast, using A126-1B2 cells that are deficient in A-kinase activity we observed no increase in forskolin-stimulated CREB phosphorylation (see Figure 5A). In light of previous work showing that A126 cells are also incapable of supporting somatostatin transcription in response to cAMP (Montminy et al., 1986), these results provide compelling evidence to support kinase-A-mediated activation of CREB.

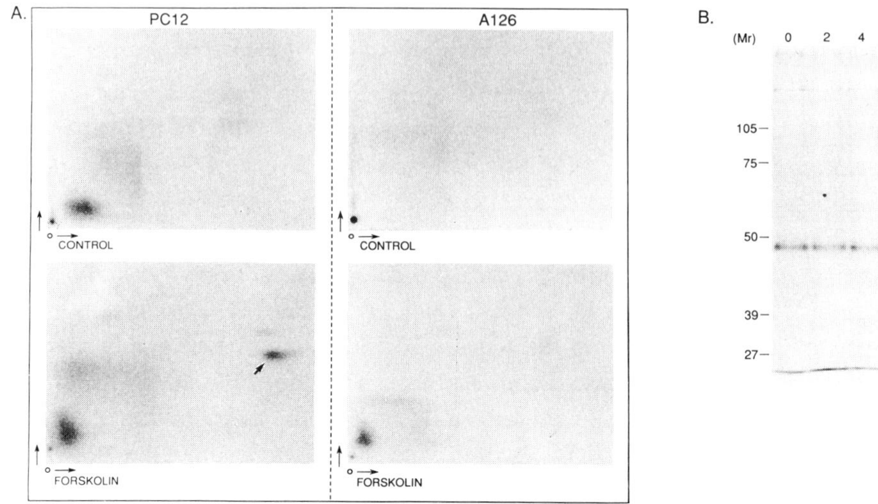

Figure 5   **A:** Phosphorylation of CREB in response to forskolin treatment of PC12 and A126 cells. Two-dimensional phosphotryptic maps of ^{32}P-labeled CREB protein after immunoprecipitation with W39 CREB antiserum. *Arrow* points to ^{35}P-labeled spot that migrates at same position as W51 CREB peptide. **B:** Biosynthesis of CREB in PC12 cells in response to forskolin, 0, 2, 4 h of forskolin treatment. *Arrow* points to ^{35}S-labeled CREB. Mr, relative molecular mass (in kilodaltons). Cells were treated with forskolin (10 μM) for times indicated and then labeled with [^{32}S]methionine (0.1mCi/ml) for 15 min. Cell extracts were immunoprecipitated with affinity-purified W39 CREB antiserum; samples were then resolved by SDS-PAGE.

To show that CREB *directly* participates in transcriptional regulation, we constructed a CREB expression vector and cotransfected this plasmid into PC12 cells with the somatostatin CRE-CAT reporter gene (Figure 6). Having observed that the RSV promoter is particularly active in PC12 cells (Montminy et al., 1986), we inserted the CREB cDNA into a plasmid containing the RSV LTR. When transfected into PC12 cells, RSV-CREB stimulated CRE-CAT activity approximately fourfold over baseline. Not surprisingly, the ability of RSV-CREB to induce CRE-CAT activity was dependent on cAMP stimulation. RSV-CREB was unable to stimulate transcription in forskolin-stimulated A126 cells (not shown), further emphasizing the requirement for A-kinase activity.

To demonstrate that cells regulate CREB bioactivity by reversible phosphorylation in response to A-kinase, we have constructed several mutants in the A-kinase site of CREB (Figure 7). Mutant M1 contains

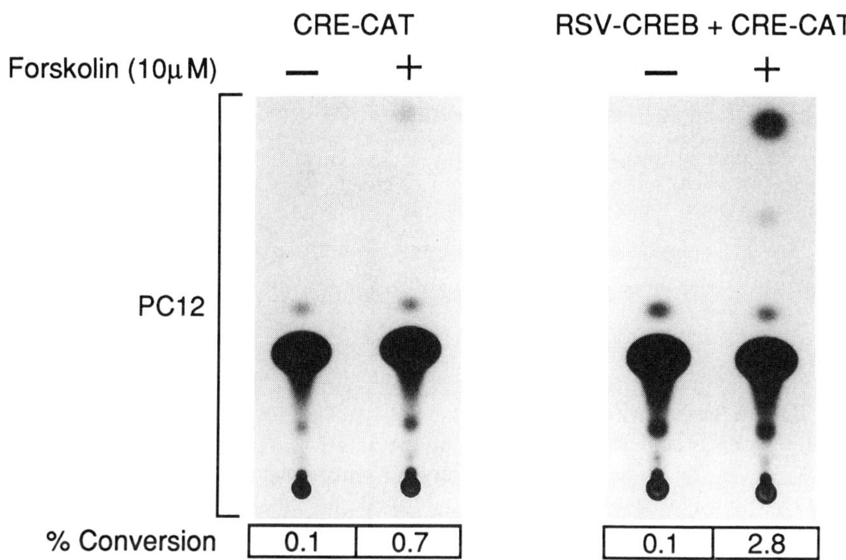

Figure 6   Effect of CREB on $\Delta(-71)$ CAT expression in PC12 cells. Cells were transfected with either $\Delta(-71)$ CAT (CRE-CAT) or $\Delta(-71)$ CAT plus CREB (RSV-CREB) expression plasmids. (−) and (+) indicate control (ethanol vehicle) or forskolin (10 μM) treatment. CAT activity, expressed as percent conversion, shown beneath each.

a conservative serine to valine substitution at position 133. This mutant would not be phosphorylated by kinase-A and should therefore not transactivate CRE-CAT in response to cAMP. Mutant M2, containing a serine-to-aspartate substitution, was designed to provide negative charge otherwise incurred by phosphorylation at this site. If phosphorylation stimulates transcription simply by making this region more acidic, we predict that M2 will be constitutively upregulated but unresponsive to cAMP. Mutant M3 was designed to test if phosphorylation by kinase-C at Ser-133 were important for activity. We therefore substituted the C-terminal basic residues Arg-135, Lys-136 with Met-135, Glu-136. Should kinase-C be critical to cAMP induction, this mutant would be completely inactive.

To assess the activity of each CREB mutant, we employed the murine teratocarcinoma cell line F9 (Figure 8). In the absence of retinoic acid-induced differentiation, these cells are largely unresponsive to cAMP. Cotransfection of eukaryotic expression plasmids encoding either CREB or the catalytic subunit (C-subunit) had only modest effects on CRE-a.

Construct		(130)			(136)		
			pK-A		pK-C		
Wild Type	. . . .AGG AGG CCT TCC			TAC AGG AAA. . . .			
	. . . .Arg Arg Pro Ser*			Tyr Arg Lys. . . .			
M1	. . . .AGG AGG CCT GCC			TAC AGG AAA. . . .			
	. . . .Arg Arg Pro **Ala**			Tyr Arg Lys. . . .			
M2	. . . .AGG AGG CCT GAC			TAC AGA AAA. . . .			
	. . . .Arg Arg Pro **Asp**			Tyr Arg Lys. . . .			
M3	. . . .AGG AGG CCT TCC			TAC ATG GAA. . . .			
	. . . .Arg Arg Pro Ser*			Tyr **Met Glu**. . . .			

b.

(Mr)   M1 M2 M3 WT

105—
75—
50—
39—
27—
17—

c.

C   P

I
II

Figure 7   Characterization of CREB point mutants. **A:** Nucleotide and corresponding amino acid sequence of point mutants M1, M2, and M3 compared to wild-type (WT) CREB cDNA near Ser-133 phosphoacceptor site. Consensus PK-A and PK-C phosphorylation sites overlined and labeled; position of amino acids shown in brackets; serine phosphoacceptor indicated by asterisks; mutated residues underlined and in bold. **B:** Immunoprecipitation of ^{35}S-labeled wild-type and mutant CREB proteins expressed *in vitro*. Mr, molecular weight standard (in kilodaltons); M1, M2, M3, WT, lanes corresponding to mutant and wild-type CREB proteins. **C:** Effect of C-subunit phosphorylation on CRE-binding activity of bacterial CREB fusion protein *in vitro*. Gel shift assay of extract using ^{32}P-labeled, double-stranded CRE oligonucleotide probes. I and II, putative dimer and monomer forms of CREB, respectively. C, control; P, CREB phosphorylated by C subunit in *in vitro*.

CAT reporter activity. When both CREB and C-subunit plasmids were introduced, however, CRE-CAT activity was induced 200 fold. This effect was specific for the CRE as neither CREB nor C-subunit plasmids had any inductive effect on an RSV-CAT reporter (which lacks a CRE and is unresponsive to cAMP).

Mutants M1, M2, and M3 were evaluated in F9 cells by cotransfection with C-subunit and CRE-CAT reporter plasmids (Figure 9). When compared to wild-type CREB, the mutant M1 (Ser-133 to Ala-133) was completely inactive. These results suggest that the Ser-133 phosphoacceptor is indeed critical for transcriptional induction by cAMP. Further, mutant M2 (Ser-133 to Asp-133) was also inactive, indicating that phosphorylation allows CREB to stimulate transcription by a mechanism other than simply providing negative charge. Mutant M3, in which the kinase-C phosphorylation component is removed, nevertheless maintains wild-type activity, demonstrating that the kinase-A component alone is sufficient for induction.

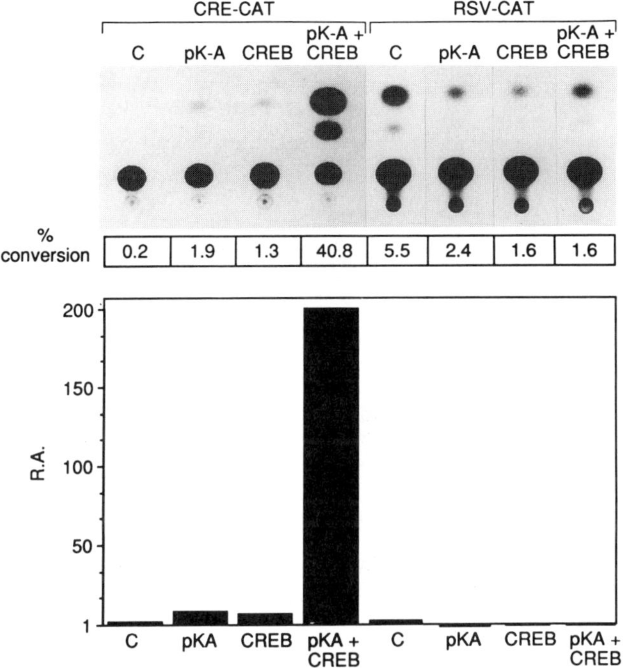

Figure 8   Effect of CREB and PK-A on $\Delta(-71)$ CAT and RSV-CAT expression in undifferentiated F9 teratocarcinoma cells. Cells were transfected with either $\Delta(-71)$ CAT or RSV-CAT reporter genes plus C-subunit (PK-A), CREB (CREB), or both (PK−A+CREB) expression plasmids. Control (C), F9 cells transfected with reporter gene alone. CAT activity is expressed as percent conversion shown below each. Relative activity (R.A.) of $\Delta(-71)$ CAT and RSV-CAT fusion genes when cotransfected with expression plasmids above depicted graphically as fold-stimulation over control.

All the mutants were expressed at levels comparable to wild-type CREB and were appropriately targeted to nuclei of transfected cells. Moreover, M1, M2, and M3 had DNA-binding activity that was indistinguishable from the wild-type protein. Taken together, these results indicate that phosphorylation of CREB specifically alters the transactivation domain so as to render the protein transcriptionally active. Future studies will focus on other structural determinants of the CREB protein that are critical for induction by cAMP.

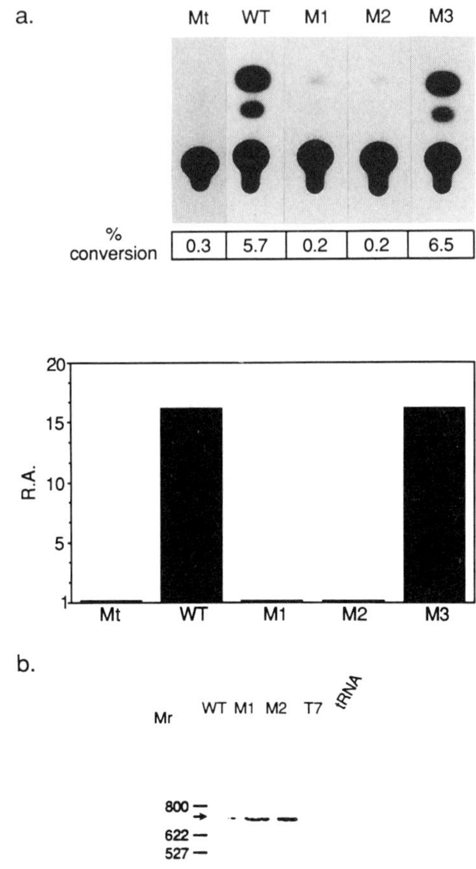

Figure 9   Analysis of wild-type and mutant CREB proteins by transient expression assay. Representative CAT assays of F9 cells transfected with wild-type (WT) or mutant (Mt, M1, M2, M3) RSV-CREB plasmids. The mutant Mt, which contains a nonsense mutation at residue 131 and does not express immunoprecipitable CREB protein, was used to control for nonspecific effects of the RSV-CREB plasmid on transcription. Each CREB plasmid was cotransfected with $\Delta(-71)$ CAT reporter, C-subunit vector MtC, and RSV-βgal as described in Figure 2. Percent conversion indicated and relative activity (R.A.) shown graphically below. Assays were normalized for β-galactosidase activity. Each mutant was tested in at least three separate assays.

Control of Cellular Proliferation by cAMP

A number of growth factors and hormones regulate the proliferation and differentiation of their target cells through the second messenger cAMP.

In the pituitary, for example, both somatotroph proliferation and function are tightly controlled by the hypothalamic growth hormone-releasing factor, GRF (Billestrup et al., 1987). The effects of GRF, including enhanced somatotroph proliferation as well as stimulated transcription of growth hormone, are mimicked by forskolin and stable analogs of cAMP. The coordinate regulation of proliferation and gene transcription in somatotrophs has prompted us to speculate that cAMP may control these processes through the same biochemical intermediates.

To test the hypothesis that cAMP may regulate somatotroph proliferation and function through a common mechanism involving transcription factor CREB, we prepared transgenic mice expressing the cAMP-unresponsive CREB mutant M1 (Struthers et al., 1991), which would not be phosphorylated in cells of the anterior pituitary. Pituitary-specific expression of the mutant transgene was conferred by inserting the CREBM1 cDNA into an eukaryotic expression vector containing 320 bases of the rat growth hormone promoter (-310 to $+8$).

After microinjection of this GH-CREBM1 fusion gene (Figure 10A), we obtained two founder animals that exhibited reduced size compared to control littermates at 8 weeks of age. Offspring from one of these founders continued to exhibit a dwarf phenotype (Figure 10B), which cosegregated with transmission of the CREBM1 transgene. The phenotypic differences between control and transgenic animals were apparent as early as 3 weeks after birth, and both sexes were affected (Figure 11A); males had a 31% reduction in body weight whereas females showed a 24% reduction at 6 weeks of age. The negative effect on body size was thus comparable with other transgenic models in which somatotroph lineages were ablated with either thymidine kinase or diptheria toxin (Behringer et al., 1988; Borelli et al., 1989).

As predicted, the pituitary glands from CREBM1 transgenic mice appeared much smaller than those of their control littermates (Figure 11B). Moreover, these differences were limited to the anterior lobes, which contain predominantly somatotroph and lactotroph lineages. Immunocytochemical studies confirmed a severe depletion of somatotroph cells in transgenic pituitaries (Figure 12). The lactotroph lineage, however, appeared to be unaffected in the CREBM1 mice despite apparent expression of the mutant transgene in these as well as somatotroph cells.

The sparing of lactotroph cells was initially surprising because other studies involving targeted disruption of somatotrophs showed cosegregation of somatotroph and lactotroph lineages. However, the preferential

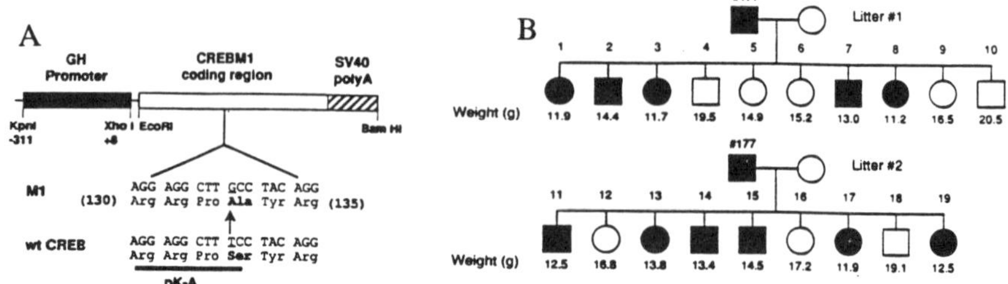

Figure 10   Production of GH-CREBM1 transgenic mice. **A:** Structure of GH-CREBM1 fusion gene employed for microinjection. The Ser-133 to Ala-133 substitution is indicated by an *arrow* with affected amino acid in **boldtype**. The PK-A recognition site is indicted by solid mar. **B:** Pedigree analysis of GH-CREBM1 F1 offspring; male offspring, square; females, circle. Animals containing transgene are shown by solid squares. Weights determined at 6 weeks of age shown below each animal.

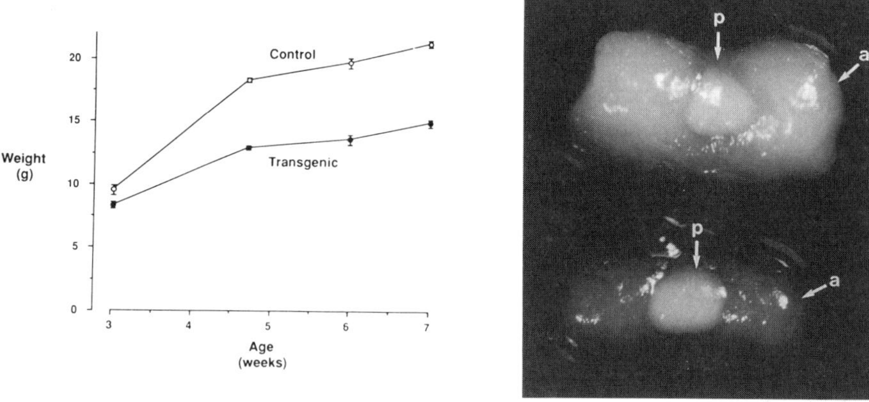

Figure 11   Effect of CREBM1 transgene expression in somatotrophs on somatic growth and pituitary development. **A:** Weight history of GH-CREBM1 transgenic mice (solid circles) versus nontransgenic littermates (open circles). Points are mean $\pm$S.E.M. for male F_1 offspring from two simultaneous litters. **B:** Macroscopic comparison of pituitary glands from control (top) and GH-CREBM1 transgenic mice (bottom).

reduction in somatotroph numbers is consistent with the observation that GRF preferentially stimulates expansion of the somatotroph population. By contrast, lactotroph proliferation appears to be regulated primarily by estrogen. Expression of an inactive CREB protein would be expected to block the cAMP response to GRF without affecting estrogen induction.

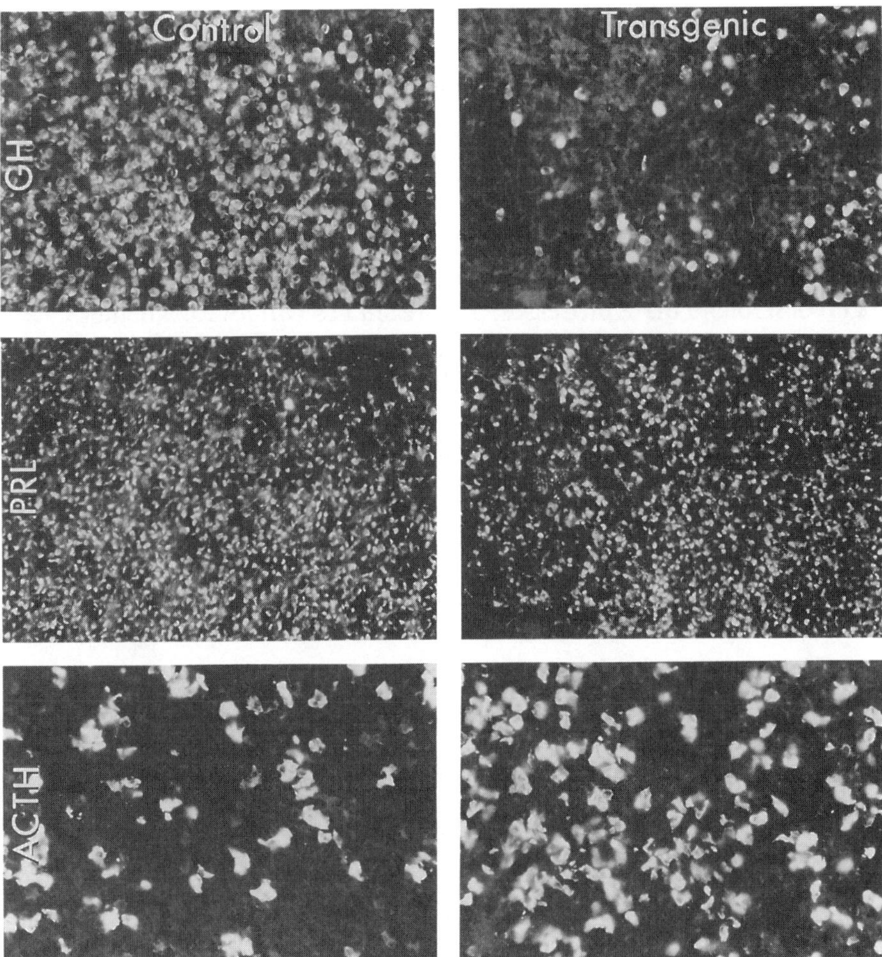

Figure 12   Expression of pituitary hormones in control and GH-CREBM1 transgenic mice. Comparison of indirect immunofluouresence staining patterns from growth hormone (GH), prolactin (PRL), and adrenocorticotropic hormone (ACTH) in control and GH-CREBM1 transgenic mice. Marked decrease in somatotroph population is evident in transgenic animals; many residual GH-immunoreactive cells are hypertrophied. By contrast, lactotroph population of anterior lobes from transgenic mice is comparable to that of controls, but density of corticotrophs is increased.

It is interesting to note that approximately one-half of all patients with acromegaly arising from growth hormone-producing pituitary tumors express constitutively active mutants of the G_s protein in these transformed cells (Landis, 1989). As such tumors contain correspondingly elevated intracellular levels of cAMP, CREB phosphorylation here may promote the proliferative process and contribute to cellular transformation.

Conclusion

Current studies on mechanisms by which cAMP regulates transcription and cellular proliferation suggest that the transcription factor CREB plays a pivotal role in these processes. Future studies will reveal whether CREB is structurally regulated by phosphorylation and whether targets of CREB action are indeed important for cellular proliferation.

Acknowledgments. The authors thank Joan Vaughan, Dr. Wylie Vale, and Dr. Jean Rivier for synthetic peptides and antisera, and Cheng Liang for expert technical assistance. This work was funded by NIH grants GM37828, CA14195 and The Foundation for Medical Research, Inc. Dr. Montminy is a Foundation investigator.

References

Andrisani OM, Hayes TE, Roos B, Dixon JE (1987): Identification of the promoter sequences involved in the cell specific expression of the rat somatostatin gene. *Nucleic Acids Res* 15:5715–5728.

Behringer RR, Mathews LS, Palmiter RD, Brinster RL (1988): *Genes & Dev* 21:453–461.

Billestrup N, Bilezikjian LM, Struthers S, Seifert H, Montminy M, Vale W (1987): *Cellular Actions of Growth Hormone Releasing Factor and Somatostatin*, pp. 53–62.

Borelli E, Heyman RA, Arias C, Sawchenko PE, Evans RM (1989): *Nature (Lond)* 339:593–597.

Comb M, Burnberg NC, Seascholtz A, Herbert E, Goodman HM (1986): A cyclic-AMP and phorbol ester-inducible DNA element. *Nature* (Lond) 323:353–356.

Delegeane AM, Ferland LH, Mellon PL (1987): Tissue-specific enhancer of the human glycoprotein hormone alpha subunit: Dependence on cyclic AMP-inducible elements. *Mol Cell Biol* 7:3994–4002.

Gonzalez GA, Montminy MR (1989): Cyclic AMP stimulates somatostatin gene transcription by phosphorylation of CREB at serine 133. *Cell* 59:675–680.

Gonzalez GA, Yamamoto KK, Fischer WH, Karr D, Menzel P, Biggs W III, Vale WW, Montminy MR (1989): A cluster of phosphorylation sites on the cyclic AMP-regulated nuclear factor CREB predicted by its sequence. *Nature* (Lond) 337:749–752.

Grove JR, Price DJ, Goodman HM, Avruch J (1987): Recombinant fragment of protein kinase inhibitor blocks cyclic AMP-dependent gene transcription. *Science* 238:530–533.

Hoeffler JP, Meyer TE, Yun Y, Jameson JL, Habener JF (1988): Cyclic-AMP-responsive DNA-binding protein: Structure based on a cloned placental cDNA. *Science* 242:1430–1432.

Horikoshi M, Hai T, Lin Y, Gree MR, Roeder RG (1988): Transcription factor ATF interacts with the TATA factor to facilitate establishment of a preinitiation complex. *Cell* 54:1033–1042.

Jameson JL, Jaffe RC, Gleason SL, Habener JF (1986): Transcriptional regulation of chorionic gonadotropin alpha- and beta-subunit gene expression by 8-boromo-adenosine 3',5'-monophosphate. *Endocrinology* 119:2560–2567.

Lamers WH, Hanson RW, Meisner HM (1982): cAMP stimulates transcription of the gene for cytosolic phosphoenolpyruvate carboxykinase in rat liver nuclei. *Proc Natl Acad Sci USA* 79:5137–5141.

Landis CA (1989): *Nature* (Lond) 340:692–696.

Landschulz WH, Johnson PF, McKnight SL (1988): The leucine zipper: A hypothetical structure common to a new class of DNA binding proteins. *Science* 240:1759–1763.

Lee KA, Green MR (1987): A cellular transcription factor E4F1 interacts with an Ela-inducible enhancer and mediates constitutive enhancer function in vitro. *EMBO J* 6:1345–1353.

Ma J, Ptashne M (1987): A new class of yeast transcriptional activators. *Cell* 51:113–119.

Montminy MR, Bilezikjian LM (1987): Binding of a nuclear protein to the cyclic-AMP response element of the somatostatin gene. *Nature* (Lond) 328:175–178.

Montminy MR, Sevarino KA, Wagner JA, Mandel G, Goodman RH (1986): Identification of a cyclic-AMP responsive element within the rat somatostatin gene. *Proc Natl Acad Sci USA* 83:6682–6686.

Ptashne M (1988): How eukaryotic transcriptional activators work. *Nature* (Lond) 335:683–689.

Riabowol KT, Fink JS, Gilman MZ, Walsh DA, Goodman RH, Feramisco JR (1988): The catalytic subunit of cAMP-dependent protein kinase induces expression of genes containing cAMP-responsive enhancer elements. *Nature* (Lond) 336:83–86.

Sasaki K, Cripe TP, Koch SR, Andreone TL, Peterson DD, Beale EB, Granner KK (1984): Multihormonal regulation of phosphoenolpyruvate carboxykinase gene transcription. *J Biol Chem* 259:15242–15251.

Short JM, Wynshaw-Boris A, Short HP, Hanson RW (1986): Characterization of the phosphoenolpyruvate carboxykinase (GTP) promoter-regulatory region.

II. Identification of cAMP and glucocorticoid regulatory domains. *J Biol Chem* 261:9721–9726.

Struthers RS, Vale WW, Arias C, Sawchenko PE, Montminy MR (1991): Somatotroph hypoplasia and dwarfism in transgenic mice expressing a non-phosphorylatable CREB mutant. *Nature* (Lond) 350:622–624.

Yamamoto KK, Gonzalez GA, Biggs WH III, Montminy MR (1988): Phosphorylation-induced binding and transcriptional efficacy of nuclear factor CREB. *Nature* (Lond) 334:494–498.

Chapter 5

Transcription Factors Controlling Muscle-Specific Gene Expression

John J. Schwarz, James F. Martin, and Eric N. Olson

Understanding the mechanisms responsible for activation of cell-type-specific gene expression during development is a fundamental problem in molecular biology. Skeletal muscle has provided an important system for investigating this problem because differentiation of skeletal myoblasts is accompanied by the coordinate induction of a battery of genetically unlinked muscle-specific genes whose products are required for the specialized functions of the mature muscle fiber. Analysis of the events associated with muscle differentiation has also been facilitated by the availability of established muscle cell lines that differentiate rapidly and synchronously and whose differentiation can be negatively regulated by individual peptide growth factors or activated oncogenes encoding proteins which transduce growth factor signals from the cell membrane to the nucleus.

The recent cloning of the muscle-specific regulatory factor MyoD (Davis et al., 1987), and the related proteins myogenin, myf-5 (Braun et al., 1989b; Edmondson and Olson, 1989; Wright et al., 1989), and MRF-4 (Rhodes and Konieczny, 1989), which can activate the entire program for muscle differentiation, has rapidly illuminated the mechanisms responsible for transcriptional control of gene expression during myogenesis. In this chapter, we consider the mechanisms through which the MyoD family of regulatory factors coordinately orchestrate the complex changes in gene expression associated with muscle differentiation. Although multiple mechanisms seem to exist for transcriptional regulation of different muscle-specific genes, many of these mechanisms are activated as a consequence of an initial regulatory step controlled by the

GENE EXPRESSION: GENERAL AND CELL-TYPE-SPECIFIC
Michael Karin, Editor
© 1993 Birkhäuser Boston

MyoD family that sets in motion a regulatory cascade culminating in activation of the complete myogenic program.

The MyoD family of transcription factors converts nonmuscle cells into skeletal muscle. By far the best characterized muscle-specific transcription factors are those of the MyoD family. This family includes MyoD (Davis et al., 1987), myogenin (Edmonson and Olson, 1989; Wright et al., 1989), myf-5 (Braun et al., 1989b), and MRF-4 (Rhodes and Konieczny, 1989), which is also known as herculin (Miner and Wold, 1990), and myf-6 (Braun et al., 1990b). These factors share extensive homology in a region encompassing a basic–amino-acid-rich sequence adjacent to a proposed helix-loop-helix motif (bHLH) that is also shared with other members of the helix-loop-helix class of transcription factors (Murre et al., 1989a).

The hallmark of the MyoD gene family members is their ability to induce the skeletal muscle differentiation program in nonmyogenic cells. This was originally demonstrated with MyoD in C3H10T1/2 fibroblasts (Davis et al., 1987) and has subsequently been repeated with the other factors (Braun et al., 1989b, 1990b; Davis et al., 1987; Edmondson and Olson, 1989; Miner and Wold, 1990; Rhodes and Konieczny, 1989; Wright et al., 1989). This ability to convert fibroblast cell lines to skeletal muscle has been extended to many different cell lines including melanoma, neuroblastoma, adipocyte, and certain types of liver cells as well as to primary cultures of fibroblasts, smooth muscle, chondroblasts, and retinal pigmented epithelial cells. Thus, it appears that at least some aspects of myogenesis can be activated in cells derived from all three germ layers (Choi et al., 1990; Weintraub et al., 1989). However, three cells lines have proved to be refractory to conversion: CV1 (African green monkey kidney cells), HeLa (human cervical carcinoma cells), and HepG2 (human hepatoma cells) (Weintraub et al., 1989). Heterokaryon experiments indicate that both positive and negative factors are involved in the susceptibility of myogenic conversion (Schafer et al., 1990; Weintraub et al., 1989). However, the identities of such factors remain to be determined.

MyoD causes exit from the cell cycle in addition to inducing myogenesis. In many systems terminal differentiation and proliferation are antagonistic processes. This is certainly the case for myogenesis of immortalized muscle cell lines in which the concentration of mitogens must be reduced in the medium for MyoD to induce differentiation (Davis et al., 1987; Weintraub et al., 1989). Further, expression of MyoD at high levels can cause withdrawal from the cell cycle of immortalized cell lines and primary cells without the necessity of reducing the mitogen concentration (Choi et al., 1990). Experiments indicate that the bHLH region of MyoD is necessary and sufficient for growth arrest as well as initiating myogenesis. However, the ability to cause growth arrest is independent of differentiation because a MyoD mutant in which the basic region is replaced with the corresponding region of the ubiquitous HLH protein E12 is incapable of inducing myogenesis, but can nevertheless arrest growth (Crescenzi et al., 1990; Sorrentino et al., 1990). The mechanism for growth inhibition remains unknown, but is likely to involve titration of a limiting cellular factor required for induction of DNA synthesis. The requirement of the bHLH for growth inhibition suggests that such a factor will also be an HLH protein, but this too remains to be demonstrated.

The myogenic factors exhibit a differential pattern of expression during embryogenesis and in muscle cell lines. The four myogenic factors exhibit different patterns of expression during embryogenesis, which suggests distinct roles for these factors during development (Bober et al., 1991; Sassoon et al., 1989). Expression of MRF-4 is in many respects the most specialized with a biphasic pattern of expression in which there is transient expression between days 9.0 and 11.0 post coitum (p.c.) in the rostral myotomes, after which there is no expression until 16 days p.c. when there is expression in fetal skeletal muscle at approximately the same time as secondary muscle fiber and neuromuscular junctions are forming (Bober et al., 1991). MRF-4 also differs from the other myogenic factors in its ability to transactivate certain muscle-specific genes, again suggesting a unique role in muscle-specific gene expression (see following).

The first myogenic factor to be expressed during development is myf-5, which has a transient expression pattern in the somites and developing limbs (Ott et al., 1991). Myf-5 is first detected at day 8 p.c. in the dermamyotome before differentiation of the dermatome and myotome, but it becomes undetectable by day 13 p.c. in the rostral myotomes. Expression of myf-5 in the limb buds is also transient with expression ceasing by day 14 p.c. in the hindlimb buds (Ott et al., 1991). Myogenin and MyoD do not exhibit these transient expression patterns, but continue to express throughout fetal development after they have initiated expression (Sassoon et al., 1989). The expression of myogenin in the myotomes precedes that of MyoD by approximately 2 days with myogenin being detectable at 8.5 days p.c. in the rostral myotome and MyoD at 10.5 days. However, both these factors are expressed at 11 days p.c. in the forelimb bud. For all these factors, the rostral-to-caudal pattern of expression mirrors that of the developing somites, suggesting that their expression is linked to the developmental stage of the somites (Bober et al., 1991).

Although the expression pattern of the myogenic factors in muscle cell lines varies with certain factors being undetectable in some cell lines (Braun et al., 1989c, 1990b; Brennan and Olson, 1990; Miller, 1990; Montarras et al., 1991), two general observations can be made with regard to the function and regulation of these factors that may be specific. The first is that MyoD and myf-5 can be expressed in proliferating myoblasts whereas other muscle-specific genes are not expressed (Davis et al., 1987; Montarras et al., 1991). This suggests their expression is under different control from that of the other muscle-specific genes and perhaps that they are the initiating factors for terminal differentiation. The second is that myogenin is expressed in all muscle cell lines at the onset of terminal differentiation (Braun et al., 1989c), possibly indicating that expression of myogenin is essential for terminal differentiation. This possibility is supported by the observation that antisense oligonucleotides for myogenin can inhibit myoblast differentiation in culture (Brunetti and Goldfine, 1990; Florini and Ewton, 1990).

The MyoD binding site is present in the transcriptional control regions of many muscle-specific genes. MyoD was originally isolated as a factor that could induce myogenesis. However, it was subsequently demonstrated that a bacterially produced glutathione S-transferase-MyoD fusion protein could bind the previously identified muscle-specific MEF1 binding site in the muscle creatine kinase (MCK) enhancers, and moreover that MyoD and MEF1 were antigenically related (Buskin and Hauschka, 1989; Lassar et al., 1989b). The bHLH region of MyoD was also shown to be sufficient to mediate this DNA binding (Lassar et al., 1989b). Interestingly, the bHLH region alone is sufficient to activate the myogenic program when expressed in stably transfected 10T1/2 fibroblasts (Tapscott et al., 1988). Thus, MyoD induces the muscle differentiation program at least in part by binding to a muscle-specific *cis*-acting element, and this activity is located in the bHLH region of the protein.

The MEF1 site possesses an E-box core, CANNTG, which is recognized by all members of the bHLH family of transcription factors (Murre et al., 1989a). Although this site is found in many muscle-specific genes, the demonstration that the E box within these genes actually binds to the myogenic factors and that it possesses functional significance has been carried out for only a few genes. A common theme from this analysis is that a single E box in itself is not functional, but requires either additional E boxes or binding sites for other transcription factors. The 3′ enhancer of the myosin light chain (MLC)–1/3 locus, for example, has three E boxes that bind the myogenic factors (Wentworth et al., 1991). One of these E boxes is required for activity of this enhancer but by itself is inactive and requires the presence of at least one of the other E boxes (Wentworth et al., 1991). The acetylcholine α-subunit gene likewise possesses two E boxes that bind MyoD homooligomers, both of which are required for maximal transcriptional activity (Piette et al., 1990), again demonstrating the need for at least two E boxes for activity. Evidently, the requirement is not solely for a minimum of two E boxes but for an E box and a site for another factor. This is clear from analysis of the troponin I (TnI) gene, which has a single E box that is insufficient for enhancer activity unless associated with two other sites that bind to factors which are not skeletal muscle specific. These sites are also incapable in themselves of activating TnI transcription (Lin et al., 1991). The human cardiac α-actin gene also has only a single E box, which appears to cooperate with the CArG and Sp1 binding sites for activity (Sartorelli et al., 1990; see also French et al., 1991). Significantly, a sequence containing the E box and CArG box of the chicken α-actin gene cannot give significant muscle-specific expression in the absence of upstream sequences, indicating that additional regulatory sequences are yet to be identified (French et al., 1991).

This requirement for multiple E boxes or cognate sites for other transcription factors suggests a high degree of cooperativity for transcriptional activation. Weintraub and coworkers addressed this issue directly by assaying transactivation by MyoD of an artificial promoter composed of the thymidine kinase TATA and one to three copies of the MEF1 site from the MCK enhancer. They determined that a significant level of *trans*-activation could only occur if at least two E boxes are present (Weintraub et al., 1990). Moreover, the ability of MyoD to cooperate with itself maps by deletion analysis to the N-terminal region, which also contains the transcriptional activating domain (Weintraub et al., 1990, 1991). Whether this cooperativity function is physically separable from transcriptional activation is unclear. However, the fact that a chimeric protein in which the transcriptional activating domain of VP16 substitutes for the N-terminal activating domain of MyoD also exhibits the same type of cooperativity (Weinstraub et al., 1991) suggests that this synergism may be a general feature of this class of acidic activating domains and not unique to that of MyoD.

The myogenic factors function in a heterooligomer with the widely expressed products of the E2A gene, E12 and E47. The proposal that the bHLH region could function as a DNA-binding and oligomerization motif provided the conceptual framework for understanding the structure and function of the MyoD family (Murre et al., 1989a). Subsequently, it was shown that MyoD or myogenin and E12 form a heterooligomer that has high affinity for the E box in the MCK enhancer (Brennan and Olson, 1990; Murre et al., 1989b). Even though a MyoD homooligomer can bind to the MCK E box *in vitro* (Lassar et al., 1989b), the actual form of MyoD and the other myogenic factors that binds to DNA *in vivo* appears to be a heterooligomer. The most direct evidence that the myogenic factors require hetero-oligomerization for efficient DNA binding was provided by Sun and Baltimore (1991), who purified E12, E47, and MyoD and determined the equilibrium constants for dimer formation and DNA binding; they demonstrated that heterodimers formed between MyoD and E12 or E47 form 400 fold more efficiently than MyoD homodimers. Lassar et al. (1991) extended this work by demonstrating that heterooligomerization with E12/E47 or similar proteins was necessary for the *in vivo* function of the myogenic factors. Similar results have also been obtained with myogenin (Brennan and Olson, 1990; Chakraborty et al., 1991a; Lassar et al., 1991).

Mutagenesis of the basic regions of MyoD and myogenin reveals that the basic region controls transcriptional activation in addition to DNA binding. A fundamental problem in understanding the specificity of transcriptional activation by the myogenic factors as well as other members of the bHLH family of transcription factors is how they can all recognize the E-box motif CANNTG but activate such diverse sets of genes without apparent activation of inappropriate genes. The problem can be partially explained by different affinities of the HLH proteins for particular E boxes. This however does not appear to be the case for the myogenic factors and the E2A gene products, which recognize the same site with equivalent affinities. Consensus binding sites have been determined for MyoD, myogenin, and the E2A gene products E12 and E47 using polymerase chain reaction (PCR) mediated binding site selection (Blackwell and Weintraub, 1990; Sun and Baltimore, 1991; Wright et al., 1991). The results of these experiments indicate that E12 and E47 have a slightly different consensus binding site than MyoD or myogenin; however, MyoD and presumably myogenin can bind well to sites selected as E12/E47 consensus binding sites.

The role of the DNA-binding domain in transcriptional specificity of bHLH proteins was originally addressed experimentally by Weintraub and coworkers who substituted the basic region of MyoD with those of E12 and the *Drosophila* protein *achaete-scute* T4 (Davis et al., 1990). As would be expected from the similar DNA-binding specificities of these proteins, these replacements did not significantly decrease binding to the E box in the MCK enhancer, but they severely diminished the ability of MyoD to transactivate an MCK reporter gene and to activate myogenesis. These results demonstrate that the basic region contributes muscle-specific transcriptional activation in addition to determining DNA-binding specificity. Subsequently, systematic replacement of residues in the myogenin (Brennan et al., 1991a) and MyoD (Weintraub et al., 1991) basic regions with the corresponding amino acids of E12 identified two adjacent amino acids, alanine and threonine, that confer muscle specificity to the basic region in that replacement of these amino acids with residues at the corresponding position of other HLH proteins severely decreased transactivation of muscle-specific genes but did not affect DNA binding. Corraboration for the importance of these amino acids for muscle-specific gene activation is that they are specific and conserved in all members of the myogenic subfamily of bHLH proteins (Brennan et al., 1991a). These amino acids are therefore considered part of a protein motif that has been termed a myogenic recognition motif (MRM) (Schwarz et al., 1992).

To further investigate the control of transcriptional activation, the transcriptional activating domains (TADs) of MyoD and myogenin were identified by means of a combination of deletion analysis and chimeras between parts of the myogenic factors and the DNA-binding domain of the *Saccharomyces cerevisiae* transcription factor GAL4. From these experiments, a TAD was identified N terminal of the MyoD bHLH and both N- and C terminal of the myogenin bHLH (Schwarz et al., 1992; Weintraub et al., 1991). In this respect, myogenin is similar to myf-5, which also has TADs in the N- and C termini (Braun et al., 1990a). In an attempt to overcome the block to transactivation and myogenesis caused by the E12 basic region substitutions, the C-terminal regions of the MyoD and myogenin substitution mutants were replaced with the powerful TAD of VP16. In both cases, the VP16 activating domain was capable of restoring transcriptional activity on a minimal promoter composed of the thymidine kinase TATA and four copies of the MCK MEF-1 site, but not on the MCK enhancer. This replacement was also unable to restore the ability of myogenin to induce myogenesis. These results are not merely the consequence of the VP16 TAD replacing the C-terminal regions of MyoD and myogenin, because the corresponding chimeras with a wild-type basic region are powerful transactivators of the MCK reporters (Schwarz et al., 1992; Weintraub et al., 1991). Thus, the basic region control of transcriptional activation is dominant to a strong TAD when the target is a complex muscle-specific enhancer.

Two nonmutually exclusive models have been advanced to explain the role of the MRM in control of gene expression (Schwarz et al., 1992; Weintraub et al., 1991). The first model proposes that the activation domains of the myogenic factors are in an inactive conformation that undergoes an allosteric rearrangement to an active conformation on binding to a muscle-specific E box. Such a rearrangement would be dependent on the MRM. MyoD or myogenin mutants containing E12 replacements in the MRM would bind to the E box but be unable to signal this rearrangement to the activating domains. This type of allosteric rearrangement on binding to the proper DNA sequence has been observed for the *S. cerevisiae* pheromone/receptor transcription factor (PRTF) (Tan and Richmond, 1990). The second model proposes that the MRM is required to recognize a "coactivator" that is necessary for muscle-specific transcriptional activation. This coactivator could conceivably have many different functions, one of which would be providing a potent transcriptional activating domain. A well characterized example of such a coactivator is the powerful viral transcription activator VP16, which recognizes the Oct-1 homeodomain in conjunction with a DNA sequence adjacent to the octomer site. Particular amino acids in the Oct-1 homeodomain that differ from those in the corresponding position in Oct-2 specify this interaction (Stern et al., 1989). Transcriptional activation by serum response factor (SRF) has similarly shown to be modulated by an accessory protein termed $p^{62^{TCF}}$ that recognizes the DNA-binding domain of SRF (Schroter et al., 1990).

Although no conclusive experiments have thus far been performed to distinguish between these two models, two pieces of evidence support the coactivator model: the first is the fact that MyoD mutants that contain the E12 basic region will transactivate in CV1 cells although they cannot transactivate in 3T3 or 10T1/2 cells, suggesting that CV1 cells possess a coactivator that is not in 3T3 or 10T1/2 cells which can recognize the E12 basic region in the context of MyoD (Weintraub et al., 1991), and the second is the inability of the VP16 chimeras containing the E12 basic to activate the MCK enhancer even though they can activate the simple E-box-containing promoter. Presumably the VP16 TAD is constitutively active and therefore the MRM must be performing another function in these chimeras, such as mediating protein–protein interactions.

Although the myogenic factors are similar, they possess target gene specificity. Given the differential pattern of expression exhibited by the myogenic factors during embryogenesis and terminal differentiation of muscle cell lines, it seems reasonable to assume that the factors perform distinct functional roles, perhaps activating transcription of different sets of target genes (Bober et al., 1991; Braun et al., 1989a; Brennan and Olson, 1990; Miller, 1990; Montarras et al., 1991; Ott et al., 1991; Sassoon et al., 1989). In addition, the synergy between the myogenic factors and other transcription factors observed in many muscle-specific regulatory regions indicates that protein–protein interactions may be essential for transcriptional activation and also suggests a mechanism by which these factors can transactivate different target genes (French et al., 1991; Lin et al., 1991; Sartorelli et al., 1990). Indeed, differential transactivation of muscle-specific regulatory regions by the myogenic factors has been observed. Specifically, MRF-4, unlike the other factors, does not transactivate the MCK, MLC1/3, or TnI genes even though it is capable of binding the MCK enhancer E box with an affinity comparable to that of the other factors (Chakraborty et al., 1991b; Rosenthal et al., 1990; Yutzey et al., 1990). Moreover, MRF-4 has virtually the same ability as myogenin to activate a minimal promoter containing four E boxes (Chakraborty and Olson, 1991). An extensive series of chimeras between myogenin and MRF-4 was made to identify the regions of these proteins conferring target gene specificity. From this analysis, it appears that this specificity resides in both the N- and C terminal regions because the only chimera between MRF-4 and myogenin that has a significant ability to transactivate the MCK enhancer is composed of the MRF-4 bHLH region and myogenin N- and C terminal regions (Chakraborty and Olson, 1991).

The expression and activity of the myogenic factors are influenced by many agents. Inhibition of myogenesis by growth factors, especially basic fibroblast growth factor (bFGF) and transforming growth factor (TGF-β), oncogenes, and high levels of serum, has been the subject of intense experimental interest for a number of years. Excellent reviews on this subject have been published by Florini et al. (Florini and Magri, 1989; Florini et al., 1991). With the discovery of the myogenic factors that initiate the process of terminal differentiation, it becomes possible to pursue the mechanism of inhibition at a more molecular level and ask if the myogenic factors are the targets of this inhibition. Our review of this subject is therefore confined to recent progress in understanding the effects of these agents on the myogenic factors.

Mechanistically, the best understood inhibitor of the myogenic factors is Id (Benezra et al., 1990). This factor is expressed in several cell lines that can be induced to differentiate and is downregulated on terminal differentiation (Benezra et al., 1990; Sun et al., 1991). Id resembles other members of the helix-loop-helix family and is capable of forming heterooligomers with E12 and MyoD. However, it lacks the basic region that is necessary for DNA binding. As a consequence, Id forms heterooligomers with MyoD or E12 that cannot bind to their cognate sites and are hence inactive. Thus, Id acts as a *trans*-dominant inhibitor of these factors. Although Id can interact with both MyoD and E12, its affinity for E12 appears to be about 5 to 10 times stronger than its affinity for MyoD. Its natural target for regulation may therefore be E12 (Benezra et al., 1990; Sun and Baltimore, 1991). However, because MyoD must heterooligomerize with E12/E47 for activity, the effect of Id is the same as if MyoD were the target.

Of the growth factors that inhibit myogenesis, basic fibroblast growth factor (bFGF) and transforming growth factor β (TGF-β) have been most thoroughly studied. Experiments investigating the effects of both these agents on the myogenic factors have produced similar results. Both growth factors can inhibit the expression of MyoD and myogenin (Brunetti and Goldfine, 1990; Vaidya et al., 1989). Moreover, because these growth factors are able to block terminal differentiation and transactivation even if MyoD or myogenin is being constitutively expressed from a viral promoter they appear to also inhibit the activity of myogenin and MyoD, (Brennan et al., 1991b; Heino and Massague, 1990; Martin et al., 1992; Vaidya et al., 1989; Yutzey et al., 1990). This inhibition of MyoD and myogenin activity may potentially explain the decrease in expression of these factors after treatment with FGF and TGF-β because inhibition of their activity would block the auto- and crossactivation of their promoters (Thayer et al., 1989). For TGF-β, the inhibition has been demonstrated to be independent of Id and to not effect DNA binding (Brennan et al., 1991b). The evidence for this is that cells constitutively expressing myogenin will show a normal decrease in Id expression and a concomitant appearance of a myogenin-containing complex able to bind DNA when shifted to a differentiation-promoting media containing TGF-β (Brennan et al., 1991b). Thus, the block to myogenesis imposed by TGF-β must be downstream of myogenin's binding to its target sequence.

Several oncogenes whose products are components of signal transduction pathways have also been shown to act as surrogates for growth factors and to block the actions of myogenic HLH proteins. Among these are *ras*, *jun*, and *fos* (Lassar et al., 1989a; Li et al., 1992), which inhibit myogenesis by decreasing expression of MyoD and can block MyoD's ability to transactivate muscle target genes. This effect on myogenesis is titratable because expression of high levels of MyoD will overcome the inhibition. The nuclear oncogene *myc*, which is also a bHLH protein, is also capable of inhibiting myogenesis (Schneider et al., 1987). Like TGF-β, this inhibition is independent of Id because Id levels decrease normally in the presence of *myc*, even though myogenesis does not occur (Miner and Wold, 1991).

The multifunctional viral oncogene E1A is a potent inhibitor of myogenesis (Webster et al., 1988). E1A interacts with the retinoblastoma gene Rb and other cellular proteins, which control the decision to enter S phase. Moreover, there exist several mutants that affect the ability of E1A to bind different sets of these protein and produce the transformed phenotype. Wild-type and mutant E1As have been used to determine that E1A inhibits expression of MyoD and myogenin, and that this inhibition depends on the conserved regions 1 and 2 of E1A, which are also required for transformation (Enkemann et al., 1990).

Several muscle-specific genes regulated by the MyoD family do not contain E boxes in their control regions. Although many muscle-specific genes appear to be regulated through direct interaction of myogenic HLH proteins with E boxes in their control regions, several genes that are induced during myogenesis do not appear to be regulated through E boxes. The fact that these genes can be induced in response to the MyoD family suggests the existence of indirect mechanisms for activation of muscle-specific transcription. The three best characterized DNA sequences that can impart muscle specificity independent of the E-box motif are the CArG box, the M-CAT motif, and the MEF-2 site. The involvement of each of these DNA sequences in the control of muscle-specific transcription is discussed next.

The CArG box confers muscle specificity independently of the E-box motif. The sarcomeric actin genes were among the first muscle-specific genes to be analyzed in detail. Muscle-specific transcription of these genes is dependent on the conserved sequence motif $CC(A/T)_6GG$, referred to as a CArG box (Minty and Kedes, 1986). In the case of the α-skeletal actin gene, which lacks an E box in its promoter, the CArG box is solely responsible for muscle specificity of transcription (Bergsma et al., 1986; Walsh and Schimmel, 1987, 1988). The α-cardiac actin gene promoter also contains a CArG box, which collaborates with an E box to activate transcription in muscle cells (French et al., 1991; Sartorelli et al., 1990). CArG boxes have also been shown to contribute to transcriptional activity of muscle-specific enhancers such as the MCK and MLC-1/3 enhancers, but they are not essential for transcriptional activity of these enhancers (Ernst et al., 1991; Sternberg et al., 1988).

The mechanism through which the CArG box directs muscle transcription remains unclear. This issue is especially confusing because CArG boxes have been shown to serve as the binding site for the serum response factor (SRF), which is widely expressed and is responsible for serum inducibility of the c-*fos* promoter, as well as the promoters of several growth factor-inducible immediate early genes (reviewed in Boxer et al., 1989; Treisman, 1990). Moreover, mutations within the CArG box sequence can abolish serum inducibility and muscle-specific transcription of the associated genes (Minty and Kedes, 1986; Treisman, 1990; Walsh and Schimmel, 1988). The possibility that serum inducibility and muscle specificity require subtle nucleotide differences in the CArG element and surrounding sequences has been considered as a potential mechanism to explain this paradox, but experimental evidence to support this model is lacking. Indeed, the ability of the α-skeletal actin CArG box (also called the myogenic response element, MRE) to function as an SRE and of the c-*fos* SRE to function as an MRE, when linked to a neutral promoter, argues against subtle nucleotide differences in the CArG motif as the basis for divergent transcriptional activities (Taylor et al., 1989; Tuil et al., 1990; Walsh, 1989). How then can a ubiquitous factor such as SRF confer disparite regulation through a common DNA sequence element? One possibility is that SRF is actually expressed at a higher level in muscle cells than in other cell types and that transcriptional activation through the CArG element thus mirrors this quantitative difference in the level of SRF. An alternate explanation is that the transcriptional activity of SRF is modulated by posttranslational mechanisms. This could occur through posttranslational modification of the protein in a cell-type-specific manner or through accessory proteins that interact with SRF to modulate its transcriptional activity (Schroter et al., 1990). In this regard, the DNA-binding domain of SRF has also been shown to serve as an interface for binding of an accessory protein termed $p62^{TCF}$ that lacks DNA-binding activity of its own. It has been suggested that $p62^{TCF}$ may mediate the final effects of diverse growth factor pathways by modulating SRF activity. Whether a related factor is responsible for activation of muscle transcription in combination with SRF remains an open question. Whatever the mechanism for activation of muscle-specific transcription through the CArG element, it is clear that this mechanism is subordinate to the more global control over the myogenic program imposed by the MyoD family because forced expression of MyoD in nonmuscle cells leads to activation of sarcomeric actin gene expression.

MEF-2 is a muscle-specific DNA binding activity related to a family of widely expressed DNA-binding proteins. Another *cis*-acting sequence that participates in muscle-specific transcription is the MEF-2 site, which is an A+T-rich sequence associated with the promoters and enhancers of numerous muscle-specific genes (Cserjesi and Olson, 1991; Gossett et al., 1989). The MEF-2 element was initially discovered and characterized in the MCK enhancer, where it is required for maximal enhancer activity. The MCK MEF-2 site shows little or no activity alone, but it is able to collaborate with heterologous regulatory elements to augment transcription. Multimerized MEF-2 sites also can direct high levels of expression in muscle cells in combination with basal promoters. MEF-2 sites are often found closely associated with E boxes, suggesting that MEF-2 and myogenic HLH proteins may cooperate to activate muscle-specific transcription. There are also examples of muscle-specific control regions that contain MEF-2 sites but lack E boxes. In these cases, MEF-2 may be responsible for directing muscle-specific transcription.

The nature of the factors that bind the MEF-2 sequence has been controversial. Gossett et al. (1989) first identified MEF-2 as a DNA-binding activity for this site that was undetectable in myoblasts and nonmuscle cells and was induced during the myoblast-to-myotube transition. Induction of MEF-2 activity is detected within 30 min after exposure of myoblasts to growth factor-deficient medium and requires a new protein synthesis. The latter observation suggests that the appearance of MEF-2 does not result from modification of a preexisting protein, although the possibility that MEF-2 has a short half-life and is rapidly degraded in the presence of cycloheximide has not been ruled out. The muscle-specific MEF-2 activity recognizes a degenerate sequence motif $CT(A/T)_4ATAG$ (Cserjesi and Olson, 1991). A variety of ubiquitous DNA-binding activities for the MEF-2 site have also been identified (Braun et al., 1989c; Buskin and Hauschka, 1989; Horlick and Bernfield, 1989; Horlick et al., 1990; Pollock and Treisman, 1991). At least some of these factors have a DNA-binding specificity subtly different from that of MEF-2.

Pollack and Triesman (1991) have described the cloning of a series of related proteins that can bind the MEF-2 site and activate transcription through that site. These factors, termed RSRFs (related to SRFs), share about 80% amino acid homology with the DNA-binding domain of SRF. In contrast to MEF-2, which appears to be muscle specific, RSRFs are widely expressed. The relationship between RSRF and MEF-2 is not defined. It is conceivable that MEF-2 is encoded by a member of the RSRF gene family or that MEF-2 represents a muscle-specific form of RSRF generated through alternate usage of tissue-specific exons. MEF-2 could also arise through binding of a muscle-specific accessory protein to a ubiquitous form of RSRF.

It appears that MEF-2 functions "downstream" of the MyoD family because MEF-2 can be induced in fibroblasts by forced expression of myogenin or MyoD (Cserjesi and Olson, 1991). Induction of MEF-2 by these myogenic regulators is dependent on withdrawal of exogenous growth factors, indicating that mitogenic signals disrupt this regulatory interaction. MEF-2 can also be induced by myogenin in CV-1 kidney cells that cannot be converted to muscle. These results demonstrate that myogenic HLH proteins retain certain functions in cells that are refractory to myogenic conversion and indicate that MEF-2 is controlled independently of muscle-specific genes such as MCK that are associated with terminal differentiation.

In addition to its importance for induction of several muscle-specific genes associated with terminal differentiation, MEF-2 appears to play a role in transcription of the mouse *myogenin* (Edmondson et al., 1992) and *MRF4* and *Xenopus MyoD* genes (M. Perry, personal communication). Because MEF-2 appears to be induced after activation of the differentiation program by the myogenic HLH proteins, the MEF-2 sites in these promoters may play a role in maintaining the expression of these genes after they have been activated by other factors.

MCBF is a myogenic transcriptional activator that acts independently of E boxes. Regulation of the cardiac troponin-T (cTnT) gene provides further evidence for an indirect mechanism for activation of muscle-specific transcription by the MyoD family. The regulatory element that confers muscle-specific expression to the cTnT gene is known as the M-CAT motif for "muscle CAT" (Mar and Ordahl, 1988). This motif, which is conserved in many muscle-specific genes, appears to bind a ubiquitous factor as well as a muscle-enriched factor.

The cTnT gene is expressed in cardiac muscle and for a short period during early development in skeletal muscle (Cooper and Ordahl, 1984, 1985; Long and Ordahl, 1988). Deletional analysis of the cTnT gene performed in embryonic skeletal muscle cells revealed a minimal promoter region at $-129/-50$ from the transcription initiation site. Fusion of the cTnT minimal promoter to a heterologous promoter showed that the $-129/-50$ element was skeletal muscle specific (Mar and Ordahl, 1988). Two tandem repeats of the conserved M-CAT sequence (CATTCCT) were identified within the minimal promoter, both of which are required for cTnT promoter activity. Also found within the minimal promoter were sequences similar to a MEF-1 site and a CArG box; however, mutagenesis of these potential recognition elements has no effect on promoter activity. DNase I footprint and gel shift analysis demonstrated that a factor (MCBF) bound both of the M-CAT motifs, and mutations in the M-CAT sequence that inhibited expression from the minimal promoter were also shown to block binding of MCBF. Competition experiments showed that the same factor bound both M-CAT motifs but that binding to each site occurred independently. Variation of the normal spacing between the M-CAT motifs resulted in diminished expression from the cTnT minimal promoter, implying that an interaction between the two sites was necessary for full activity of the promoter (Mar and Ordahl, 1990). Study of the cTnT promoter in cardiac cells has also demonstrated the importance of the tandem M-CAT motifs. Deletional analysis of the cTnT promoter in cardiac cells identified an upstream element necessary for expression. Interestingly, this cardiac-specific enhancer element contains a MEF-2 recognition site. Cardiac expression from the cTnT promoter required the upstream MEF-2-containing element in conjunction with the downstream tandem M-CAT motifs. Deletion of either of these elements abolished cardiac-specific expression from the cTnT promoter (Iannello et al., 1991).

Both cardiac muscle- and skeletal muscle-specific expression of the cTnT gene requires cooperative interactions of the M-CAT motif. Gel shift analysis revealed that the M-CAT binding activity was not tissue specific because the activity was present in liver and brain extracts. Interestingly, the MCAT motif is similar to the binding site for the transcription factor TEF-1, which is widely expressed (Xiao et al., 1991). The relationship between MCBF and TEF-1 remains to be determined.

Analysis of MCAT binding activities in MyoD and myogenin-transfected 10T1/2 cells suggests that MCAT lies "downstream" of myogenic HLH proteins in a myogenic regulatory cascade because forced expression of the myogenic regulators in fibroblasts can activate a muscle-specific MCAT binding activity (Lassar et al., 1991). The relationship between the muscle-specific and ubiquitous MCAT binding activities is presently unclear. The reason that ubiquitous MCAT binding activities are seemingly unable to activate the cTnT promoter in nonmyogenic cells is also unknown.

Conclusions

Current knowledge of the mechanisms involved in activation of muscle-specific transcription is consistent with a model in which members of the MyoD family act within a regulatory pathway that leads to induction of the complete program for muscle-specific gene expression. A schematic representation of this myogenic regulatory circuit is shown in Figure 1. It is likely that additional regulatory factors which exist "upstream" of the MyoD family initiate MyoD expression in early embryogenesis. We imagine that activation of muscle-specific gene expression results from a cascade of regulatory steps that can be entered at many levels. Thus, forced expression of myogenic HLH proteins in certain nonmuscle cell backgrounds can result in activation of all downstream events leading to establishment of the muscle differentiation program. These factors

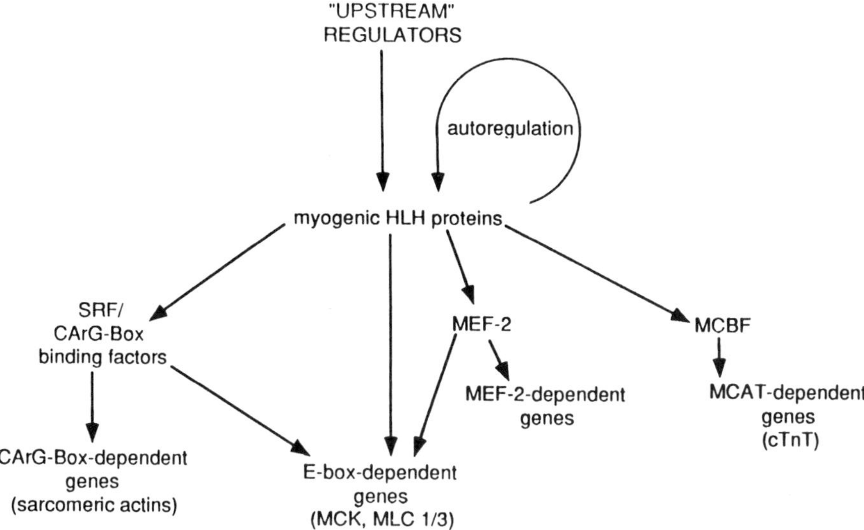

Figure 1   Schematic diagram of regulatory circuit of transcription factors involved in muscle-specific gene expression. Genes encoding myogenic HLH proteins become activated in early development through actions of "upstream" regulatory factors that are not defined. Myogenic HLH proteins autoregulate their own expression and activate "downstream" muscle-specific genes by binding to E boxes in their control regions. Myogenic HLH proteins can also activate muscle-specific genes indirectly by inducing expression of MEF-2 and M-CAT binding factor (MCBF), which act independently or in conjunction with myogenic HLH proteins to induce muscle transcription. Sarcomeric actin genes, under control of the CArG box that binds SRF, can also be induced by myogenic HLH proteins, through an undefined mechanism.

need not function in cell determination but could simply result in activation of the final stages of the muscle differentiation program. Many muscle-specific genes are regulated directly by MyoD whereas others are controlled indirectly through activation of intermediate regulatory factors. In the future, it is important to identify the regulatory factors that control the MyoD family and to further elucidate the mechanisms for indirect regulation of muscle-specific genes.

References

Bergsma D, Grichnik J, Gossett L, Schwartz R (1986): Delimitation and characterization of *cis*-acting DNA sequences required for the regulated expression and transcriptional control of the chicken skeletal α-actin gene. *Mol Cell Biol* 6:2462–2475.

Benezra R, Davis RL, Lockshon D, Turner DL, Weintraub H (1990): The protein Id: A negative regulatory of helix-loop-helix DNA binding proteins. *Cell* 61:49–59.

Blackwell TK, Weintraub H (1990): Differences and similarities in DNA-binding preferences of MyoD and E2A protein complexes revealed by binding site selection. *Science* 250:1104–1110.

Bober E, Lyons GE, Braun T, Cossu G, Buckingham M, Arnold H (1991): The muscle regulatory gene, myf-6, has a biphasic pattern of expression during early mouse development. *J Cell Biol* 113:1255–1265.

Boxer LM, Prywes R, Roeder RG, Kedes L (1989): The sarcomeric actin CArG-binding factor is indistinguishable from the c-*fos* serum response factor. *Mol Cell Biol* 9:515–522.

Braun T, Tannich E, Buschhausen-Denker G, Arnold HH (1989a): Promoter upstream elements of the chicken cardiac myosin light-chain 2A gene interact with *trans*-acting regulatory factors for muscle-specific transcription. *Mol Cell Biol* 9:2513–2525.

Braun T, Buschhausen-Denker G, Bober E, Tannich E, Arnold HH (1989b): A novel human muscle factor related to but distinct from MyoD1 induces myogenic conversion in 10T1/2 fibroblasts. *EMBO J* 8:701–709.

Braun T, Bober E, Buschhausen-Denker G, Kotz S, Grzeschik K, Arnold HH (1989c): Differential expression of myogenic determination genes in muscle cells: Possible autoactivation by the myf gene products. *EMBO J* 8:3617–3625.

Braun T, Winter B, Bober E, Arnold HH (1990a): Transcriptional activation domain of the muscle-specific gene-regulatory protein myf-5. *Nature* (Lond) 346:663–665.

Braun T, Bober B, Winter B, Rosenthal N, Arnold HH (1990b): Myf-6, a new member of the human gene family of myogenic determination factors: Evidence for a gene cluster on chromosome 12. *EMBO J* 9:821–831.

Brennan TJ, Chakraborty T, Olson EN (1991a): Mutagenesis of the myogenin basic region identifies an ancient protein motif critical for activation of myogenesis. *Proc Natl Acad Sci USA* 88:5675–5679.

Brennan TJ, Edmondson DG, Li L, Olson EN (1991b): Transforming growth factor β represses the actions of myogenin through a mechanism independent of DNA binding. 88:3822–3826.

Brennan TJ, Olson EN (1990): Myogenin resides in the nucleus and acquires high affinity for a conserved enhancer element on heterodimerization. *Genes & Dev* 4:582–595.

Brunetti A, Goldfine ID (1990): Role of myogenin in myoblast differentiation and its regulation by fibroblast growth factor. *J Biol Chem* 265:5960–5963.

Buskin JN, Hauschka SD (1989): Identification of a myocyte nuclear factor which binds to the muscle-specific enhancer of the mouse muscle creatine kinase gene. *Mol Cell Biol* 9:2627–2640.

Chakraborty T, Brennan TJ, Li L, Edmondson D, Olson EN (1991a): Inefficient homodimerization contributes to the dependence of myogenin on E2A products for efficient DNA binding. *Mol Cell Biol* 11:3633–3641.

Chakraborty T, Brennan TJ, Olson EN (1991b): Differential *trans*-activation of muscle-specific enhancer by myogenic helix-loop-helix proteins is separable from DNA binding. *J Biol Chem* 266:2878–2882.

Chakraborty T, Olson EN (1991): Domains outside of the DNA-binding domain impart target gene specificity to myogenin and MRF4. *Mol Cell Biol* 11:6103–6108.

Choi J, Costa ML, Mermelstein CS, Chagas C, Holtzer S, Holtzer H (1990): MyoD converts primary dermal fibroblasts, chondroblasts, smooth muscle, and retinal pigmented epithelial cells into striated mononucleated myoblasts and multinucleated myotubes. *Proc Natl Acad Sci USA* 87:7988–7992.

Cooper T, Ordahl C (1984): A single troponin T gene regulated by different programs in cardiac and skeletal muscle development. *Science* 226:979–982.

Cooper T, Ordahl C (1985): A single troponin-T gene generates embryonic and adult isoforms via developmentally regulated alternate splicing. *J Biol Chem* 260:11140–11148.

Crescenzi M, Fleming TP, Lassar AB, Weintraub H, Aaronson SA (1990): MyoD induces growth arrest independent of differentiation in normal and transformed cells. *Proc Natl Acad Sci USA* 87:8442–8446.

Cserjesi P, Olson EN (1991): Myogenin induces the myocyte-specific enhancer binding factor MEF-2 independently of other muscle-specific gene products. *Mol Cell Biol* 11:4854–4862.

Davis RL, Weintraub H, Lassar AB (1987): Expression of a single transfected cDNA converts fibroblasts to myoblasts. *Cell* 51:987–1000.

Davis RL, Cheng P, Lassar AB, Weintraub H (1990): The MyoD DNA binding domain contains a recognition code for muscle-specific gene activation. *Cell* 60:733–746.

Edmondson DG, Olson EN (1989): A gene with homology to the myc similarity region of MyoD1 is expressed during myogenesis and is sufficient to activate the muscle differentiation program. *Genes & Dev* 3:628–640.

Edmondson DG, Cheng TC, Cserjesi P, Chakraborty T, Olson EN (1992): Analysis of the myogenin promoter reveals an indirect pathway for positive autoregulation mediated by the muscle-specific enhancer factor MEF-2. *Mol Cell Biol* 12:3665–3677.

Enkemann SA, Konieczny SF, Taparowsky EJ (1990): Adenovirus 5 E1A represses muscle-specific enhancers and inhibits expression of the myogenic regulatory factor genes, *MyoD1* and *myogenin. Cell Growth & Differ* 1:375–382.

Ernst H, Walsh K, Harrison CA, Rosenthal N (1991): The myosin light chain enhancer and the skeletal actin promoter share a binding site for factors involved in muscle-specific gene expression. *Mol Cell Biol* 11:3735–3744.

Florini JR, Ewton DZ (1990): Highly specific inhibition of IGF-I-stimulated

112   J. J. Schwarz et al.

differentiation by an antisense oligodeoxyribonucleotide to myogenin mRNA. *J Biol Chem* 265:13435–13437.

Florini JR, Ewton DZ, Magri KA (1991): Hormones, growth factors, and myogenic differentiation. *Annu Rev Physiol* 53:201–216.

Florini JR, Magri KA (1989): Effects of growth factors on myogenic differentiation. *Am J Physiol* 256:c701–711.

French BA, Chow K, Olson EN, Schwartz RJ (1991): Heterodimers of myogenic helix-loop-helix regulatory factors and E12 bind a complex element governing myogenic induction of the avian cardiac α-actin promoter. *Mol Cell Biol* 11:2439–2450.

Gossett LA, Kelvin DJ, Sternberg EA, Olson EN (1989): A new myocyte-specific enhancer-binding factor that recognizes a conserved element associated with multiple muscle-specific genes. *Mol Cell Biol* 9:5022–5033.

Heino J, Massague J (1990): Cell adhesion to collagen and decreased myogenic gene expression implicated in the control of myogenesis by transforming growth factor-β. *J Biol Chem* 265:10181–10184.

Horlick RA, Benfield PA (1989): The upstream muscle-specific enhancer of the rat muscle creatine kinase gene is composed of multiple elements. *Mol Cell Biol* 9:2396–2413.

Horlick RA, Hobson GM, Patterson JH, Mitchell MT, Benfield PA (1990): Brain and muscle creatine kinase genes contain common TA-rich recognition protein-binding regulatory elements. *Mol Cell Biol* 10:4826–4836.

Ianello RC, Mar JH, Ordahl CP (1991): Characterization of a promoter element required for transcription in myocardial cells. *J Biol Chem* 266:3309–3316.

Lassar AB, Thayer MJ, Overell RW, Weintraub H (1989a): Transformation by activated *ras* or *fos* prevents myogenesis by inhibiting expression of MyoD1. *Cell* 58:659–667.

Lassar AB, Buskin JN, Lockshon D, Davis RL, Apone S, Hauschka SD, Weintraub H (1989b): MyoD is a sequence-specific DNA binding protein requiring a region of *myc* homology to bind to the muscle creatine kinase enhancer. *Cell* 58:823–831.

Lassar AB, Davis RL, Wright WE, Kadesch T, Murre C, Voronova A, Baltimore D, Weintraub H (1991): Functional activity of myogenic HLH proteins requires heterooligomerization with E12/E47-like proteins in vivo. *Cell* 66:305–315.

Li L, Chambard J-C, Karin M, Olson EN (1992): Fos and Jun repress transcriptional activation by myogenin and MyoD: The amino terminus of Jun mediates repression. *Genes & Dev* 6:676–689.

Lin H, Yutzey K, Koniecny SF (1991): Muscle-specific expression of the troponin I gene requires interactions between helix-loop-helix muscle regulatory factors and uniquitous transcription factors. *Mol Cell Biol* 11:267–280.

Long C, Ordahl C (1988): Transcriptional repression of an embryo-specific muscle gene. *Dev Biol* 127:228–234.

Mar J, Ordahl C (1988): A conserved CATTCCT motif is required for skeletal muscle-specific expression of the cardiac troponin T gene promoter. *Proc Natl Acad Sci USA* 85:6404–6408.

Mar J, Ordahl CP (1990): M-CAT binding factor, a novel trans-acting factor governing muscle-specific transcription. *Mol Cell Biol* 10:4271–4283.

Martin J, Li L, Olson EN (1992): Repression of myogenin function by TGF-β is targeted at the basic-helix-loop-helix motif and is independent of E2A products *J Biol Chem* 267:10956–10960.

Miller JB (1990): Myogenic programs of mouse muscle cell lines: Expression of myosin heavy chain isoforms, MyoD1, and myogenin. *J Cell Biol* 111:1149–1159.

Miner JH, Wold B (1990): Herculin, a fourth member of the MyoD family of myogenic regulatory genes. *Proc Natl Acad Sci USA* 87:1089–1093.

Miner JH, Wold B (1991): c-*myc* inhibition of MyoD and myogenin-initiated myogenic differentiation. *Mol Cell Biol* 11:2842–2851.

Minty A, Kedes L (1986): Upstream regions of the human cardiac actin gene that modulate its transcription in muscle cells: Presence of an evolutionarily conserved repeated motif. *Mol Cell Biol* 6:2125–2136.

Montarras D, Chelly J, Bober E, Arnold H, Ott M, Gros F, Pinset C (1991): Developmental patterns in the expression of Myf5, MyoD, myogenin, and MRF4 during myogenesis. *New Biol* 3:592–600.

Murre C, McCaw PS, Baltimore D (1989a): A new DNA binding and dimerization motif in immunoglobulin enhancer binding, daughterless, MyoD, and myc proteins. *Cell* 56:777–783.

Murre C, McCaw PS, Vaessin H, Caudy M, Jan LY, Jan YN, Cabrera CV, Buskin JN, Hauschka SD, Lassar AB, Weintraub H, Baltimore D (1989b): Interactions between heterologous helix-loop-helix proteins generate complexes that bind specifically to a common DNA sequence. *Cell* 58:537–544.

Ott M-O, Bober E, Lyons G, Arnold H, Buckingham M (1991): Early expression of the myogenic regulatory gene, myf-5, in precursor cells of skeletal muscle in the mouse embryo. *Development* 111:1097–1107.

Piette J, Bessereau J, Huchet M, Changeux J (1990): Two adjacent MyoD1-binding sites regulate expression of the acetylcholine receptor α-subunit gene. *Nature* (Lond) 345:353–355.

Pollock R, Treisman R (1991): Human SRF-related proteins: DNA-binding properties and potential regulatory targets. *Genes & Dev* 5:2327–2341.

Rhodes SJ, Konieczny SF (1989): Identification of MRF4: A new member of the muscle regulatory factor gene family. *Genes & Dev* 3:2050–2061.

Rosenthal N, Berglund EB, Wentworth BM, Donoghue M, Winter B, Bober E, Braun T, Arnold HH (1990): A highly conserved enhancer downstream of the human MLCY$_3$ locus is a target for multiple myogenic determination factors. *Nuc Acid Res* 18:6239–6246.

Sartorelli V, Webster KA, Kedes L (1990): Muscle-specific expression of the cardiac α-actin gene requires MyoD1, CArG-box binding factor, and Sp1. *Genes & Dev* 4:1811–1822.

Sassoon D, Lyons G, Wright WE, Lin V, Lassar A, Weintraub H, Buckingham M (1989): Expression of two myogenic regulatory factors myogenin and MyoD1 during mouse embryogenesis. *Nature* (Lond) 341:303–307.

Schafer BW, Blakely BT, Darlington GJ, Blau HM (1990): Effect of cell history on response to helix-loop-helix family of myogenic regulators. *Nature* (Lond) 344:454–458.

Schneider MD, Perryman MB, Payne PA, Spizz G, Roberts R, Olson EN (1987): Autonomous expression of c-*myc* in BC₃H1 cells partially inhibits but does not prevent myogenic differentiation. *Mol Cell Biol* 7:1973–1977.

Schroter H, Mueller CG, Meese K, Nordheim A (1990): Synergism in ternary complex formation between the dimeric glycoprotein p67SRF, polypeptide p62TCF and the c-*fos* serum response element. *EMBO J* 9:1123–1130.

Schwarz JJ, Chakraborty T, Martin J, Zhou J, Olson EN (1992): The basic region of myogenin cooperates with two transcription activation domains to induce muscle-specific transcription. *Mol Cell Biol* 12:266–275.

Sorrentino V, Pepperkok R, Davis RL, Ansorge W, Philipson L (1990): Cell proliferation inhibited by MyoD1 independently of myogenic differentiation. *Nature* (Lond) 345:813–815.

Stern S, Tanaka M, Herr W (1989): The Oct-1 homoeodomain directs formation of a multiprotein-DNA complex with the HSV transactivator VP16. *Nature* (Lond) 341:624–630.

Sternberg EA, Spizz G, Perry WM, Vizard D, Weil T, Olson EN (1988): Identification of upstream and intragenic regulatory elements that confer cell-type-restricted and differentiation-specific expression on the muscle creatine kinase gene. *Mol Cell Biol* 8:2896–2909.

Sun X-H, Baltimore D (1991): An inhibitory domain of E12 transcription factor prevents DNA binding in E12 homodimers but not in E12 heterodimers. *Cell* 64:459–470.

Tan S, Richmond TJ (1990): DNA binding-induced conformational change of the yeast transcriptional activator PRTF. *Cell* 62:367–377.

Tapscott SJ, Davis RL, Thayer MJ, Cheng P, Weintraub H, Lassar AB (1988): MyoD1: A nuclear phosphoprotein requiring a *myc* homology region to convert fibroblasts to myoblasts. *Science* 242:405–411.

Taylor M, Treisman R, Garrett N, Mohun T (1989): Muscle-specific (CArG) and serum-responsive (SRE) promoter elements are functionally interchangeable in *Xenopus* embryos and mouse fibroblasts. *Development* 106:67–78.

Thayer MJ, Tapscott SJ, Davis RL, Wright WE, Lassar AB, Weintraub H (1989): Positive autoregulation of the myogenic determination gene MyoD1. *Cell* 58:241–248.

Treisman R (1990): The SRF: A growth factor responsive transcriptional regulator. *Semin Cancer Biol* 1:47–58.

Tuil D, Clergue N, Montarras D, Pinset C, Kahn A, Phan-Dinh-Tuy F (1990): CC Ar GG boxes, *cis*-acting elements with a dual specificity. Muscle-specific transcriptional activation and serum responsiveness. *J Mol Biol* 213:677–686.

Vaidya TB, Rhodes SJ, Taparowsky EJ, Konieczny SF (1989): Fibroblast growth factor and transforming growth factor β repress transcription of the myogenic regulatory gene MyoD1. *Mol Cell Biol* 9:3576–3579.

Walsh K (1989): Cross-binding of factors of functionally different promoter elements in c-*fos* and skeletal actin genes. *Mol Cell Biol* 9:2191–2201.

Walsh K, Schimmel P (1987): Two nuclear factors compete for the skeletal muscle actin promoter. *J Biol Chem* 262:9429–9432.

Walsh K, Schimmel P (1988): DNA-binding site for two skeletal actin promoter factors is important for expression in muscle cells. *Mol Cell Biol* 8:1800–1802.

Webster KA, Muscat GEO, Kedes L (1988): Adenovirus E1A products suppress myogenic differentiation and inhibit transcription from muscle-specific promoters. *Nature* (Lond) 332:553–557.

Weintraub H, Davis R, Lockshon D, Lassar A (1990): MyoD binds cooperatively to two sites in a target enhancer sequence: Occupancy of two sites is required for activation. *Proc Natl Acad Sci USA* 87:5623–5627.

Weintraub H, Dwarki VJ, Verma I, Davis R, Hollenberg S, Snider L, Lassar A, Tapscott SJ (1991): Muscle-specific transcriptional activation by MyoD. *Genes & Dev* 5:1377–1386.

Weintraub H, Tapscott SJ, Davis RL, Thayer MJ, Adam MA, Lassar AB, Miller AD (1989): Activation of muscle-specific genes in pigment, nerve, fat, liver, and fibroblast cell lines by forced expression of MyoD. *Proc Natl Acad Sci USA* 86:5434–5438.

Wentworth BM, Donoghue M, Engert JC, Berglund EB, Rosenthal N (1991): Paired MyoD binding sites regulate myosin light chain gene expression. *Proc Natl Acad Sci USA* 88:1242–1246.

Wright WE, Binder M, Funk W (1991): Cyclic amplification and selection of targets (CASTing) for the myogenin consensus binding site. *Mol Cell Biol* 11:4104–4110.

Wright WE, Sassoon DA, Lin VK (1989): Myogenin, a factor regulating myogenesis, has a domain homologous to MyoD1. *Cell* 56:607–617.

Xiao JH, Davidson I, Mathes H, Garnier JM, Chambon P (1991): Cloning, expression and transcriptional properties of the human enhancer factor TEF-1. *Cell* 65:551–568.

Yutzey KE, Rhodes SJ, Konieczny SF (1990): Differential *trans* activation associated with the muscle regulatory factors MyoD1, myogenin and MRF4. *Mol Cell Biol* 10:3934–3944.

Chapter 6

Gene Expression in Hematopoietic Cells: The β-Globin Gene

Beverly M. Emerson

Study of the regulation of gene expression in hematopoietic cells has primarily focused on the globin gene families. These genes are very well characterized in terms of their physical arrangement within the genomes of many different species and in their involvement in hemoglobinopathies (see reviews by Bunn and Forget, 1986; Collins and Weissman, 1984; Nienhuis and Maniatis, 1987; Stamatoyannopoulos and Nienhuis, 1987; Weatherall et al., 1989). The β-globin genes are quite interesting from a transcriptional point of view because their expression is highly erythroid specific and individual members of the multigene family are expressed at defined developmental stages. Transcriptional control is regulated primarily by chromatin structure and the availability of multiple, tissue-specific, *trans*-acting proteins (see review by Evans et al., 1990). Chromatin structural studies have shown that the entire β-globin gene locus is maintained in an "active" nucleosomal conformation throughout erythroid development although the locus is packaged in an "inactive" conformation in nonexpressing cells. Once the chromosomal locus is activated, individual globin genes are differentially regulated by *trans*-acting factors operating on specific promoters and enhancers that are associated with each gene. The important early step of activating the chromosomal locus prior to gene expression is thought to be regulated by sequences called the locus control regions (LCRs), which are located quite distal to globin structural genes. Thus, globin expression is transcriptionally controlled by the *global activation* of a large chromosomal locus and the *local activation* of individual genes.

GENE EXPRESSION: GENERAL AND CELL-TYPE-SPECIFIC
Michael Karin, Editor
© 1993 Birkhäuser Boston

The *cis*-acting elements that regulate globin transcription, such as promoters, enhancers, and LCRs, interact with a combination of erythroid-specific and ubiquitous proteins. A current focus of research is to understand how precise combinations of multiple proteins function with eath other when bound to transcriptional control regions. Often these regions contain consensus protein-binding sites that are overlapping and mutually exclusive. This implies that transcriptional factors must interact at different times to regulate gene activation in discrete stages. The availability of recently cloned erythroid-specific regulatory proteins should be instrumental in addressing these questions. Another important area of interest concerns the mechanism by which distal regulatory regions interact with each other. Most β-globin genes are activated by distal enhancers, and specific protein complexes within individual promoters that mediate enhancer-dependent transcription have been identified. Finally, deciphering the mechanism of LCR action in terms of its ability to activate large chromosomal domains and to regulate individual globin genes within a multigene family remains a very crucial area of research.

The complex regulatory pathways that determine the tissue- and development-specific activation of globin genes undoubtedly have many features in common with other genes that are under transcriptional control; for example, tissue-specific members of the GATA family of transcription factors have been discovered. Thus, laboratories studying hematopoietic gene regulation should continue to provide valuable insight into the broader principles governing development and differentiation.

Organization and Chromatin Structure of the β-Globin Gene Locus

The globin genes of the α and β families are contained in separate chromosomal loci in an arrangement that usually reflects the developmental order of gene expression (Bunn and Forget, 1986). For example, the human β-globin locus is approximately 90 kb and consists of five genes that are arranged in the 5' to 3' order in which they are expressed during erythroid maturation: embryonic (ε), fetal ($^{G}\gamma$, $^{A}\gamma$), and adult (δ, β). The ε gene is expressed in the embryonic yolk sac, γ in the fetal liver, and δ and β predominantly in the adult bone marrow. The human α-globin locus contains three genes that are also expressed in the same order: embryonic (ζ) and fetal/adult (α_2, α_1). However, the chick β-globin locus (approximately 30 kb) has a gene order that is nonlinear in its developmental expression: embryonic (ρ), adult (β^{Hatching}, β^{A}), and embryonic (ε). A diagram of the β-globin gene chromosomal locus in human and chick is shown in Figure 1.

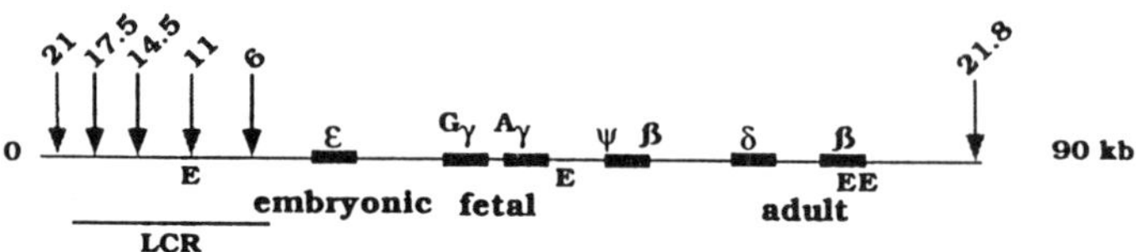

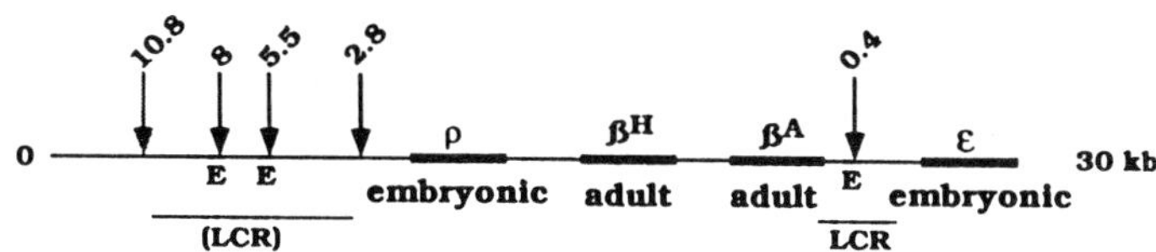

Figure 1   Organization of the β-globin gene locus in human and chick; the developmental stage (embryonic, fetal, adult) at which they are expressed is indicated underneath. Enhancer elements and locus control regions that have been identified in the globin gene cluster are designated E and LCR, respectively; (LCR) denotes potential locus control regions. *Arrows* flanking genes represent DNase-hypersensitive sites mapped in chromatin of globin locus in erythroid cells (*stippled arrows* indicate sites that are not erythroid specific). In the human locus, the 5′ sites are labeled by distance (kb) from the ε-globin gene, and the 3′ site is labeled by distance from the β-globin gene. The 5′ sites in the chick locus are labeled by approximate distance from the ρ-globin gene; the 3′ site in the β-ε enhancer is 0.4 kb from the end of the β-globin gene. (Adapted from Evans et al., 1990).

Much of the early work defining the chromatin structure of transcriptionally active and inactive genes was accomplished using the chick globin genes as a model system (McGhee et al.,1981; Stalder et al., 1980). These studies demonstrate that both the α- and β-globin gene loci are maintained in a DNase-sensitive structure throughout erythroid development. By contrast, the α- and β-globin loci in nonerythroid cells exist in a DNase-insensitive structure that is correlated with permanently repressed genes. Similar findings have been determined for the human β-globin (Forrester et al., 1986) and α-globin loci (Yagi et al., 1986). In erythroid cells, only a subset of β-globin genes within the locus is expressed at any given time (Brown and Ingram, 1974; Bruns and Ingram, 1973; Chapman and Tobin, 1979). This indicates that the DNase-sensitive chromatin structure must allow the transient activation and repression of globin genes within its active domain. An "activated" chromatin structure is therefore necessary for transcription but is insufficient without the regulation imposed by local control elements that are associated with individual genes. In erythroid cells the chromatin structure of the human (Forrester et al., 1986; Tuan et al., 1985) and chick β-globin genes (Reitman and Felsenfeld, 1990) is characterized by multiple nuclease hypersensitive (HS) sites that are distributed throughout the locus but are mostly located far upstream of the globin structural genes (see Figure 1). Hypersensitive sites usually represent a localized disruption in the normal nucleosomal array near transcriptional control regions, and their appearance is often correlated with gene activity. The HS sites in the β-globin locus can be divided into several classes: ubiquitous, erythroid-specific, or erythroid stage-specific. The ubiquitous and erythroid-specific sites that remain throughout red cell development are especially sensitive to digestion by DNase I and are the major hypersensitive sites in the gene cluster. In the human β-globin locus, the HS sites at -21 and -11 kb upstream of the ε-globin gene and at 21.8 kb downstream of the β-globin gene are ubiquitous whereas the HS sites at -17.5, -14.5, and -6 kb are erythroid specific (Dhar et al., 1990; Forrester et al., 1989). The four HS sites at -17.5, -14.5, -11, and -6 kb reside within sequences that comprise the locus control region (LCR). This transcriptional control element is functionally distinct from local control regions, although an LCR can act as an enhancer, and is required for the tissue-specific activation of the β-globin genes within its chromosomal domain (Grosveld et al., 1987; Forrester et al., 1987). In the chick β-globin locus, a single HS site at the far 5$'$ end of the gene cluster is present in many tissues whereas four HS sites are restricted to erythroid cells and are present throughout development (Reitman and Felsenfeld, 1990). Three of the erythroid-specific sites are present in DNA sequences having transcriptional enhancer activity, and at least one site has been shown to have LCR activity (Evans et al., 1990). The available evidence on the role of the ubiquitous and erythroid-specific HS sites within the β-globin locus suggests that these sites may serve to generate and define the boundaries of large transcriptionally active chromatin domains.

In addition to HS sites that persist throughout erythroid development, another set of HS sites appear only in an erythroid stage-specific manner.

This class of developmentally regulated chromatin structural changes is most clearly illustrated in the chick β-globin gene locus (Reitman and Felsenfeld, 1990). By 5 days of chick embryonic development, the circulating erythroid cell population consists of the primitive red cell lineage (defined as the unsustained erythroid cell line in young embryos) (Bruns and Ingram, 1973), and HS sites are present in the 5'- and 3'-flanking regions of the two embryonic globin genes, ρ and ε, which are expressed at this stage. However, by 10 days of development, the primitive red cell lineage has been replaced by the definitive lineage (the sustained erythroid cell line) in which only the two adult globin genes, β^{Hatching} and β^{Adult}, are expressed. At this stage, the HS sites flanking the embryonic globin genes have disappeared and new HS sites are present in the 5' promoter regions of the two adult β-globin genes (Figure 2). In the case of the β^{Adult} gene, the 5' HS site reflects a developmentally regulated nucleosome-free region in the promoter whose appearance is correlated with gene expression (McGhee et al., 1981). The removal of this nucleosome creates an accessible region of approximately 260 base pairs to which multiple transcription factors can bind (Emerson et al., 1985). Because the 3' enhancer of the β^{A}-globin gene is hypersensitive and presumably accessible at all stages of erythroid development, the removal of a nucleosome from the promoter is the last major regulated chromatin structural change before transcription. Thus, chromatin activation appears to proceed in stages: initially the major HS sites that remain throughout erythroid development in regions of LCR activity may function to activate a large chromosomal domain while the stage-specific HS sites regulate the accessibility of local control elements to activate individual genes at a precise time. Several lines of evidence indicate that the generation of both the major and stage-specific HS sites is caused by the binding of diffusible protein factors present in erythroid cells. In one set of experiments, the fusion of mouse erythroleukemia (MEL) cells with human nonerythroid cells resulted in the generation of the complete set of major HS sites in the normally inactive human β-globin gene locus, suggesting that erythroid proteins derived from MEL cells could alter the chromatin structure of the repressed globin locus in human cells (Forrester et al., 1987). Moreover, in chromatin reconstitution experiments, the generation of HS sites in the 5' promoter and 3' enhancer of the chick β^{Adult} globin gene required the preincubation of stage-specific erythroid proteins before nucleosome assembly (Emerson and Felsenfeld, 1984). The identification of proteins that regulate the tissue-specific activation

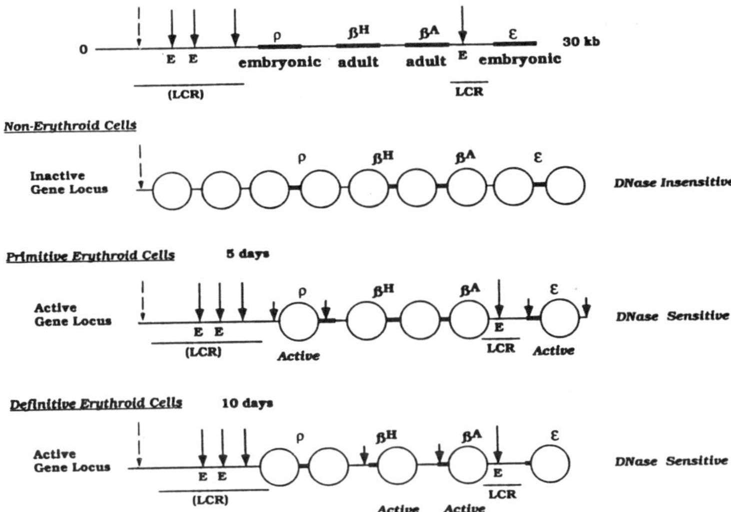

Figure 2   Chromatin structure of the chick β-globin locus is modulated during erythroid development. The chick β-globin gene locus is assembled in a DNase-insensitive chromatin structure in nonerythroid cells, indicated by closely spaced nucleosomes (*circles*), and is transcriptionally quiescent. By contrast, in erythroid cells the β-globin locus is assembled in a DNase-sensitive chromatin structure, indicated by loosely spaced nucleosomes (*circles*). Throughout erythroid development this structure is characterized by the presence of five major nuclease hypersensitive sites (*large arrows*): four are erythroid specific, three of which occur near enhancer regions (E), and one is present in both erythroid and nonerythroid cells (*stippled arrow*). In primitive erythroid cells, four hypersensitive sites transiently appear (*small arrows*) near the embryonic globin genes (ρ and ε) that are expressed at this time. In definitive erythroid cells, embryonic globin genes are inactive and the associated hypersensitive sites have disappeared. Adult globin genes (β^H and β^A) are expressed at this stage, and two new hypersensitive sites appear in the promoters of these genes. LCR, locus control region; (LCR), potential LCR. (Adapted from Reitman and Felsenfeld, 1990.)

of chromatin structure over the entire β-globin locus or that affect the structure of individual genes is an area of active investigation.

Regulation of β-Globin Gene Transcription by Trans-Acting Factors

Most of the information about the regulation of hematopoietic genes comes from studies on the human and chick β-globin genes. The local activation of globin genes within the gene family is regulated by the interaction of multiple *trans*-acting factors with promoter and enhancer DNA elements that are associated with individual genes. These proteins function as both direct transcriptional activators and repressors as well as modifiers of chromatin structure.

Chick β-Globin Promoter and Enhancer Regulation: Multiple Activators Regulate Stage-Specific Chick β-Globin Promoter Activity

The chick β-globin gene is transcriptionally regulated by a 5′ promoter and a distal 3′ enhancer. Chromatin structural studies have shown that the promoter is blocked by a nucleosome until the β^A-globin gene is expressed at about day 9 (McGhee et al., 1981) but the enhancer is nucleosome free throughout erythroid development (Emerson et al., 1987). This indicates that the enhancer region is accessible to the binding of transcription factors before β^A-globin expression occurs. However, the promoter remains inaccessible until the correct stage of development when it undergoes a pronounced chromatin structural change that enables regulatory proteins to bind and activate transcription. Protein–DNA interaction studies have shown that at least eight proteins bind to the promoter region (Figure 3) in definitive red cells and can prevent a nucleosome from forming over this control element (Emerson and Felsenfeld, 1984). Most of these proteins, such as GATA-1, NF-E4, GC, and BGP1, are erythroid specific (Emerson et al., 1985; Jackson et al., 1989; Plumb et al., 1986). At least one, NF-E4, which is present in highest abundance in definitive red cells, has been shown to be erythroid stage-specific (Gallarda et al., 1989). To determine the role of these proteins in β^A-globin regulation, functional studies have been carried out by DNA transfection into chick erythroid cells (Jackson et al., 1989) and by *in vitro* transcription using erythroid extracts (Emerson et al., 1989). The results from both types of analyses, which are almost identical, are summarized in Figure 3. Deletion to −165 removes the binding sites for both the erythroid-specific factor BGP1 and the ubiquitous factor Sp1, which together bind a sequence of 18 contiguous guanine residues that is flanked by a GC box (Emerson et al., 1985; Plumb et al., 1986) but has no effect on promoter activity. However, some evidence exists that these two proteins may have a role in generating a developmentally regulated nuclease-hypersensitive chromatin structure in this region. BGP1 has been purified and shown to be related to Sp1 by immunological cross-reactivity (Lewis et al., 1988).

Promoters deleted to −122, which abolishes protein binding to the CACC box, as well as promoters deleted to −58, which removes the

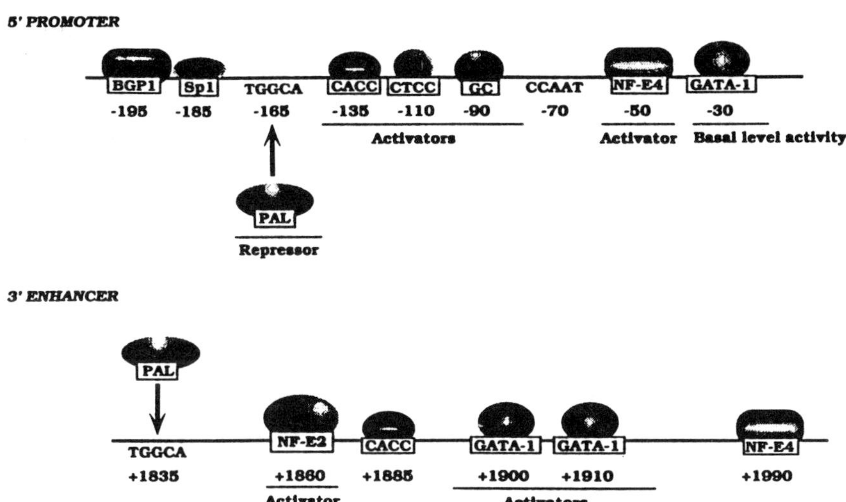

Figure 3   Chick β-globin promoter and enhancer regions. Specific protein–DNA complexes on the chick β-globin promoter and enhancer regions were mapped using extracts from 11-day-old chick erythroid cells that actively express this gene. The protein composition changes during red cell development and consists of both erythroid-specific (GATA-1, NF-E4, GC, BGP1, NF-E2) and general transcription factors. Tissue-specific members of complex gene families, such as CACC, may bind to the β-globin gene in erythroid cells. Interaction of the PAL repressor protein with the promoter and enhancer is favored in mature, inactive red cells where it is present in high concentrations.

CTCC, GC, and CCAAT sites, show a marked decrease in β^{A}-globin expression. There is no detectable binding activity to the CCAAT box in chick erythroid extracts, and it is unclear what functional significance this site may have in chick β^{A}-globin expression. The protein binding to the CTCC site may be a ubiquitous factor, but the GC box interacts with a protein that appears to be unrelated to Sp1 and is present in high abundance in transcribing red cells. The CACC box is conserved among all β-globin promoters (Dierks et al., 1983) and is also present in the regulatory regions of nonerythroid genes. Proteins binding to this recognition sequence are related to the transcription factor Sp1 (Kadonaga et al., 1987) but may exist in tissue-specific forms. For example, the nuclear protein TEF-2 recognizes CACC sequences in the SV40 enhancer but forms a complex that is distinct from that of Sp1 (Davidson et al., 1988; Nomiyama et al., 1987; Xiao et al., 1987). In erythroid cells, there are several different proteins that bind CACC boxes and consist of ubiquitous Sp1 as well as a form(s) that is highly abundant in red cells (Catala et al., 1989; Emerson et al., 1989; Jackson et al., 1989; Mantovani et al., 1988). The erythroid-enriched form of the CACC binding protein was affinity purified and shown to be a potent activator of β^A-globin expression when added back to erythroid transcription extracts depleted of this factor (Emerson et al., 1989).

The promoter region from -58 to -20 contains binding sites for the erythroid-specific proteins NF-E4 and GATA-1. NF-E4 is present in high concentrations in definitive but not primitive chick red cells, and has been implicated in regulating the switch in expression from embryonic ε-globin to adult β^A-globin through a DNA site at -50 called the stage-selector element or SSE (Choi and Engel, 1988; Gallarda et al., 1989). This protein is related to, but distinct from, Sp1 and functions as an activator of the β^A-globin promoter in both genetic (Choi and Engel, 1988) and biochemical (Emerson et al., 1989) analyses. NF-E4 binds near the transcriptional initiation site at -30, which in each member of the chick β-globin gene family except β^H, contains a noncanonical TATA box that is 'GATA.' This site represents the core recognition sequence for the erythroid-specific protein GATA-1. Binding sites for GATA-1 have been found in the transcriptional control regions of all erythroid-expressed genes and are distributed throughout the β-globin gene locus (Evans et al., 1988; Plumb et al., 1989). Mutation of the -30 GATA site to a sequence that eliminates the binding of both GATA-1 and the TATA-binding protein, TFIID, completely abolishes β^A-globin transcription *in vivo* and *in vitro* (Fong and Emerson, 1992).

The functional studies conducted on the chick β^A-globin promoter demonstrate that the CACC, GC, and NF-E4 proteins are the most important activators of transcription. These proteins most likely function by upregulating the activity of the basal initiation complex formed at the specialized TATA box by GATA-1 and TFIID. The concentrations of these erythroid-specific proteins are developmentally regulated during chick erythropoiesis such that their promoter occupancy is highest in definitive red cells at approximately 10–14 days of embryogenesis when β^A-globin transcription is maximal.

Chick β-Globin Promoter Inactivation is Controlled by the PAL Protein in Vitro

In addition to multiple activators regulating the β^A-globin gene promoter, there is compelling evidence for the existence of a potent transcriptional repressor that is also developmentally regulated. Removal of the palindromic sequence at -165 results in a severalfold increase in promoter activity, indicating that the PAL protein which recognizes this site may be functioning as a transcriptional repressor. This was substantiated by the observation that PAL is present in high abundance in nonerythroid cells and in mature, transcriptionally quiescent red cells but in very low abundance in expressing red cells (Emerson et al., 1989; Jackson et al., 1989). Moreover, when affinity-purified PAL is added back to transcribing red cell extracts, β^A-globin gene expression is effectively repressed unless the palindromic site is mutated to prevent binding (Emerson et al., 1989). Interestingly, the PAL protein appears to be a member of the CTF/NF-1 protein family whose recognition sequence is TGGCA (Jones et al., 1987), because purified human CTF binds to identical sequences within the β^A-globin promoter as the PAL protein (Emerson et al., 1989). However, PAL and CTF are functionally distinct because CTF is unable to repress β^A-globin transcription when bound to the promoter. This indicates that mere occupancy of the TGGCA binding site is insufficient for repression to occur and implies that the non-DNA-binding portion of the PAL protein, which presumably differs from that of CTF, may interfere with protein–protein contacts between the activator proteins and the basal transcriptional machinery resulting in promoter inactivation. DNase protection experiments indicate that repression does not appear to result from a displacement of CACC or other activator proteins bound to the promoter (Emerson et al., 1989), suggesting that PAL does not function in the manner of the CCAAT displacement protein (Barberis et al., 1987). However, other studies employing a gel mobility shift analysis indicate that some displacement may occur between PAL and the CACC protein (Jackson et al., 1989). In addition to binding the β^A-globin promoter, PAL also interacts with the distal 3′ enhancer in adult red cells and may have a role in repressing enhancer function as well. Interestingly, DNase protection studies in intact nuclei ("*in vivo* footprinting") have demonstrated that the PAL binding site in the β^A-globin promoter is occupied in chick oviduct and mature erythrocytes but not in expressing erythroid cells (Jackson and Felsenfeld, 1985). Moreover, nuclease digestion studies reveal that nucleosomes appear to be phased on the

β^A-globin promoter in nonerythroid cells, in contrast to the hypersensitive chromatin structure found in erythroid cells. This suggests that the PAL protein may influence nucleosome distribution in permanently repressed β^A-globin genes. A diagram depicting the effects of the PAL protein on β^A-globin gene expression and chromatin structure is shown in Figure 4.

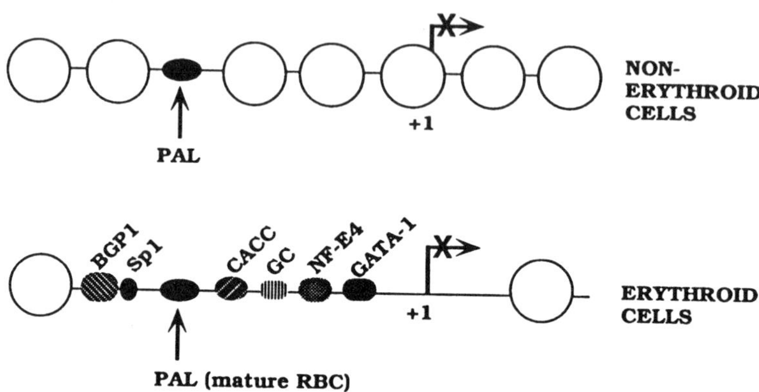

Figure 4   PAL protein, a transcriptional repressor of the chick β-globin promoter *in vitro*, is developmentally regulated in chick red cells; it is present in low concentrations in transcriptionally active erythrocytes and in high concentrations in mature erythrocytes and nonerythroid cells. PAL is a potent repressor of chick β^A-globin transcription *in vitro* and interacts with the β^A-globin promoter in nonerythroid cells to potentially affect nucleosomal phasing. PAL may repress chick β^A-globin transcription by two distinct mechanisms: as a direct transcriptional repressor by interfering with multiple activator proteins in erythroid cells, and as an indirect repressor by altering β^A-globin chromatin structure in nonerythroid cells. *Circles*, nucleosomes; X, repressed transcription.

The PAL protein binds to the chick β^A-globin promoter in mature erythrocytes and in nonerythroid cells and may function as a transcriptional repressor in both cell types by different mechanisms of action. For example, in nonerythroid cells, where β-globin genes are permanently inactive, PAL presumably binds to the promoter before chromatin assembly and may affect nucleosome distribution to produce phasing. The functional consequence of this may be a more stringent form of repression by ensuring that a nucleosome is tightly positioned over the transcriptional initiation and cap sites. In mature erythrocytes, where β-globin genes have been expressed but are now inactive, the β^A-globin promoter already exists in a hypersensitive chromatin structure because of the binding of multiple activator proteins at an earlier stage of red cell development. PAL can freely interact with the promoter at this stage, because its binding site is accessible, but cannot affect chromatin structure. Thus, PAL acts as a direct transcriptional repressor in erythroid cells by interfering with the activity of the activator proteins; in nonerythroid cells, however, PAL may function as an indirect repressor by modulating chromatin structure to produce a more tightly repressed gene. Because PAL also binds to the β-globin enhancer, it will be interesting to determine whether it contributes to the inactivation of this important control region.

In summary, tissue-specific β^A-globin promoter function is dependent on at least two conditions: a combination of activators whose concentrations are regulated to be highest in red cells of the appropriate developmental stage, and the absence in expressing erythroid cells of the potent, sequence-specific repressor PAL and the presence of high levels of this repressor in nonexpressing cells. A balance of these regulated factors determines the net activity of the promoter. Thus, the β^A-globin promoter in expressing definitive red cells is transcriptionally activated as well as derepressed.

Chick β-Globin Enhancer Activity is Positively Regulated by Two Erythroid-Specific Proteins, GATA-1 and NF-E2

In addition to regulation by a complex promoter, β^A-globin gene expression is also dependent on a distal, erythroid-specific enhancer. The 3' enhancer is located between the β^A and ε-globin genes at approximately 1800 bp from the β^A-globin cap site (Choi and Engel, 1986; Hesse et al., 1986). Interestingly, the enhancer can regulate both the β^A and ε-globin genes at different times in development (Choi and Engel, 1988; Nickol and Felsenfeld, 1988), consistent with the observation that this control region, unlike the β^A-promoter, is accessible in chromatin and presumably bound by protein at all stages of erythropoiesis. The $\beta - \varepsilon$ enhancer stimulates transcription of the chick β^A-globin gene approximately 100 fold on transfection into primary chick red blood cells (Hesse et al., 1986) or into a chick erythroid cell line, HD3 (Choi and Engel, 1986). Protein–DNA interaction studies have shown that at least five proteins bind to the enhancer region in definitive red cells (see Figure 3), many of which also interact with the promoter: PAL (in mature erythrocytes), NF-E2, an erythroid-specific member of the AP1 transcription factor family (Angel et al., 1987; Lee et al., 1987), CACC, GATA-1, and NF-E4. An extensive mutational analysis of individual protein binding sites indicates that NF-E2 and GATA-1 are the most important positive regulators of $\beta - \varepsilon$ enhancer activation of the β^A-globin promoter. Mutations of the CACC, PAL, and NF-E4 sites in the $\beta - \varepsilon$ enhancer have little effect (Reitman and Felsenfeld, 1988). However, NF-E4 may be required to mediate the switch in enhancer activation between the β^A and ε-globin promoters when both genes are linked in *cis* (Gallarda et al., 1989). The GATA-1 and NF-E2 proteins were previously shown to be required for the erythroid-specific induction of the human porphobilinogen deaminase gene in MEL cells (Mignotte et al., 1989a, 1989b).

Human β-Globin Promoter and Enhancer Regulation: Erythroid-Specific Activation of a Minimal Human β-Globin Promoter

The human β-globin gene is transcriptionally regulated by three local control elements: a $5'$ promoter, an intragenic enhancer, and a $3'$ enhancer located just past the poly A site (Behringer et al., 1987; Kollias et al., 1987; Trudel and Costantini, 1987). Both the promoter and enhancer elements interact with multiple proteins, which consist of erythroid-specific and general transcription factors (Figure 5). Functional studies on the human and mouse β-globin promoter indicate that a limited region extending from -30 to -100 is required for minimal promoter activity (Dierks et al., 1983; Myers et al., 1986). This region interacts with at least three proteins through the TATA (-30), CCAAT (-70), and CACC (-90) DNA-binding sites. Each of these sites is required for β-globin expression in transfected erythroid and nonerythroid cells (Cowie and Myers, 1988). Proteins interacting with the TATA and CCAAT sites are most likely general transcription factors whereas the CACC box interacts with a family of related proteins, some of which are erythroid-specific (Mantovani et al., 1988). Mutations in the TATA and CACC sequences have been found in human thalassemias that are characterized by reduced levels of β-globin expression (reviewed by Weatherall et al., 1989). Sequences upstream of the minimal promoter may be important for developmental regulation because they are required for the induction of β-globin transcription in differentiated MEL (mouse erythroleukemia) cells (Antoniou et al., 1988). In particular, a combination of the -150 site with either the -200 or -120 site is capable of inducing the minimal promoter, presumably through the CACC site, in an erythroid-specific manner in the absence of an enhancer. The proteins thought to be responsible for this induction are the CCAAT box protein, CP1 (Chodosh et al., 1988), which interacts at −150, and GATA-1 (Martin et al., 1989; Wall et al., 1988), which binds with high and low affinity at −200 and −120, respectively (deBoer et al., 1988). Thus, a general transcription factor, CP1, must act in conjunction with an erythroid-specific factor, GATA-1, to regulate the basal promoter in induced MEL cells. In the presence of an erythroid enhancer, however, only 48 bp of the promoter (consisting of the TATA box) is required for erythroid-specific β-globin gene expression in both MEL cells (Wright et al., 1984) and transgenic mice (Townes et al., 1985). This suggests that the relative importance of particular promoter control regions in determining erythroid specificity may be modulated by a functional enhancer.

Human β-Globin Enhancer Activity is Regulated by Multiple GATA-1 Proteins

The 3′ enhancer region is located about 2200 bp from the human β-globin transcriptional start site. This enhancer stimulates β-globin expression about 10 fold (Trudel and Costantini, 1987) and, in combination with the intragenic enhancer, can regulate a permissive promoter (SV40 early) in a tissue-specific and erythroid stage-specific manner (Magram et al., 1989) similar to its functional counterpart, the chick $\beta - \varepsilon$ enhancer. Protein–DNA interaction studies have shown that the 260-bp minimal enhancer contains recognition sequences for four GATA-1 sites (see Figure 5). These sites are bound by GATA-1 in erythroid extracts from β-globin expressing (induced MEL) and nonexpressing cells (uninduced MEL and K562), indicating that other factors must act in conjunction with GATA-1 to regulate stage-specific transcription. Proteins present in HeLa cell extracts also bind to regions containing three of the four GATA sites within the minimal enhancer (Wall et al., 1988). Interestingly, GATA-1 consensus binding sites in both the human β-globin enhancer and promoter appear to overlap, or be very close to recognition sequences for general factors such as CP1 and OCTA (deBoer et al., 1988; Wall et al., 1988). Previous studies have shown that the octamer binding protein OCTA can be either ubiquitous or tissue-specific (Rosales et al., 1987), and at least one form is regulated in erythroid K562 cells (Mantovani et al., 1987). Because these factors are present with GATA-1 in erythroid cells, protein competition may exist to determine the occupancy of overlapping regulatory sequences. A recent study describing overlapping binding sites for the GATA-1, CP2, and Sp1 proteins in the human ζ-globin promoter demonstrated that GATA-1 can displace bound Sp1 molecules (Yu et al.,

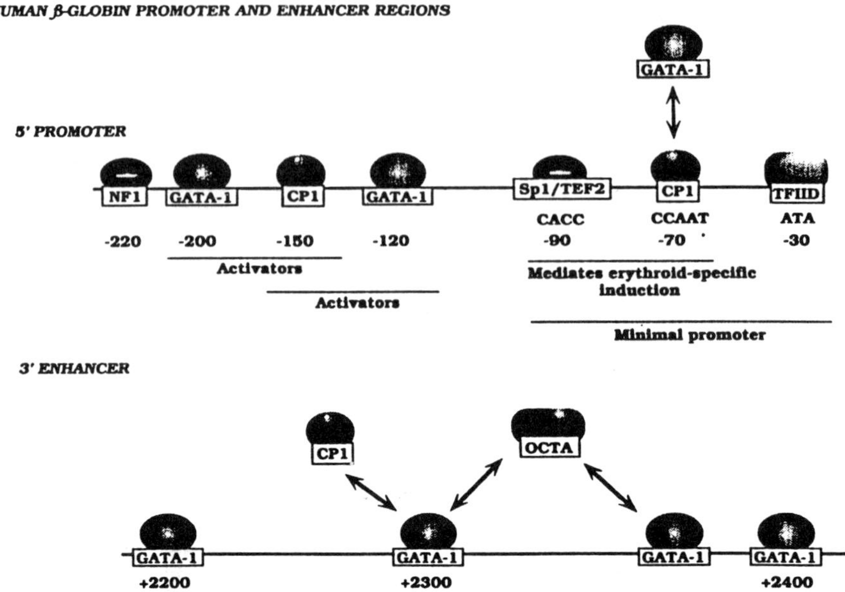

Figure 5   Positions of specific protein–DNA complexes on the human β-globin promoter and enhancer regions were determined using extracts from mammalian erythroid cells that actively express β-globin genes. Proteins consist of erythroid-specific (GATA-1) and general transcription factors. Tissue-specific members of complex gene families, such as CACC, may bind in erythroid cells. Proteins with overlapping consensus binding sites are indicated by *arrows*. (Adapted from Wall et al., 1988 and deBoer et al., 1988.)

1990). Thus, a common mechanism that may emerge for the globin gene family is the ability to regulate transcription by protein competition or displacement from overlapping and mutually exclusive binding sites. In this way, temporal gene activation could be achieved by two proteins occupying the same site and having distinct functions at different times.

Regulation of β-Globin Promoter–Enhancer Interactions

Identification of β-Globin Promoter Sequences that are
Enhancer Responsive

In both human and chick, erythroid-specific β-globin gene regulation depends on the functional communication of promoter sequences with a distal enhancer. In each system, promoter sequences that are required for enhancer-dependent β-globin expression have been identified. For example, the region extending from -112 to -20 in the chick β^A-globin promoter is necessary for developmental stage-specific activation of the β^A-globin gene and repression of the embryonic ε-globin gene (Choi and Engel, 1988). This sequence interacts with several proteins, one of which is an erythroid stage-specific factor, NF-E4 (see Figure 3). This factor has been implicated in causing the regulated switch in transcription from the ε- to the β^A-globin gene by altering the interaction of the $\beta - \varepsilon$ globin enhancer with the promoters for each of these templates (Gallarda et al., 1989). Promoter–enhancer interaction studies on the human β-globin gene have utilized a "microlocus" cassette that contains the active regions of the LCR condensed into 6.5 kb and placed directly upstream of the minimal β-globin promoter linked to a reporter gene. In these constructs, the condensed LCR functions as an erythroid-specific enhancer and can replace the two downstream β-globin enhancers (Collis et al., 1990). Sequences within the human β-globin promoter that are responsive to the LCR enhancer have been localized to the TATA, CACC, and CCAAT consensus binding sites (Antoniou and Grosveld, 1990), which interact with TFIID, Sp1/TEF-2, and CP1, respectively (see Figure 5). The LCR stimulates basal promoter activity through the initiation complex at the TATA box but requires either the CACC or CCAAT box to mediate an erythroid-specific activation of β-globin transcription in induced MEL cells. The CACC and CCAAT elements can also function synergistically with the LCR, and with each other, to stimulate basal or induced β-globin expression in MEL cells at levels higher than obtained with either element alone (Antoniou and Grosveld, 1990). The importance of these sequences in mediating LCR function is substantiated by the observation that the LCR can direct erythroid-specific expression from heterologous promoters that contain functional CACC or CCAAT boxes. These promoters include the herpes simplex virus thymidine kinase (Talbot et al., 1989) and the murine Thy-1 (Blom van Assendelft et al., 1989). Moreover, the murine histone H4 promoter, which does not contain CACC or CCAAT boxes, is not responsive to the LCR even though an Sp1 site is present. This suggests that a tissue-specific member of the Sp1 family, such as TEF-2, may be required to mediate activation by the LCR.

These studies on the chick and human β-globin genes suggest that enhancer-dependent promoter activation is regulated by protein–protein contact through complexes bound to each control region. Erythroid-specific proteins bound to the CACC and CCAAT boxes or within the

−112 to −20 region in the human and chick β-globin promoters, respectively, appear to confer tissue specificity to the enhancer response. The model most often used to explain distal regulatory effects involves DNA looping, which brings activator proteins bound to enhancers into close proximity with promoters (Ptashne, 1988). Because many of the proteins that bind to the β-globin promoter, enhancer, and LCR are identical (see Figures 3 and 5), models have been proposed in which similar proteins bound to distal sites come into contact with each other through DNA looping (Gallarda et al., 1989).

In an effort to explore this mechanism further, a series of experiments were conducted on the chick β^A-globin gene (Fong and Emerson, 1992). Initially it was noticed that all members of the chick β-globin gene family, with the exception of β^{Hatching}, have a noncanonical TATA box that is 'GATA' at −30. Because this is the consensus recognition sequence for the erythroid-specific protein GATA-1, experiments were undertaken to determine whether this protein could regulate transcription through the initiation site as well as through the distal enhancer.

GATA-1 Binding to the Chick β-Globin −30 GATA Box is Required for Efficient Enhancer-Dependent Transcription

A series of point mutations were made at the −30 site in the β^A-globin promoter that were designed to eliminate the binding of GATA-1, TFIID, or both, and their effects were measured on transcription. Protein–DNA interaction studies demonstrate that the GATA-1 protein can interact at the −30 GATA site with very high affinity, as expected, but cloned human TFIID binds very weakly. If this site is mutated to a canonical TATA box, the affinities of these two proteins are reversed: TFIID interacts with high affinity whereas the binding of GATA-1 is undetectable. A mutation of the −30 GATA box to GGGA completely eliminates the interaction of both TFIID and GATA-1. To examine the functional consequence of the interaction of GATA-1 at the initiation site, the activity of β^A-globin promoters containing the GATA, TATA, and GG mutations at −30 was assessed. For this analysis, transcription levels were determined both *in vitro*, using erythroid extracts from the correct developmental stage, and *in vivo*, by transfection into primary chick embryonic red cells. In primary red cells, expression of chick β^A-globin genes is almost completely enhancer-dependent because a deletion of this element reduces transcription to a negligible amount even in the presence of a functional promoter (Hesse et al., 1986). A comparison of β^A-globin promoters bearing mutations at the -30 site revealed that the GG mutation, which eliminates the binding of both GATA-1 and TFIID, severely reduces β^A-globin expression relative to the unmodified gene. Interestingly, when the -30 GATA site is converted to a canonical TATA box, which binds TFIID but not GATA-1, a marked decrease in enhancer-dependent transcription is also observed. The protein binding and transcription data are summarized in Figure 6.

−30 Initiation Site	Protein Binding GATA-1	Transcription TFIID	Transcription *in vitro* Promoter (%)	*in vivo* Promoter + Enhancer (%)
GATA	++++	+	100	100
TATA	−	++++	100	29
GG	−	−	0	16
GATA (-enh)	++++	+	100	6

Figure 6   Summary of protein binding and transcriptional activity of chick β-globin promoters; the binding affinities of GATA-1 and TFIID for β^A-globin promoters containing various mutations at the -30 initiation site and the effect of different protein–DNA complexes on the regulation of β^A-globin gene expression. Transcription levels of β^A-globin genes having these initiation site mutations were observed *in vitro* using chick erythroid extracts to measure promoter activity and *in vivo* by transfection into primary chick erythrocytes to measure enhancer dependence. (GATA(-enh)), β^A-globin genes containing a normal promoter (-30 GATA) and a deleted enhancer.

TFIID but not GATA-1 is Required for Chick β-Globin Promoter Activity in Vitro

Promoter mutations that decrease β^A-globin expression in transfected erythroid cells may do so by either lowering promoter activity itself or by affecting the ability of this element to respond to the distal enhancer. To discriminate between these two possibilities, β^A-globin templates containing mutations in the -30 GATA box were examined for transcriptional activity *in vitro*. In most *in vitro* systems, gene expression levels are determined solely by promoter strength because distal enhancer elements are not functional. In this case, the chick β^A-globin gene is transcribed in erythroid extracts with equal efficiency in the presence or absence of its 3′ enhancer. Therefore, the *in vitro* transcription levels obtained from β^A-globin genes mutated at the −30 GATA box are a measure of promoter activity rather than promoter–enhancer interactions. By contrast, expression levels obtained from genes transfected into erythroid cells are dependent on both promoter strength and enhancer activity. As summarized in Figure 6, mutations that abolish the binding of both TFIID and GATA-1 at −30 (GG) result in completely inactive promoters when measured *in vitro* whereas promoters mutated to a canonical TATA box are as efficient as the unmodified form (GATA).

A comparison of transcriptional levels obtained *in vivo* and *in vitro* indicates that the inactivity of β^A-globin genes containing the −30 GG mutation is most likely because TFIID is unable to bind the promoter and establish an initiation complex. *In vivo*, low-level transcription is observed with the GG mutation, but there is no activity *in vitro*, presumably because the enhancer does not function under these conditions and expression is entirely promoter-dependent. Thus the GG mutational effects can be attributed to an inactive promoter. By contrast, β^A-globin genes containing the TATA mutation are much more active *in vitro* than *in vivo* relative to the normal −30 GATA template. Because this sequence change does not lower the activity of the promoter, the decreased expression *in vivo* must result from a disruption in promoter–enhancer interaction that requires the binding of GATA-1 at −30.

Other *in vitro* transcription/complementation experiments have demonstrated that the GATA-1 protein cannot functionally replace TFIID or influence its activity within the initiation complex (Fong and Emerson, 1992). This further substantiates the conclusion that GATA-1 does not function as a transcriptional activator when bound to the −30 GATA box but instead may have a structural role by mediating distal enhancer activation of the promoter. This presumably occurs through the formation of a DNA loop between the two control regions.

Interaction of GATA-1 at the Chick β-Globin Promoter and Enhancer May Facilitate Interaction Between the Two Regions

Because GATA-1 is present at all stages of erythroid development, DNA loop formation between the β^A promoter and $\beta - \varepsilon$ enhancer may occur in primitive red cells long before the β^A-globin gene is expressed. Alternatively, in definitive red cells, this connection between the two control regions could be stabilized by the binding of other proteins, such as the erythroid stage-specific factor, NF-E4 (Gallarda et al., 1989), to the promoter and enhancer. Interestingly, the GATA-1–enhancer complex is situated between the β^A- and ε-globin genes and controls their expression at different stages of erythroid development (Choi and Engel, 1988; Nickol and Felsenfeld, 1988). The NF-E4 protein is thought to function by regulating the differential interaction of the $\beta - \varepsilon$ enhancer with the β^A- and ε-globin promoters in definitive red cells (Gallarda et al., 1989). A model depicting the proposed DNA loop structure between the chick β^A-globin promoter and the $\beta - \varepsilon$ enhancer is shown in Figure 7.

TFIID Requires Adaptor Proteins to Interact with the Chick β-Globin Specialized TATA Box and to Displace GATA-1 from this Site

In erythroid cells, TFIID and GATA-1 must function through the same -30 GATA site to regulate enhancer-dependent chick β^A-globin transcription in erythroid cells. For this to occur, the two proteins must form a complex together or occupy the specialized TATA box at different times during β^A-globin gene activation. To distinguish between these two mechanisms, protein–DNA–binding studies were conducted using affinity purified chick GATA-1 and bacterially expressed human TFIID (Fong and Emerson, 1992). These experiments demonstrate that no

Figure 7   Proposed stage-specific interaction between chick β-globin promoter and $\beta - \varepsilon$ enhancer mediated by erythroid proteins: the $\beta - \varepsilon$ enhancer regulates both the β^A- and ε-globin genes at separate stages of red cell development, presumably by differential interaction with their respective promoters. In early definitive red cells, the stage-specific transcription factor NF-E4 may, in combination with GATA-1, stabilize DNA loop formation between $\beta - \varepsilon$ enhancer and β^A-globin promoter as an initial step in β^A-globin gene activation. The $\beta - \varepsilon$ enhancer is hypersensitive throughout erythroid development whereas the β^A promoter becomes hypersensitive only in definitive red cells when proteins upstream of GATA-1 and NF-E4 have bound. The β^A-globin gene is maximally expressed when the full complement of activator proteins are associated with the promoter and enhancer regions. In late definitive red cells, where transcription is becoming quiescent, the concentration of the PAL repressor protein increases relative to that of the activator proteins. At this time, PAL may interact with both the β^A promoter and the $\beta - \varepsilon$ enhancer to inactive transcription of the β^A-globin gene.

See Figure 7 on next page $\longrightarrow$

CHICK β- AND ε -GLOBIN GENES
5' β Promoter
β-GLOBIN
β- ε Enhancer
5' ε Promoter
ε -GLOBIN
GATA-1
-30
GATA-1
+1904
GATA-1
-30
DEFINITIVE ERYTHROID
(ACTIVE)
β PROMOTER
5' Non-hypersensitive
β-ε ENHANCER
3' Hypersensitive
GATA-1
NF-E4
GATA-1
NF-E4
β PROMOTER
5' Hypersensitive
β-ε ENHANCER
3' Hypersensitive
GATA-1
GATA-1
DEFINITIVE ERYTHROID
ADULT
(INACTIVE)
PAL
β PROMOTER
5' Hypersensitive
GATA-1
GATA-1
PAL
β-ε ENHANCER
3' Hypersensitive

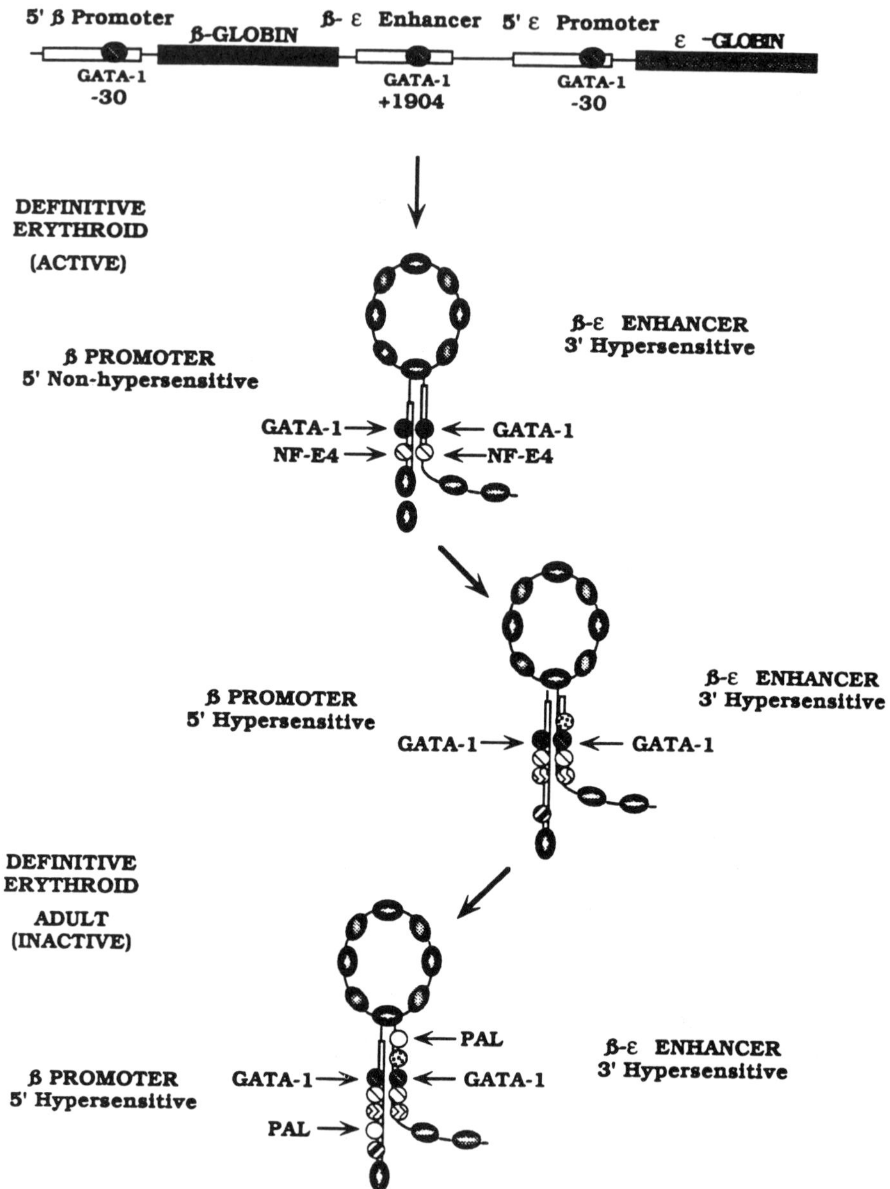

Figure 7

stable association occurs when both purified proteins are combined. Purified TFIID interacts very weakly with GATA DNA sites and is unable to affect the binding of the GATA-1 protein. However, when TFIID is combined with non-DNA-binding proteins from definitive chick red cells, which are likely to contain factors that mediate protein–protein interactions, it is able to efficiently interact with GATA sites. The TFIID/GATA DNA complex formed in the presence of these "adaptor" proteins is higher in mobility than the complex formed with TFIID alone, indicating that at least one protein in the adaptor fraction is associated with TFIID and greatly increases its affinity for GATA sites. By contrast, the GATA-1 protein is not required for TFIID to bind efficiently to GATA sequences, nor does it appear to function in a manner similar to the adenovirus protein E1A, which can directly interact with TFIID and mediate activation by other transcription factors (Horikoshi et al., 1991; Lee et al., 1991). The TFIID–adaptor complex is disrupted by an antibody to TFIID and the association between the adaptor proteins and TFIID on GATA sites is specific because other DNA-binding proteins, such as Sp1 or Jun, fail to form a high molecular weight complex when combined with the adaptor fraction.

Interestingly, when the adaptor fraction is combined with the GATA-1 protein, no change in mobility of the GATA-1/GATA DNA complex is observed but there is a significant decrease in GATA-1 binding. This decrease is even more pronounced in the presence of the TFIID-adaptor complex whose mobility is unaffected by the GATA-1 protein. This indicates that GATA-1 is not a component of the TFIID–adaptor complex and that this complex is a potent inhibitor of GATA-1 binding. Because the decrease of GATA-1 binding in the presence of the TFIID–adaptor complex was so striking, the concentration of GATA-1 was varied to examine the dynamics of interaction between these two complexes. At low GATA-1 concentrations, the TFIID–adaptor complex is present in high abundance while GATA-1 binding is effectively eliminated. As the concentration of GATA-1 increases, a change in the relative abundance of the TFIID–adaptor complex and the GATA-1/GATA DNA complex is observed such that at the highest GATA-1 levels the TFIID–adaptor complex is negligible while the GATA-1/GATA DNA complex is efficiently formed. These results suggest that there is an active displacement of GATA-1 or TFIID binding and that the protein occupancy of GATA DNA sites is determined by the relative concentrations of GATA-1, TFIID, and adaptor proteins. This displacement is more likely caused by competition for adaptor proteins between GATA-1 and TFIID rather than a competition for GATA DNA-binding sites because the binding reactions are performed in DNA excess. This indicates that the protein occupancy of the −30 GATA box in the chick β^A-globin promoter is carefully regulated by the relative concentrations of GATA-1, TFIID, and adaptor proteins. Whether the adaptors are similar to the coactivator or adaptor proteins that mediate activation of the basal transcriptional machinery by other regulatory proteins (Berger et al., 1990; Kelleher et al., 1990; Meisterernst et al., 1991; Pugh and Tjian, 1990) or represent the class of factors known as TAFs (Dynlacht et al., 1991) remains to be determined.

These results suggest that the −30 GATA box is occupied by GATA-1 and TFIID at different times to first establish an interaction between the promoter and enhancer and then to form an active transcriptional initiation complex that is enhancer-dependent. The purpose of incorporating GATA-1 into the initiation site through the specialized TATA box may be to enable the TFIID complex to be regulated by the distal enhancer. By this mechanism, erythroid-specific and enhancer-dependent control of the initiation complex is imparted through a specialized TATA box. If the −30 GATA box is modified to a canonical TATA sequence, the binding affinity of the TFIID–adaptor complex increases but GATA-1 cannot interact with this site. This does not alter the efficiency of the promoter but enhancer-dependent expression is markedly decreased, suggesting that the high-affinity interaction of GATA-1 at −30 is critical for establishing promoter–enhancer interaction before formation of an active initiation complex. Moreover, the timing of these two distinct steps in β^A-globin gene activation may be determined by the relative concentration of adaptor proteins that regulate whether GATA-1 or TFIID is functional on −30 GATA sites. A model depicting these interactions is shown in Figure 8.

Regulation of β-Globin Gene Expression by the Locus Control Region

The β-globin locus control region (LCR) was originally identified by the presence of several erythroid-specific nuclease hypersensitive (HS) sites located 6–18 kb upstream of the human ε-globin gene and one HS site located 3′ of the β-globin gene (see Figure 1; Forrester et al., 1986; Tuan et al., 1985). These sequences were analyzed functionally by the construc-

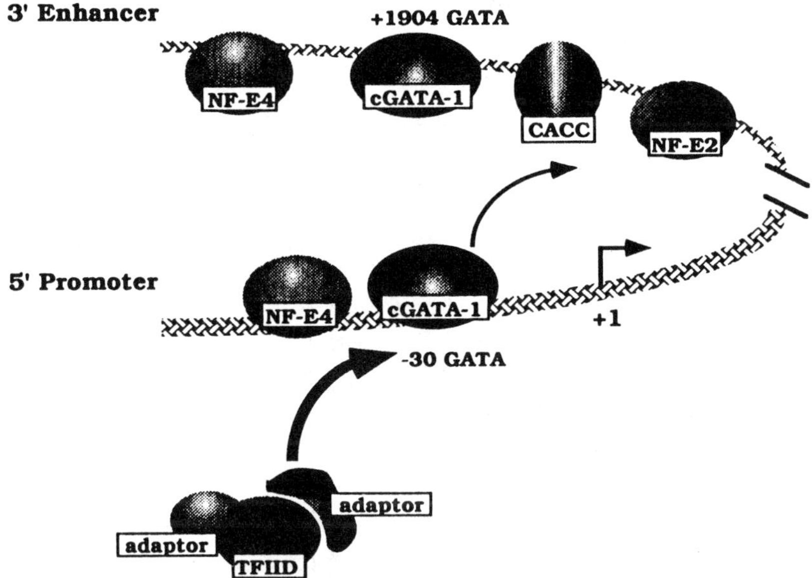

Figure 8   Regulation of chick β-globin enhancer-dependent transcription by GATA-1, TFIID, and adaptor proteins. Diagram of proposed relationship between GATA-1, TFIID, and adaptor proteins at specialized -30 GATA initiation site. This model predicts that regulation of the transcriptional initiation complex by the erythroid-specific $\beta - \varepsilon$ enhancer occurs in discrete stages. First, GATA-1 at the -30 GATA box interacts with proteins bound to the distal enhancer to establish a stable interaction between these two control regions. Second, TFIID and adaptor proteins displace GATA-1 from the -30 box to form an active initiation complex that is now regulated by the erythroid-specific enhancer. Proposed structure between promoter and enhancer at the initiation site results in a highly active promoter and an efficient mechanism by which the enhancer can activate expression. Mutation of the -30 site to a canonical TATA box, which cannot bind GATA-1, significantly reduces ability of the TFIID initiation complex to respond to the tissue-specific enhancer.

tion of a mini- or microlocus containing the four 5' HS sites in a 20-kb or 6.5-kb fragment, respectively, linked to the human β-globin gene. When these constructs were tested in either transgenic mice or erythroid cells, the LCR was found to have the unique property of conferring high-level, erythroid-specific β-globin expression that is dependent on gene copy number but not on integration site (Blom van Assendelft et al., 1989; Forrester et al., 1989; Grosveld et al., 1987; Talbot et al., 1989). By contrast, human β-globin genes containing only local control elements, the 5' promoter and two 3' enhancers, are expressed at very low levels in erythroid cells relative to the endogenous gene. This low-level expression appears to be independent of gene copy number and very dependent on the site of integration. These studies indicate that sequences quite distal from the human β-globin gene can exert a far more profound regulatory effect than the local control regions associated with this gene. Moreover, the LCR has the novel capacity of insulating genes from chromosomal position effects and acting as a powerful transcriptional enhancer. The human β-globin LCR can also direct high-level erythroid-specific expression of α-globin message (Hanscombe et al., 1989; Ryan et al., 1989) as well as heterologous genes such as murine Thy-1 and the herpes thymidine kinase promoter when linked to a reporter gene, *tk-neo* (Blom van Assendelft et al., 1989; Talbot et al., 1989). Thus, the LCR can override local regulatory elements to confer erythroid-specific expression on genes that are not normally active in this tissue. Interestingly, a sequence that is analogous in function, structure, and position to the β-globin LCR has been identified upstream of the human α-globin gene family and is also characterized by the presence of erythroid-specific hypersensitive sites (Higgs et al., 1990).

Function of Distinct Hypersensitive Sites Within the Human β-Globin LCR

In an effort to functionally dissect the LCR, each of the 5' hypersensitive sites (−6, −11, −14.5, −17.5 kb; see Figure 1) have been tested individually and in various combinations for their effects on globin gene expression. These studies demonstrate that HS sites at −6 and −17.5 kb are 10% as efficient as a microlocus containing the four erythroid-specific HS sites and HS sites at −11 and −14.5 kb are 50% as efficient when linked individually to β-globin genes in MEL cells (Collis et al., 1990). However, in transgenic mice HS sites at −11 and −17.5 are each 30% as efficient as the full LCR whereas the site at −14.5 is 70% as efficient. HS sites at −11, −14.5, and −17.5 can each confer position-independent, copy number-dependent expression and deletion of either site −11 or −14.5 from the complete LCR severely decreases transcription (Fraser et al., 1990). An analysis of multiple combinations of the four HS sites indicates that the HS −11 and −14.5 sites are critical for transcriptional activation but require HS sites −6 or −17.5 for fully regulated function. The enhancer activities of HS sites −11 and −14.5 do not cooperatively interact with each other but appear to be synergetic with either site −6 or −17.5. If only one of the enhancers is present, both HS sites −6 and −17.5 are required for high transcription levels whereas if both enhancers are together, either the HS site at -6 or -17.5 kb is sufficient for full LCR activity (Collis et al., 1990). In another study, the combination of single enhancers $(-11, -14.5)$ with nonenhancers $(-6, -17.5)$ results in significant levels of activity relative to the complete LCR; 30% activity with HS sites -6 and -11 kb and 40% activity with HS sites -14.5 and -17.5 in MEL cells (Forrester et al., 1989). Taken together, these results demonstrate that distinct transcriptional activities are associated with the multiple regions of the human β-globin LCR. These distinct activities are functionally redundant and can be synergetic with each other in various combinations to give full inducibility of β-globin expression.

Previous experiments indicated that classic enhancer activity (the ability of a DNA sequence to activate transcription in a position-independent manner) is associated only with an 800-bp DNA fragment containing the HS site at -11 kb when assayed in erythroid K562 cells (Tuan et al., 1989) and in transgenic mice (Curtin et al., 1989). By contrast, the enhancer function of the HS -14.5 kb site could be detected only in stable transfectants, indicating that this region of the LCR must be integrated into chromatin for proper function (Collis et al., 1990). Because the HS -11 kb site is active in stable and transient assays as well as in transgenic mice, it must consist of sequences that confer both enhancer and LCR activity (integration site independence and gene copy number dependence). A detailed analysis of this region reveals consensus binding sites for the CACC, NF-E2/AP-1, and GATA-1 proteins (Figure 9), which are fully occupied in erythroid nuclei, as determined by *in vivo* footprinting (Reddy and Shen, 1991). The -11 kb site is hypersensitive in both erythroid and nonerythroid cells (Dhar et al., 1990). In nonerythroid tissues, hypersensitivity could result from the interaction of the ubiquitous factor AP-1 at the NF-E2 consensus sequence. A careful mutational study localized enhancer ability to the two closely spaced NF-E2/AP-1 sites (Ney et al., 1990). Selective mutation demonstrates that the binding of NF-E2, rather than AP-1, to each site is responsible for enhancer function but the two sites are not equivalent: the 5′ site is much more critical for activation than the 3′ site. This indicates that interaction between NF-E2 and flanking proteins, such as CACC or GATA-1, may be required for enhancer activity (Talbot and Grosveld, 1991). Indeed, multimerized copies of the NF-E2 sites are not sufficient for transcriptional activation (Talbot et al., 1990). Interestingly, mutations in the NF-E2 sites that significantly reduce transcription have no effect on the ability of the remaining sequences to confer position-independent, copy

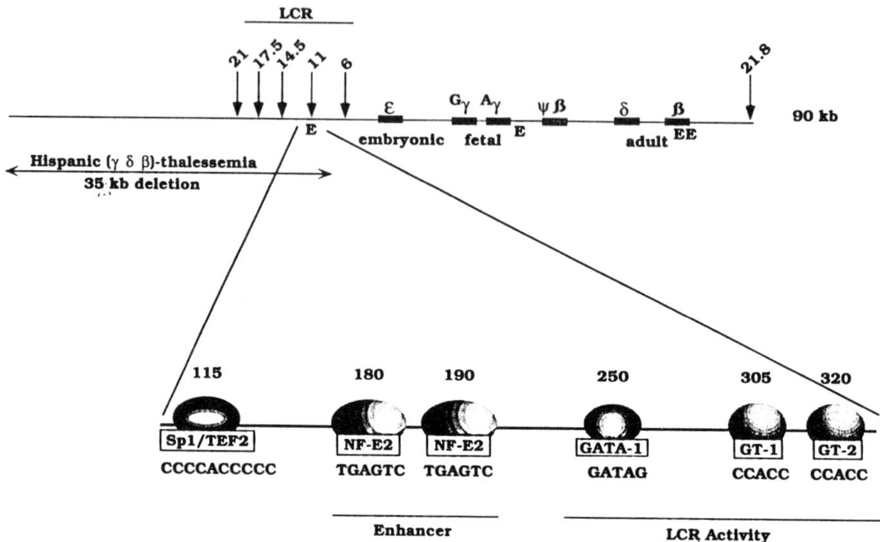

Figure 9   Specific protein–DNA interactions regulate the human β-globin locus control region at -11 kb: expanded region of hypersensitive site at -11 kb within human β-globin LCR. NF-E2 complexes are critical for enhancer activity and function cooperatively with downstream proteins to mediate this effect. However, deletion of NF-E2 binding sites does not affect ability of the remaining proteins to confer position-independent and copy number-dependent expression. CCACC motifs at 3′ end of region may bind proteins distinct from Sp1 or TEF2. Protein–DNA interactions were determined by *in vivo* footprinting in erythroid nuclei; numbers refer to distance within a 375 bp *Hind*III-Xba 1 fragment isolated from -11 kb hypersensitive site. (Adapted from Reddy and Shen, 1991.)

number-dependent expression in transgenic mice. Thus, the HS -11 kb element can be divided into regions that confer either enhancer or LCR function, both of which appear to be independent activities (Talbot and Grosveld, 1991).

The Human β-Globin LCR Mediates Globin Gene Switching During Erythroid Development

The role of the human β-globin LCR in gene switching has been examined in several ways. Initially, the effect of the LCR on the expression of individual genes during erythroid development was monitored. Some members of the human β-globin gene family that are linked to the β-globin LCR are expressed in transgenic mice with tissue specificity but lose their ability to be developmentally regulated. For example, the human $^A\gamma$- and β-globin genes each maintain their correct pattern of expression, albeit at low levels, in transgenic mice in the absence of the LCR (Chada et al., 1986; Kollias et al., 1986). However, when these genes are individually linked to the LCR, they are expressed erythroid-specifically but at all stages of development: yolk sac, fetal liver, and adult bone marrow (Enver et al., 1989). The LCR is apparently dominant over any stage-specific control sequences that may reside within the $^A\gamma$- or β-globin promoter and enhancer regions. Interestingly, if both the $^A\gamma$- and β-globin genes are linked in *cis* to the LCR, rather than individually attached, proper developmental regulation of each gene is restored (Behringer et al., 1990; Enver et al., 1990). One explanation for this effect is that both the $^A\gamma$- and β-globin promoters are in competition for interaction with the LCR. If only one promoter is present, it can stably bind to the LCR and remain active throughout erythroid development. However, in the presence of two promoters, the LCR may preferentially interact with the promoter that is bound by stage-specific proteins. As red cell maturation proceeds, the composition of these proteins changes and promoter-LCR affinities are affected. This model is similar to that proposed for the embryonic- to adult-globin gene switch in the chick. In this case, switching is thought to occur by a competition between the ε- and β-promoters and the enhancer that is situated between them. Promoter-specific interaction with the enhancer may be controlled by the binding of proteins that are developmentally regulated (Choi and Engel, 1988; Nickol and Felsenfeld, 1988). This model could also explain the transcriptional patterns of some nondeletion hereditary persistence of fetal hemoglobin (HPFH) mutations. This condition is characterized by individuals who continue to express high levels of fetal, rather than adult, hemoglobin even though the adult β-globin genes are structurally intact. Instead, point mutations are found within the promoters of the fetal hemoglobin genes that could alter their affinity for transcription factors and for the LCR. According to the competition model, if the affinity is increased, the LCR would favor interaction with fetal globin promoters resulting in the continued expression of these genes and the underexpression of adult hemoglobin.

In addition to promoter competition, there is evidence for the existence of transcriptional silencers in the regulation of certain members of the β-globin gene family. For example, when the human ε-globin gene is linked to the LCR in the absence of any other genes, it shows correct developmental expression in transgenic mice (Raich et al., 1990). This indicates that the ε-globin gene itself contains sufficient information for erythroid stage-specific regulation. In fact, recent studies have identified a transcriptional silencer within the ε promoter (Cao et al., 1989) that is, apparently, dominant over the LCR in definitive red cells. A similar result has been found for the $^A\gamma$-globin gene. By contrast with studies described earlier (Behringer et al., 1990; Enver et al., 1990), one group has found that the $^A\gamma$-globin gene is highly expressed in embryonic erythroid cells but inactive in adult cells. This occurs in the presence of the LCR and does not require linkage to the β-globin gene (Dillon and Grosveld, 1991). This apparent contradiction may result from differences in DNA constructs used in these experiments, which change the distance of the $^A\gamma$-globin gene to the LCR, and in the number of gene copies integrated. Thus, there is evidence for at least two types of control mechanisms involved in globin gene switching: a silencing mechanism in the ε- and γ-globin genes that overrides the positive effects of the LCR and effectively represses these genes in definitive red cells, and a competitive mechanism that regulates β-globin expression by differential promoter affinity with the LCR. The ε- and γ-globin genes are, therefore, autonomously controlled while the β-globin gene is controlled by its competition with other promoters in the globin family. Without this competition, the β-globin gene would be expressed throughout development because it appears not to have an independent silencing mechanism that is dominant over the LCR.

Another important parameter in globin gene switching is the effect of gene order in the β-globin family relative to the LCR. Recent studies have shown that when the position of the β-globin gene is altered within the gene cluster such that it is the closest promoter to the LCR, premature expression occurs in embryonic erythroid cells. Premature expression of the β-globin gene can normally be blocked by the insertion of either the γ-globin or an α-globin gene proximal to the LCR, but no blockage occurs if these genes are placed distal to the LCR. Moreover, when the position of the γ-globin gene is changed such that it is distal rather than proximal to the β-globin gene, both genes are expressed at the same rate, unlike the situation *in vivo*. Thus, linkage to other genes is not sufficient to generate a regulated developmental pattern of expression unless these genes are in the correct order. This supports the notion that competition between the LCR and multiple promoters is strongly influenced by polarity. Promoters situated at a closer distance to the LCR, such as ε-globin, will have a greater frequency of interaction and should be preferentially activated early in red cell development. Promoters at a greater distance from the LCR, such as β-globin, will interact less frequently and will not be expressed at this stage. LCR activation of the β-globin promoter is favored later in red cell development when the proximal ε- and γ-globin genes are silenced, possibly by stage-specific repressors that weaken LCR interaction (Hanscombe et al., 1991). A model summarizing these studies is shown in Figure 10.

Intact Human β-Globin LCR is Required for Active Chromatin Structure and Proper Timing of Replication

An analysis of naturally occurring mutations within or distal to the human β-globin gene locus has indicated that globin expression can be deregulated or inactivated by sequence changes that affect the control regions of specific genes or the whole locus while leaving individual genes intact. For example, mutations that result in the hereditary persistence of fetal hemoglobin (HPFH) inactivate β-globin transcription but allow the fetal globin genes to be expressed in adult red cells (Collins et al., 1987; Tuan et al., 1983). By contrast, deletions in the $(\gamma\delta\beta)^0$-thalassemias occur in the 5' end of the β-globin locus and inactivate all *cis*-linked genes at every stage of erythroid development despite the fact that these genes are structurally intact (Taramelli et al., 1986; Wood et al., 1979). These studies suggest that distinct mechanisms may exist for the tissue-specific activation of the β-globin gene locus, which is affected by the $(\gamma\delta\beta)^0$-thalassemic deletions, and for the developmental expression of specific genes within the locus, which can be deregulated by the HPFH class of mutations. Thus far, the three types of $(\gamma\delta\beta)^0$-thalassemia contain deletions in all or part of the LCR that result in β-globin gene inactivation. The Dutch, English, and Hispanic $(\gamma\delta\beta)^0$-thalassemias have breakpoints that are 2.5, 25, and 60 kb 5' of the β-globin gene and delete 100, 90, and 35 kb, respectively (Curtin et al., 1985; Driscoll et al., 1989; Van der Ploeg et al., 1980). Characterization of the Dutch $(\gamma\delta\beta)^0$-thalassemia demonstrates that the entire β-globin chromosomal locus is maintained in a DNase-resistant chromatin structure in fetal liver cells where the structure is normally DNase sensitive. In these thalassemic cells the β-globin gene is inactive even though it contains no mutations and can be expressed when transfected into erythroid cells (Kioussis et al., 1983). The Hispanic $(\gamma\delta\beta)^0$-thalassemia has a much smaller deletion that removes all the LCR, except the HS site at -6 kb, and is located at a greater

embryonic

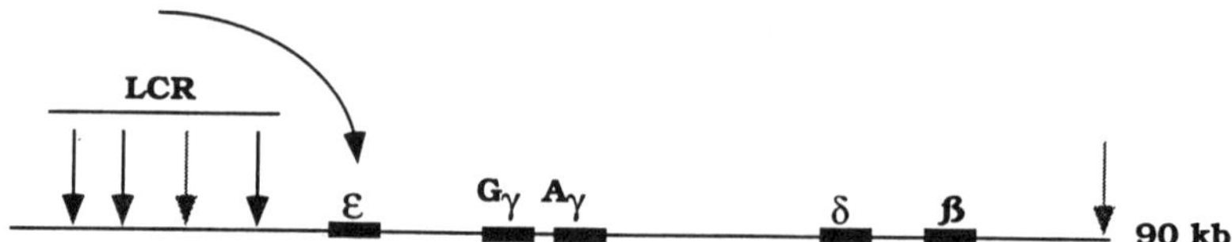

fetal

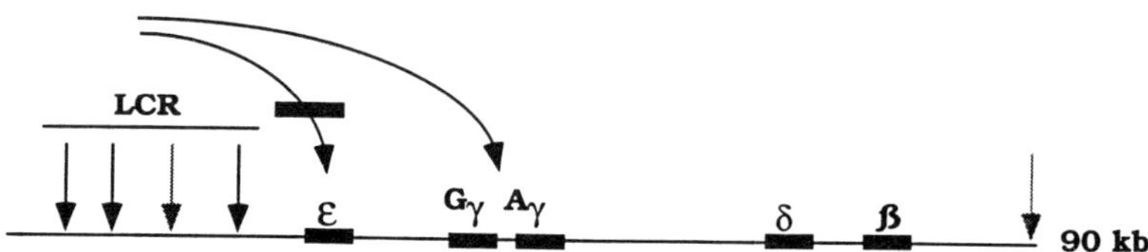

adult

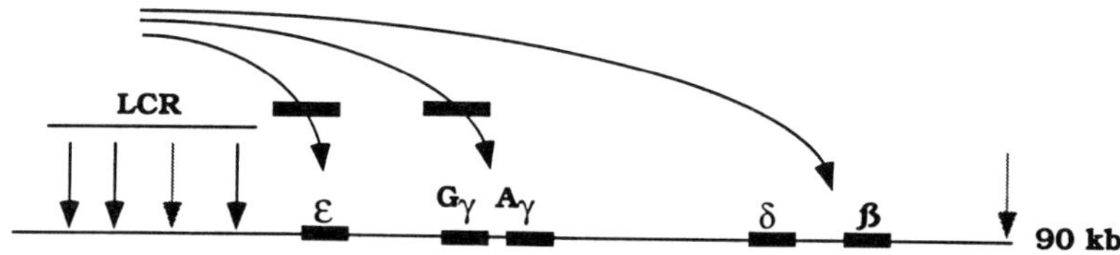

Figure 10   A Model for LCR regulation of erythroid stage-specific gene expression within the β-globin family suggests that the LCR regulates the timed expression of each member of the β-globin gene family by interacting (*solid arrows*) with individual promoters at different stages of erythroid development. Frequency and affinity of LCR–promoter contact is determined by proximity or gene order and the presence of stage-specific transcription factors that may act positively to strengthen LCR interaction or negatively to abolish the interaction and silence the gene. There is evidence that silencer elements (*solid bars*) exist in the ε-globin and possibly γ-globin genes that could alter the affinities of these promoters for the LCR to favor interaction with the δ- and β-globin promoters in definitive red cells of the adult bone marrow. (Adapted from Hanscombe et al., 1991.)

distance from the β-globin gene (see Figure 9). Cells containing this defect were analyzed in somatic cell hybrids in which nonerythroid human thalassemic cells are fused to mouse erythroleukemia (MEL) cells to activate the human erythroid program. The effect of the Hispanic $(\gamma\delta\beta)^0$ deletion in these hybrids is to transcriptionally inactivate the β-globin gene locus and establish a DNase-resistant chromatin structure extending over a region of more than 200 kb (Forrester et al., 1990). The timing of DNA replication was also examined in these hybrids because previous work had shown that active genes, in general, replicate early in S phase (Goldman et al., 1984; Holmquist, 1987) and the β-globin gene locus, in particular, replicates early in erythroid cells (Dhar et al., 1988; Epner et al., 1981). These studies demonstrate that the timing of DNA synthesis is changed in cells containing the Hispanic $(\gamma\delta\beta)^0$ deletion such that the β-globin gene locus is replicated late in S phase rather than early (Forrester et al., 1990). Thus, results obtained with both the Dutch and Hispanic $(\gamma\delta\beta)^0$ thalassemias suggest that sequences upstream of the β-globin gene family, which comprise the LCR, regulate the activation of chromatin structure and the timing of DNA replication over the entire β-globin gene locus and flanking regions extending several hundred kilobases. Because a DNase-sensitive chromatin structure is established before the onset of transcription (Forrester et al., 1987) and early replication occurs in regions that are DNase-resistant (Epner et al., 1988), these important aspects of gene activation appear to be mechanistically separable but dependent on elements in the LCR. It will be of great interest to analyze targeted mutations within the LCR to ascertain its role in tissue-specific chromatin activation and to determine how the timing of DNA replication may influence this process.

The GATA Family of Transcription Factors

Functional Domains and Tissue Restriction of the GATA-1 Protein

The transcription factor GATA-1 is a basic nuclear protein that is present throughout red cell development and may have a central role in the regulation of all erythroid-specific genes. The cDNA encoding this protein has been cloned and characterized in chick (Evans and Felsenfeld, 1989), mouse (Tsai et al., 1989), human (Trainor et al., 1990; Zon et al., 1990), and *Xenopus* (Zon et al., 1991b). The murine and human GATA-1 proteins are each 49 kDa (413 amino acids) in size and 86% identical in sequence. The GATA-1 protein from chick and *Xenopus* are smaller, 37 kDa (304 amino acids) and 39 kDa, respectively, and differ significantly from their mammalian counterparts in the amino- and carboxy-terminal portions of the molecule. All four proteins contain a highly conserved central region composed of two cysteine-rich zinc finger repeats with the motif: Cys-Xaa-Asn-Cys-Xaa$_4$-Thr-Xaa-Leu-Trp-Arg-Arg-Xaa$_3$-Gly-Xaa-Cys-Asn-Ala-Cys. GATA-1 proteins are regulated at the RNA level (Evans and Felsenfeld, 1989; Tsai et al., 1989) and transcripts have been detected in mast cells and megakaryocytes, indicating that these factors are not exclusively erythroid-specific (Martin et al., 1990; Romeo et al., 1990). In addition to its important role in globin gene regulation, GATA-1 has also been shown to activate the promoters of the human porphobilinogen deaminase gene (Mignotte et al., 1989a) and the murine erythropoietin receptor gene (Zon et al., 1991a) in erythroid cells as well as the murine carboxypeptidase A gene in mast cells (Zon et al., 1991c).

Functional studies demonstrate that GATA-1 cDNA can transactivate expression of minimal promoters containing single or multiple GATA consensus binding sites (T/A GATA A/G) in a nonerythroid background such as COS or 3T3 cells (Evans and Felsenfeld, 1991; Martin and Orkin, 1990). This effect is not apparent in transfected erythroid cells where oligomerized GATA binding sites linked to minimal globin promoters are insufficient for high-level expression, presumably because other erythroid transcription factors are required. In COS cells, mouse GATA-1 function is dependent on multiple domains within the protein that separately regulate DNA binding and transactivation (Martin and Orkin, 1990). The first activation domain consists of residues within the amino terminus (amino acids 1–63), the second domain is defined by internal deletional mutations (64–193 and 149–188), and the third domain resides in the carboxy terminus (deletions 308–413, 328–382, 383–413, and 349–413). These activation domains appear to be functionally distinct because the acidic, serine-rich amino-terminal domain (2–66) can act independently as a potent activator when fused to a heterologous DNA-binding domain, GAL 4, whereas the internal (2–193) or carboxy-terminal domains (300–413) cannot. Sequence-specific DNA binding to the consensus site is conferred by two zinc finger domains that act cooperatively but are functionally distinct, like the activation domains. GATA-1 proteins deleted of the amino finger (200–248) still bind DNA and are competent for transactivation whereas GATA-1 proteins deleted of the carboxy finger (249–290) fail to bind DNA or activate transcription. Protein–DNA binding experiments indicate that mutations affecting the amino-terminal finger decrease the stability and specificity of DNA interaction while those affecting the carboxy-terminal finger severely diminish binding affinity which contributes most significantly to transactivation (Figure 11).

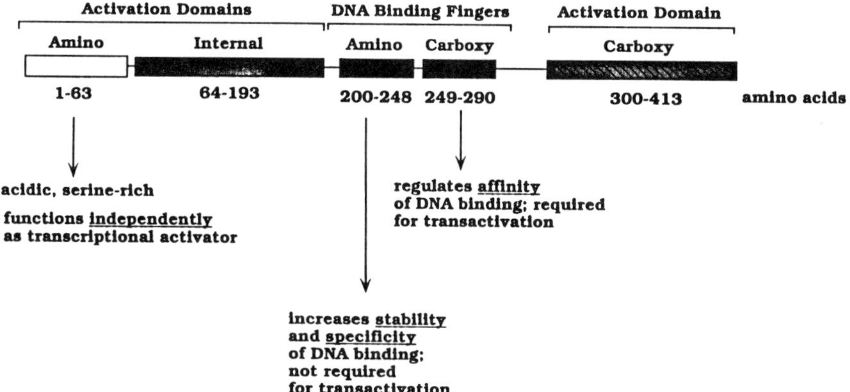

Figure 11   Functional domains of the GATA-1 protein. Murine GATA-1 factor contains 413 amino acids and consists of two DNA binding fingers and three transcriptional activation domains. The two fingers act cooperatively with each other but are functionally distinct: the carboxy-terminal finger regulates the affinity of DNA interaction and is essential for transactivation; the amino-terminal finger contributes to the stability and specificity of DNA binding. The three activation domains are also functionally distinct: the acidic amino terminal domain can transactivate independently whereas the internal and carboxy domains cannot. (Adapted from Martin and Orkin, 1990.)

The structural features of the GATA-1 protein may enable it to achieve a diversity of regulatory potential in hematopoietic cells. This flexibility is necessary in view of the changing context of GATA binding sites among various promoters, enhancers, and LCRs of hematopoieticly expressed genes that require interaction with different sets of transcription factors. The multiple and distinct activation domains of GATA-1 could allow these differential interactions with specific regulatory proteins to occur while the cooperativity of the two DNA-binding motifs may enable GATA-1 to respond to subtle changes in DNA-binding sites.

Structure and Promoter Activity of the Gene Encoding GATA-1

The gene encoding GATA-1 has been cloned from mouse to chick (Hannon et al., 1991; Tsai et al., 1991). The GATA-1 gene from both species consists of six exons that span a region of approximately 8 kb. The first exon is noncoding while the second exon contains the initiating start site. Exons IV and V separately encode the two DNA-binding finger domains. Characterization of the promoter region revealed the presence of consensus binding sites for Sp1, CACC, and GATA but no discernible TATA motif, which may account for the observed 5' heterogeneity of transcripts from this gene. Protein-DNA interaction studies by *in vivo* footprinting of the mouse GATA-1 promoter demonstrated that two CACC motifs (−218 to −192) and a double GATA element (−687 to −673) are bound in erythroid but not in nonerythroid cells (Tsai et al., 1991). A functional analysis of the GATA-1 promoter indicated that it is preferentially active in erythroid cells (Hannon et al., 1991; Tsai et al., 1991) and mutation of either the CACC or GATA sites severely decreases transcription of a linked reporter gene (Tsai et al., 1991). The presence of GATA binding sites within its own promoter raises the interesting possibility that the GATA-1 gene is autoregulatory. Because the concentration of GATA-1 protein increases throughout erythroid development (Whitelaw et al., 1990), autoregulation would most likely be in the form of a positive feedback loop (Hannon et al., 1991; Tsai et al., 1991). This is substantiated by the observation that the chick GATA-1 promoter can be activated in primary chick embryo fibroblasts by cotransfection with a GATA-1 cDNA expression plasmid (Hannon et al., 1991), although this was not observed with the mouse GATA-1 gene in NIH-3T3 fibroblasts (Tsai et al., 1991). Other tissue-specific transcription factors that are under positive autoregulation include MyoD1 (Lassar et al., 1986; Thayer et al., 1989) and GHF-1/Pit 1 (Chen et al., 1990; McCormick et al., 1990).

Characterization of Other Members of the GATA Gene Family

GATA-1 is part of a multigene family whose members are tissue-restricted in activity and regulate a variety of genes within those cell types (Wilson et al., 1990; Yamamoto et al., 1990). For example, GATA-1 is present in erythroid, megakaryocyte, and mast cells (Martin et al., 1990; Romeo et al., 1990), whereas GATA-2 has a wide tissue distribution including erythrocytes, fibroblasts, embryonic brain, liver, cardiac muscle, and adult kidney and has recently been shown to regulate the human preproendothelin-1 gene in endothelial cells (Dorfman et al., 1992; Wilson et al., 1990). GATA-3 is most abundantly expressed in mature erythrocytes, T lymphocytes, embryonic brain, and adult kidney (Yamamoto et al., 1990). This protein is implicated in the regulation of T-cell-specific genes because the GATA consensus binding site is present in the enhancer regions of the α, β, and δ chains of the T-cell receptor (Gottschalk and Leiden, 1990; Ho et al., 1989; Redondo et al., 1990). Recent functional studies have shown that human and mouse GATA-3 can transactivate expression of the human T-cell receptor (TCR) α and δ genes through their respective enhancers (Ho et al., 1991; Joulin et al., 1991; Ko et al., 1991). Other expression studies indicate that all GATA proteins can transactivate synthetic and native promoters containing GATA sites, although to different extents and with little tissue specificity in these particular assays.

Chromosomal localization studies have demonstrated that the GATA gene family is widely dispersed in the human genome (Dorfman et al., 1992; Joulin et al., 1991; Zon et al., 1990). Individual members are found on the X chromosome (hGATA-1), chromosome 3 (hGATA-2), and chromosome 10 (hGATA-3). All members of the GATA gene family contain a highly conserved DNA-binding domain and recognize the GATA consensus sequence with similar affinities (Yamamoto et al., 1990). Regions apart from the two finger domains show no significant homology between the three proteins suggesting that their activation domains may have evolved differently to interact with distinct sets of tissue-restricted transcription factors. The three proteins are similar in size (50 kDa), except for chick GATA-1 (38 kDa), and their mRNAs vary between 1.3 (cGATA-1), 4.3 (cGATA-2), and 3.3 (cGATA-3) kb (Yamamoto et al., 1990). Extensive sequence homology exists between the human and chick GATA family: hGATA-3 is 93% homologous to cGATA-3, hGATA-2 is 84% homologous to cGATA-2, but hGATA-1 shares only 41% homology with cGATA-1, which differs significantly in its activation domains (Dorfman et al., 1992; Ho et al., 1991; Ko et al., 1991; Trainor et al., 1990). In fact, similar types of GATA proteins among different species (e.g. hGATA-3, cGATA-3) share greater homology than members of the GATA family within the same species (e.g. hGATA-1, hGATA-2, hGATA-3) with the exception of GATA-1 (Ko et al., 1991).

The mRNA and protein concentrations of GATA-1, GATA-2, and GATA-3 increase on maturation of erythrocytes, retinoic acid treatment of endothelial cells, and phytohemagglutin activation of T lymphocytes, respectively, suggesting a function for these factors during cellular differentiation (Dorfman et al., 1992; Joulin et al., 1991; Whitelaw et al., 1990). The role of GATA-1 in erythroid differentiation has been demonstrated in an elegant study. In this experiment, the GATA-1 gene on the X chromosome was specifically disrupted by homologous recombination in mouse embryonic stem cells. These cells were then analyzed for their ability to contribute to various tissues in chimeric mice. The GATA-1 deficient cells were found in all nonhematopoietic tissues examined as well as in white blood cells but were absent in mature erythrocytes. This clearly indicates the requirement for a specific member of the GATA multigene family in red cell differentiation because GATA-2 and GATA-3 could not functionally replace GATA-1 in this process (Pevny et al., 1991). This experimental system should be very useful in future studies designed to analyze the role of individual members of the GATA gene family as well as other hematopoietic factors in developmental regulation.

Acknowledgments. I wish to express my appreciation to Drs. Michelle Barton, Navid Madani, and Katherine Jones for their useful scientific comments. B.M.E. is a Pew Scholar in the Biomedical Sciences. This chapter is dedicated to the memory of my mother, Marie M. Emerson (1920–1989).

References

Angel P, Imagawa M, Chiu R, Stein B, Imbra RJ, Rahmsdorf HJ, Jonat C, Herrlich P, Karin M (1987): Phorbol ester-inducible genes contain a common *cis* element recognized by a TPA-modulated *trans*-acting factor. *Cell* 49:729–739.

Antoniou M, deBoer E, Habets G, Grosveld F (1988): The human β-globin gene contains multiple regulatory regions: Identification of one promoter and two downstream enhancers. *EMBO J* 7:377–384.

Antoniou M, Grosveld F (1990): β-globin dominant control region interacts differently with distal and proximal promoter elements. *Genes & Dev* 4:1007–1013.

Barberis A, Superti-Furga G, Busslinger M (1987): Mutually exclusive interaction of the CCAAT-binding factor and of a displacement protein with overlapping sequences of a histone gene promoter. *Cell* 50:347–359.

Behringer RR, Hammer RE, Brinster RL, Palmiter RD, Townes TM (1987): Two $3'$ sequences direct adult erythroid-specific expression of human β-globin genes in transgenic mice. *Proc Natl Acad Sci USA* 84:7056–7060.

Behringer RR, Ryan TM, Palmiter RD, Brinster RL, Townes TM (1990): Human γ- to β-globin gene switching in transgenic mice. *Genes & Dev* 4:380–389.

Berger SL, Cress WD, Cress A, Triezenberg SJ, Guarente L (1990): Selective inhibition of activated but not basal transcription by the acidic activation domain of VP-16: Evidence for transcriptional adaptors. *Cell* 61:1199–1208.

Blom van Assendelft M, Hanscombe O, Grosveld F, Greaves DR (1989): The β-globin dominant control region activates homologous and heterologous promoters in a tissue-specific manner. *Cell* 56:969–977.

Brown JL, Ingram VM (1974): Structural studies on chick embryonic hemoglobins. *J Biol Chem* 249:3960–3972.

Bruns GA, Ingram VM (1973): The erythroid cells and hemoglobins of the chick embryo. *Philos Trans R Soc London (Biol)* 266:225–305.

Bunn HF, Forget BG (1986): *Hemoglobin: Molecular, Genetic and Clinical Aspects.* Philadelphia: WB Saunders.

Cao SX, Gutman PD, Dave HPG, Schechter AN (1989): Identification of a transcriptional silencer in the $5'$ flanking region of the human ε-globin gene. *Proc Natl Acad Sci USA* 86:5306–5309.

Catala F, deBoer E, Habets G, Grosveld F (1989): Nuclear protein factors and erythroid transcription of the human gamma-globin gene. *Nucleic Acids Res* 17:3811–3827.

Chada K, Magram J, Costantini F (1986): An embryonic pattern of expression of a human fetal globin gene in transgenic mice. *Nature* (Lond) 319:685–689.

Chapman BS, Tobin AJ (1979): Distribution of developmentally regulated hemoglobins in embryonic erythroid populations. *Dev Biol* 69:375–387.

Chen R, Ingraham HA, Treacy MN, Albert VR, Wilson L, Rosenfeld MG (1990): Autoregulation of pit-1 gene expression mediated by two *cis*-active promoter elements. *Nature (Lond)* 346:583–586.

Chodosh LA, Baldwin AS, Carthew RW, Sharp PA (1988): Human CCAAT-binding proteins have heterologous subunits. *Cell* 55:11–24.

Choi O-R, Engel JD (1986): A $3'$ enhancer is required for temporal and tissue-specific transcriptional activation of the chicken adult β-globin gene. *Nature (Lond)* 323:731–734.

Choi O-RB, Engel JD (1988): Developmental regulation of β-globin gene switching. *Cell* 55:17–26.

Collins FS, Cole JL, Lockwood WK, Iannuzzi MC (1987): The deletion in both common types of hereditary persistence of fetal hemoglobin is approximately 105 kb. *Blood* 70:1797–1803.

Collins FS, Weisman SM (1984): The molecular genetics of human hemoglobin. *Prog Nucleic Acid Res Mol Biol* 31:315–462.

Collis P, Antoniou M, Grosveld F (1990): Definition of the minimal requirements within the human β-globin gene and the dominant control region for high level expression. *EMBO J* 9:233–240.

Cowie A, Myers RM (1988): DNA sequences involved in transcriptional regulation of the mouse β-globin promoter in murine erythroleukemia cells. *Mol Cell Biol* 8:3122–3128.

Curtin PT, Lui D, Lui W, Chang JC, Kan YW (1989): Human β-globin gene expression in transgenic mice is enhanced by a distant DNase-I hypersensitive site. *Proc Natl Acad Sci USA* 86:7082–7086.

Curtin PT, Pirastu M, Kan YW, Gobert-Jones JA, Stephens AD, Lehmann H (1985): A distant deletion affects β-globin gene function in an atypical $\gamma\delta\beta$-thalassemia. *J Clin Invest* 76:1554–1558.

Davidson I, Xiao JH, Rosales R, Staub A, Chambon P (1988): The HeLa cell protein TEF-1 binds specifically and cooperatively to two SV40 enhancer motifs of unrelated sequence. *Cell* 54:931–942.

deBoer E, Antoniou M, Mignotte V, Wall L, Grosveld F (1988): The human β-globin promoter: Nuclear protein factors and erythroid-specific induction of transcription. *EMBO J* 7:4203–4212.

Dhar V, Mager D, Iqbal A, Schildkraut CL (1988): The coordinate replication of the human β-globin gene domain reflects its transcriptional activity and nuclease hypersensitivity. *Mol Cell Biol* 8:4958–4965.

Dhar V, Nandi A, Schildkraut C, Skoultchi AI (1990): Erythroid-specific nuclease hypersensitive sites flanking the human β-globin domain. *Mol Cell Biol* 10:4324–4333.

Dierks P, van Ooyen A, Cochran MD, Dobkin C, Reiser J, Weissman C (1983): Three regions upstream from the cap site are required for efficient and accurate transcription of the rabbit β-globin gene in mouse 3T6 cells. *Cell* 32:695–706.

Dillon N, Grosveld F (1991): Human γ-globin genes silenced independently of other genes in the β-globin locus. *Nature* (Lond) 350:252–254.

Dorfman DM, Wilson DB, Bruns GAP, Orkin SH (1992): Human transcription factor GATA-2. *J Biol Chem* 267:1279–1285.

Driscoll C, Dobkin CS, Alter BP (1989): $\gamma\delta\beta$-Thalassemia due to a de novo mutation deleting the 5′ β-globin locus activating region hypersensitive sites. *Proc Natl Acad Sci USA* 86:7470–7474.

Dynlacht BD, Hoey T, Tjian R (1991): Isolation of coactivators associated with the TATA-binding protein that mediate transcriptional activation. *Cell* 66:563–576.

Emerson BM, Felsenfeld G (1984): Specific factor conferring nuclease hypersensitivity at the 5′ end of the chicken adult β-globin gene. *Proc Natl Acad Sci USA* 81:95–99.

Emerson BM, Lewis CD, Felsenfeld G (1985): Interaction of specific nuclear factors with the nuclease-hypersensitive region of the chicken adult β-globin gene: Nature of the binding domain. *Cell* 41:21–30.

Emerson BM, Nickol JM, Fong TC (1989): Erythroid-specific activation and derepression of the chick β-globin promoter in vitro. *Cell* 57:1189–1200.

Enver T, Ebens AJ, Forrester WC, Stamatoyannopoulos S (1989): The human β-globin locus activation region alters the developmental fate of a human fetal globin gene in transgenic mice. *Proc Natl Acad Sci USA* 86:7033–7037.

Enver T, Raich N, Ebens AJ, Papayannopoulou T, Costantini F, Stamatoyannopoulos G (1990): Developmental regulation of human fetal-to-adult globin gene switching in transgenic mice. *Nature* (Lond) 344:309–313.

Epner E, Forrester WC, Groudine M (1988): Asynchronous DNA replication within the human β-globin gene locus. *Proc Natl Acad Sci USA* 85:8081–8085.

Epner E, Rifkind R, Marks PA (1981): Replication of α- and β-globin DNA sequencesoccurs during early S phase in murine erythroleukemia cells. *Proc Natl Acad Sci USA* 78:3058–3062.

Evans T, Felsenfeld G (1989): The erythroid-specific transcription factor eryf1: A new finger protein. *Cell* 5:877–885.

Evans T, Felsenfeld G (1991): Trans-activation of a globin promoter in non-erythroid cells. *Mol Cell Biol* 11:843–853.

Evans T, Felsenfeld G, Reitman M (1990): Control of globin gene transcription. *Annu Rev Cell Biol* 6:95–124.

Evans T, Reitman M, Felsenfeld G (1988): An erythroid-specific DNA-binding factor recognizes a regulatory sequence common to all chicken global genes. *Proc Natl Acad Sci USA* 85:5976–5980.

Fong TC, Emerson BM: The erythroid-specific protein, cGATA-1, mediates distal enhancer activity through a specialized β-globin TATA box. *Genes & Dev* (in press).

Forrester WC, Epner E, Driscoll MC, Enver T, Brice M, Papayannopoulou T, Groudine M (1990): A deletion of the human β-globin locus activation region causes a major alteration in chromatin structure and replication across the entire β-globin gene locus. *Genes & Dev* 4:1637–1649.

Forrester WC, Novak U, Gelinas R, Groudine M (1989): Molecular analysis of the human β-globin locus activation region. *Proc Natl Acad Sci USA* 86:5439–5443.

Forrester WC, Takegawa S, Papayannopoulou T, Stamatoyannopoulos G, Groudine M (1987): Evidence for a locus activation region: The formation of developmentally stable hypersensitive sites in globin-expressing hybrids. *Nucleic Acids Res* 15:10159–10177.

Forrester WC, Thompson C, Elder JT, Groudine M (1986): A developmentally stable chromatin structure in the human β-globin gene cluster. *Proc Natl Acad Sci USA* 83:1359–1363.

Fraser P, Hurst J, Collis P, Grosveld F (1990): DNase I hypersensitive sites 1, 2, and 3 of the human β-globin dominant control region direct position-independent expression. *Nucleic Acids Res* 18:3503–3508.

Gallarda JL, Foley KP, Yang Z, Engel JD (1989): The β-globin stage selector element factor is erythroid-specific promoter/enhancer binding protein NF-E4. *Genes & Dev* 3:1845–1859.

Goldman MA, Holmquist GP, Gray MC, Caston LA, Nag A (1984): Replication timing of genes and middle repetitive sequences. *Science* 224:686–692.

Gottschalk LR, Leiden JM (1990): Identification and functional characterization of the human T-cell receptor β gene transcriptional enhancer: Common nuclear proteins interact with the transcriptional regulatory elements of the T-cell receptor α and β genes. *Mol Cell Biol* 10:5486–5495.

Grosveld F, van Assendelft GB, Greaves DR, Kollias G (1987): Position-independent, high-level expression of the human β-globin gene in transgenic mice. *Cell* 51:975–985.

Hannon R, Evans T, Felsenfeld G, Gould H (1991): Structure and promoter activity of the gene for the erythroid transcription factor GATA-1. *Proc Natl Acad Sci USA* 88:3004–3008.

Hanscombe O, Vidal M, Kaeda J, Luzzatto L, Greaves DR, Grosveld F (1989): High-level, erythroid-specific expression of the human α-globin gene in transgenic mice and the production of human hemoglobin in murine erythrocytes. *Genes & Dev* 3:1572–1581.

Hanscombe O, Whyatt D, Fraser P, Yannoutsos N, Greaves DR, Dillon N, Grosveld F (1991): Importance of globin gene order for correct developmental expression. *Genes & Dev* 5:1387–1394.

Hesse JE, Nickol JM, Lieber MR, Felsenfeld G (1986): Regulated gene expression in transfected primary chicken erythrocytes. *Proc Natl Acad Sci USA* 83:4312–4316.

Higgs DR, Wood WG, Jarman AP, Sharpe J, Lida J, Pretorius IM, Ayyub H (1990): A major positive regulatory region located far upstream of the human α-globin gene locus. *Genes & Dev* 4:1588–1601.

Ho IC, Vorhees P, Marin N, Oakley BK, Tsai SF, Orkin SH, Leiden JM (1991): Human GATA-3: A lineage-restricted transcription factor that regulates the expression of the T-cell receptor α gene. *EMBO J* 10:1187–1192.

Ho IC, Yang LH, Morle G, Leiden JM (1989): A T-cell specific transcriptional enhancer element $3'$ of Cα in the human T cell receptor α locus. *Proc Natl Acad Sci USA* 86:6714–6718.

Holmquist GP (1987): Role of replication time in the control of tissue-specific gene expression. *Am J Hum Genet* 40:151–173.

Horikoshi N, Maguire K, Kralli A, Maldonado E, Weinmann R (1991): Direct interaction between adenovirus E1A protein and the TATA box binding transcription factor IID. *Proc Natl Acad Sci USA* 88:5124–5128.

Jackson PD, Evans T, Nickol JM, Felsenfeld G (1989): Developmental modulation of protein binding to β-globin gene regulatory sites within chicken erythrocyte nuclei. *Genes & Dev* 3:1860–1873.

Jackson PD, Felsenfeld G (1985): A method for mapping intranuclear protein-DNA interactions and its application to a nuclease hypersensitive site. *Proc Natl Acad Sci USA* 82:2296–2300.

Joulin V, Bories D, Eleouet JF, Labastie MC, Chretien S, Mattei MG, Romeo PH (1991): A T-cell specific TCR δ DNA binding protein is a member of the human GATA family. *EMBO J* 10:1809–1816.

Kadonaga JT, Jones KA, Tjian R (1987): Promoter-specific activation of RNA polymerase II transcription by Sp1. *Trends Biochem* 11:20–23.

Kelleher RJ III, Flanagan PM, Kornberg RD (1990): A novel mediator between activator proteins and the RNA polymerase II transcription apparatus. *Cell* 61:1209–1215.

Kioussis D, Vanin E, deLange T, Flavell RA, Grosveld FG (1983): β-globin gene inactivation by DNA translocation in $\gamma\beta$-thalassemia. *Nature (Lond)* 306:662–666.

Ko LJ, Yamamoto M, Leonard MW, George KM, Ting P, Engel JD (1991): Murine and human T-lymphocyte GATA-3 factors mediate transcription through a *cis*-regulatory element within the human T-cell receptor δ gene enhancer. *Mol Cell Biol* 11:2778–2784.

Kollias G, Hurst J, deBoer E, Grosveld F (1987): The human β-globin gene contains a downstream developmental specific enhancer. *Nucleic Acids Res* 15:5739–5747.

Kollias G, Wrighton N, Hurst J, Grosveld F (1986): Regulated expression of human $^A\gamma$-,β- and hybrid γ/β-globin genes in transgenic mice: Manipulation of the developmental expression patterns. *Cell* 46:89–94.

Lassar AB, Patterson BM, Weintraub H (1986): Transfection of a DNA locus that mediates the conversion of 10T1/2 fibroblasts to myoblasts. *Cell* 47:649–656.

Lee W, Mitchell P, Tjian R (1987): Purified transcription factor AP-1 interacts with TPA-inducible enhancer elements. *Cell* 49:741–752.

Lee WS, Kao CC, Bryant GO, Liu X, Berk AJ (1991): Andenovirus E1A activation domain binds the basic repeat in the TATA box transcription factor. *Cell* 67:365–376.

Lewis CD, Clark SP, Felsenfeld G, Gould H (1988): An erythrocyte-specific protein that binds to the poly (dG) region of the chicken β-globin gene promoter. *Genes & Dev* 2:863–873.

Magram J, Niederreither K, Costantini F (1989): β-Globin enhancers target expression of a heterologous gene to erythroid tissues of transgenic mice. *Mol Cell Biol* 9:4581–4584.

Mantovani R, Malgaretti N, Nicolis S, Giglioni B, Comi P (1988): An erythroid specific nuclear factor binding to the proximal CACCC box of the β-globin gene promoter. *Nucleic Acids Res* 16:4299–4313.

Mantovani R, Malgaretti N, Giglioni B, Comi P, Cappelini S, Nicolis S, Ottolenghi S (1987): A protein factor binding to an octamer motif in the γ-globin promoter disappears upon induction of differentiation and hemoglobin synthesis in K562 cells. *Nucleic Acids Res* 15:9349–9364.

Martin DIK, Orkin SH (1990): Transcriptional activation and DNA-binding by the erythroid factor GF-1/NF-E1/Eryf 1. *Genes & Dev* 4:1886–1898.

Martin DIK, Tsai S-F, Orkin SH (1989): Increased gamma-globin expression in a nondeletion HPFH mediated by an erythroid-specific DNA-binding factor. *Nature* (Lond) 338:435–438.

Martin DIK, Zon LI, Mutter G, Orkin SH (1990): Expression of an erythroid transcription factor in megakaryocytic and mast cell lineages. *Nature* (Lond) 344:444–447.

McCormick A, Brady H, Theill LE, Karin M (1990): Regulation of the pituitary-specific homeobox gene GHF-1 by cell-autonomous and environmental cues. *Nature* (Lond) 345:829–832.

McGhee JD, Wood WI, Dolan M, Engel JD, Felsenfeld G (1981): A 200 base pair region at the 5$'$ end of the chicken adult β-globin gene is accessible to nuclease digestion. *Cell* 27:45–55.

Meisterernst M, Roy AL, Lieu HM, Roeder RG (1991): Activation of class II gene transcription by regulatory factors is potentiated by a novel activity. *Cell* 66:981–993.

Mignotte V, Eleouet JF, Raich N, Romeo PH (1989a): *Cis-* and *trans*-acting elements involved in the regulation of the erythroid promoter of the human

158   B. M. Emerson

porphobilinogen deaminase gene. *Proc Natl Acad Sci USA* 86:6548–6552.

Mignotte V, Wall L, deBoer E, Grosveld F, Romeo PH (1989b): Two tissue-specific factors bind the erythroid promoter of the human porphobilinogen deaminase gene. *Nucleic Acids Res* 17:37–54.

Myers RM, Tilly K, Maniatis T (1986): Fine structure genetic analysis of a β-globin promoter. *Science* 232:613–618.

Ney PA, Sorrentino BP, McDonagh KT, Nienhuis AW (1990): Tandem AP-1 binding sites within the human β-globin dominant control region function as an inducible enhancer in erythroid cells. *Genes & Dev* 4:993–1006.

Nickol JM, Felsenfeld G (1988): Bidirectional control of the chicken β- and ε-globin genes by a shared enhancer. *Proc Natl Acad Sci USA* 85:2548–2552.

Nienhuis AW, Maniatis T (1987): Structure and expression of globin genes in erythroid cells. In: *The Molecular Basis of Blood Diseases*, Stamoyannopoulos G, Nienhuis AW, Leder P, Majerus PW, eds., pp. 28–65. Philadelphia: WB Saunders.

Nomiyama H, Fromental C, Xiao JH, Chambon P (1987): Cell-specific activity of the constituent elements of the simian virus 40 enhancer. *Proc Natl Acad Sci USA* 84:7881–7885.

Pevny L, Simon MC, Robertson E, Klein WH, Tsai SF, D'Agati V, Orkin SH, Costantini F (1991): Erythroid differentiation in chimaeric mice blocked by a targeted mutation in the gene for transcription factor GATA-1. *Nature* (Lond) 349:257–260.

Plumb, MA, Frampton J, Wainwright H, Walker M, Macleod K (1989): GATA AG: A *cis*-control region binding an erythroid-specific nuclear factor with a role in globin and nonglobin gene expression. *Nucleic Acids Res* 17:73–91.

Plumb, MA, Lobanenkov VV, Nicolas RH, Wright CA, Zavou S, Goodwin GH (1986): Characterization of chicken erythroid nuclear proteins which bind to the nuclease hypersensitive regions upstream of the β^A- and β^H-globin genes. *Nucleic Acids Res* 14:7675–7693.

Ptashne M (1988): How eukaryotic transcriptional activators work. *Nature* (Lond) 335:683–689.

Pugh BF, Tjian R (1990): Mechanism of transcriptional activation by Sp1: Evidence for coactivators. *Cell* 61:1187–1197.

Raich N, Enver T, Nakamoto B, Josephson B, Papayannopoulou T, Stamatoyannopoulos G (1990): Autonomous developmental control of human embryonic globin gene switching in transgenic mice. *Science* 250:1147–1149.

Reddy PMS, Shen CKJ (1991): Protein–DNA interactions in vivo of an erythroid-specific, human β-globin locus enhancer. *Proc Natl Acad Sci USA* 88:8676–8680.

Reitman M, Felsenfeld G (1988): Mutational analysis of the chicken β-globin enhancer reveals two positive-acting domains. *Proc Natl Acad Sci USA* 85:6264–6271.

Reitman M, Felsenfeld G (1990): Developmental regulation of topoisomerase II sites and DNase-I hypersensitive sites in the chicken β-globin locus. *Mol*

Cell Biol 10:2774–2786.

Redondo JM, Hata S, Brocklehurst C, Krangel MS (1990): A T-cell specific transcriptional enhancer within the human T-cell receptor δ locus. *Science* 247:1225–1229.

Romeo PH, Prandini MH, Joulin V, Mignotte V, Prenant M, Vainchenker G, Marguerie G, Uzan G (1990): Megakaryocytic and erythroid lineages share specific transcription factors. *Nature (Lond)* 344:447–449.

Rosales R, Vigneron M, Macchi M, Davidson I, Xiao J, Chambon P (1987): In vitro binding of cell-specific and ubiquitous nuclear proteins to the octamer motif of the SV40 enhancer and related motifs present in other promoters. *EMBO J* 6:3015–3025.

Ryan TM, Behringer RR, Townes TM, Palmiter RD, Brinster RL (1989): High-level erythroid expression of human α-globin genes in transgenic mice. *Proc Natl Acad Sci USA* 86:37–41.

Stamatoyannopoulos G, Nienhuis AW (1987): Hemoglobin switching. In: *The Molecular Basis of Blood Diseases*, Stamoyannopoulos G, Nienhuis AW, Leder P, Majerus PW, eds., pp. 66–105. Philadelphia: WB Saunders.

Stalder J, Larsen A, Engel JD, Dolan M, Groudine M, Weintraub H (1980): Tissue-specific DNA cleavages in the globin chromatin domain introduced by DNase I. *Cell* 20:451–460.

Talbot D, Collis P, Antoniou M, Vidal M, Grosveld F, Greaves DR (1989): A dominant control region from the human β-globin locus conferring integration site-independent gene expression. *Nature* (Lond) 338:352–355.

Talbot D, Grosveld F (1991): The 5' HS2 of the globin locus control region enhances transcription through the interaction of a multimeric complex binding at two functionally distinct NF-E2 binding sites. *EMBO J* 10:1391–1398.

Talbot D, Philipsen S, Fraser P, Grosveld F (1990): Detailed analysis of the site 3 region of the human β-globin dominant control region. *EMBO J* 9:2169–2178.

Taramelli R, Kioussis D, Vanin E, Bartram K, Groffen J, Hurst J, Grosveld F (1986): $\gamma\beta$-thalassemias 1 and 2 are the result of a 100-kb deletion in the human β-globin gene cluster. *Nucleic Acids Res* 14:7017–7029.

Thayer MJ, Tapscott SJ, Davis RL, Wright WE, Lassar AB, Weintraub H (1989): Positive autoregulation of the myogenic determination gene MyoD1. *Cell* 58:241–248.

Townes TM, Lingrel JB, Chen HY, Brinster RL, Palmiter RD (1985): Erythroid-specific expression of human β-globin genes in transgenic mice. *EMBO J* 4:1715–1723.

Trainor CD, Evans T, Felsenfeld G, Boguski MS (1990): Structure and evolution of a human erythroid transcription factor. *Nature* (Lond) 343:92–96.

Trudel M, Costantini F (1987): A 3' enhancer contributes to the stage-specific expression of the human β-globin gene. *Genes & Dev* 1:954–961.

Tsai SF, Martin DI, Zon LI, D'Andrea AD, Wong GG, Orkin SH (1989): Cloning of cDNA for the major DNA-binding protein of the erythroid lineage through expression in mammalian cells. *Nature* (Lond) 339:446–451.

Tsai SF, Strauss E, Orkin SH (1991): Functional analysis and in vivo footprinting implicate the erythroid transcription factor GATA-1 as a positive regulator of its own promoter. *Genes & Dev* 5:919–931.

Tuan D, Feingold E, Newman M, Weissman SM, Forget BG (1983): Different 3′ endpoints of deletions causing $\delta\beta$-thalassemia and hereditary persistence of fetal hemoglobin: Implications for the control of γ-globin gene expression in man. *Proc Natl Acad Sci USA* 80:6937–6941.

Tuan D, Solomon W, Li Q, London IM (1985): The "β-like globin" gene domain in human erythroid cells. *Proc Natl Acad Sci USA* 82:6384–6388.

Tuan D, Solomon W, London IM, Lee DP (1989): An erythroid-specific, developmental stage-independent enhancer far upstream of the human "β-like globin" genes. *Proc Natl Acad Sci USA* 86:2554–2558.

Van der Ploeg LHT, Konings A, Oort M, Roos D, Bernini L, Flavell RA (1980): $\gamma\beta$-Thalassemia studies showing that deletion of the γ- and δ-genes influences β-globin gene expression in man. *Nature* (Lond) 283:637–642.

Wall L, deBoer E, Grosveld F (1988): The human β-globin gene 3′ enhancer contains multiple binding sites for an erythroid-specific protein. *Genes & Dev* 2:1089–1100.

Weatherall DJ, Clegg JB, Higgs D, Wood WG (1989): The hemoglobinopathies. In: *The Metabolic Basis of Inherited Disease*, 6th Ed., Scriver CR, Beaudet AL, Sly WS, Valle D, eds., pp. 2281–2339. New York: McGraw-Hill.

Whitelaw E, Tsai SF, Hogben P, Orkin SH (1990): Regulated expression of globin chains and the erythroid transcription factor (GF-1/NF-E1/Eryf 1) during erythropoiesis in the developing mouse. *Mol Cell Biol* 10:6596–6606.

Wilson DB, Dorfman DM, Orkin SH (1990): A non-erythroid GATA-binding protein is required for function of the human preproendothelin-1 promoter in endothelial cells. *Mol Cell Biol* 10:4854–4862.

Wood WG, Clegg JB, Weatherall DJ (1979): Hereditary persistence of fetal hemoglobin (HPFH) and $\delta\beta$-thalassemia. *Br J Haematol* 43:509–520.

Wright S, Rosenthal A, Flavell RA, Grosveld F (1984): DNA sequences required for regulated expression of β-globin genes in murine erythroleukemia cells. *Cell* 38:265–273.

Xiao JH, Davidson I, Macchi M, Rosales R, Vigneron M (1987): In vitro binding of several cell-specific and ubiquitous nuclear proteins to the GT-1 motif of the SV40 enhancer. *Genes & Dev* 1:794–807.

Yagi M, Gelinas R, Elder JT, Peretz M, Papayanopoulou T, Stamatoyannopoulos G, Groudine M (1986): Chromatin structure and developmental expression of the human α-globin cluster. *Mol Cell Biol* 6:1108–1116.

Yamamoto M, Ko LJ, Leonard MW, Beug H, Orkin SH, Engel JD (1990): Activity and tissue-specific expression of the transcription factor NF-E1 multigene family. *Genes & Dev* 4:1650–1662.

Yu C-Y, Chen J, Lin L-I, Tam M, Shen C-KJ (1990): Cell type-specific protein-DNA interactions in the human ζ-globin upstream promoter region: Displacement of Sp1 by the erythroid cell-specific factor NF-E1. *Mol Cell Biol* 10:282–294.

Zon LI, Youssoufian H, Mather C, Lodish HF, Orkin SH (1991a): Activation of the erythropoietin receptor promoter by transcription factor GATA-1. *Proc. Natl Acad Sci USA* 88:10638–10641.

Zon LI, Mather C, Burgess S, Bolce ME, Harland RM, Orkin SH (1991b): Expression of GATA-binding proteins during embryonic development in *Xenopus laevis. Proc Natl Acad Sci USA* 88:10642–10646.

Zon LI, Gurish MF, Stevens RL, Mather C, Reynolds DS, Austen KF, Orkin SH (1991c): GATA-binding transcription factors in mast cells regulate the promoter of the mast cell carboxypeptidase A gene. *J Biol Chem* 266:22948–22953.

Zon LI, Tsai SF, Burgess S, Matsudaira P, Bruns GAP, Orkin SH (1990): The major human erythroid DNA-binding protein (GF-1): Primary sequence and localization of the gene to the X chromosome. *Proc Natl Acad Sci USA* 87:668–672.

Chapter 7

Transcriptional Control of Gene Expression in Hepatic Cells

Gennaro Ciliberto, Vittorio Colantuoni,
Raffaele De Francesco, Vincenzo De Simone,
Paolo Monaci, Alfredo Nicosia, Dipak P. Ramji,
Carlo Toniatti, and Riccardo Cortese

In this chapter, we review the regulation of gene expression in hepatic cells, with emphasis on the *cis*-acting elements located in the proximal promoter region and on the properties of the transcriptional factors active in liver.

Developmental and Spatial Regulation of Liver Gene Expression

The Development of the Liver and Expression of Developmentally Regulated Liver Genes

In vertebrate embryos, the liver originates from an endodermal diverticulum of the foregut, which in the mouse develops between days 8.5 and 9.5 of gestation. The liver bud becomes recognizable microscopically at day 10.5 (Le Douarin, 1975; Theiler, 1989), and by day 12.5 it is already a relatively large and differentiated organ. These morphogenetic events have been studied mainly in chickens (Le Douarin, 1975) and in mice (Houssaint, 1980). In the mouse, the mRNA encoding the typical marker of liver differentiation albumin is already detectable at day 9.5 (Cascio and Zaret, 1991), a stage corresponding to the first contacts between the endodermal cells that will give rise to hepatocytes and the surrounding mesodermal cells. This stage precedes cell aggregation

GENE EXPRESSION: GENERAL AND CELL-TYPE-SPECIFIC
Michael Karin, Editor
© 1993 Birkhäuser Boston

and liver plate formation. A further sharp increase in albumin mRNA takes place shortly after the definitive formation of the liver (Cascio and Zaret, 1991), which is perhaps an indication that the cell–cell interaction essential for 164–165).

There are many morphological and functional differences between fetal and adult liver, including differences in the pattern of gene expression. However, in this chapter we consider only adult liver gene expression.

The Architecture of the Adult Liver and Its Role in the
Expression of Liver-Specific Genes

The integrity of the liver tissue architecture plays an important role in sustaining high levels of liver-specific gene expression. Comparisons between liver-specific gene expression in the intact liver, in liver slices, and in disaggregated hepatocytes have shown that disrupting the tissutar organization causes a 20- to 100 fold drop in the quantities of several liver-specific mRNAs. This decrease reaches its maximum 24 h after tissue disaggregation (Clayton and Darnell, 1983) and is prevented by *in vitro* tissue reassociation (Clayton and Darnell, 1983; Clayton et al., 1985a; Jefferson et al., 1984). Moreover, in most differentiated hepatoma cell lines the level of liver-specific gene expression is also similar to that of disaggregated hepatocytes (Clayton et al., 1985b). It has further been shown that coculturing hepatocytes with nonparenchymal liver epithelial cells derived from terminal biliary ductular cells increases both transcription rate and stability of liver-specific mRNAs (Fraslin et al., 1985). Thus, it seems that one important mechanism for regulating liver-specific gene expression works through cell–cell contacts in the hepatic tissue.

Another important factor is the interaction between hepatocytes and the extracellular matrix. Hepatocytes cultured on hydrated extracellular matrix (ECM) or collagen matrix show elevated liver-specific mRNAs and albumin production when compared to controls grown on plastic dishes or dried ECM (Ben-Ze'ev et al., 1988). The cell shape and cytoskeletal organization are dictated by the extracellular matrix; thus, it is conceivable that these properties of the hepatocytes are important for high levels of specific gene expression. In fact, it has been recently reported that the ECM modulates in a coordinate fashion the activity of liver transcription factors and hepatocyte morphology, selectively increasing the level of the HNF-3α and eH-TF transcriptional activators.

This increase, which occurs at transcriptional level for HNF-3α, leads to an ECM-dependent activation of the distal enhancer of the albumin promoter and results in increased albumin production (Di Persio et al., 1991; Liu et al., 1991).

Metabolic Zonation of Liver Parenchyma

All the specialized functions of the liver, such as catabolite modifications and plasmaproteins synthesis, take place in the hepatocytes. However, different rates of liver enzyme production, relating to enzyme position in the hepatic lobulus, have been observed among different hepatocytes (Gleiberman and Abelev, 1985). Hydroxymethylglutaryl-CoA reductase (HMGR), carbamoyl phosphatase synthase II (CPS), ornitine carbamoyl transferase (OCT), phosphoenolpyruvate carboxykinase (PEPCK), and other enzymes are produced at higher levels in periportal hepatocytes; others, such as glucokinase (GK), carbonic anidrase III (CAIII), and glutamine synthetase (GS), are expressed at higher levels in perivenous hepatocytes. In particular, GS is exclusively expressed in a two- to three-cell-thick layer of hepatocytes surrounding the centrolobular complex. It has been proposed that periportal hepatocytes predominantly catalyze beta oxidation of fatty acids, catabolism of amino acids, ureagenesis, gluconeogenesis for glycogen synthesis and glucose release, bile formation with cholesterol synthesis, and protective metabolism. Perivenous hepatocytes preferentially mediate glucose uptake for glycogen synthesis, glycolysis, liponeogenesis, ketogenesis, glutamine formation, and xenobiotic metabolism (Jungermann, 1988; Quistorff, 1990).

The molecular mechanisms responsible for the differences between the periportal and perivenous regions are at present unknown. Zonal gradients of pH, oxygen, hormones, or even nerve density between these two regions have been suggested as possible causes of the metabolic zonation of liver parenchyma. However, studies carried out on GS by partial hepatectomy suggest that a more profound difference exists between periportal and perivenous hepatocytes; it is as if they belonged to two distinct subpopulations (Gebhardt, 1990).

Model Systems for Liver Gene Expression

A great deal of information on liver gene expression has been obtained through studies on the entire organism, taking advantage of the fact that many liver-specific products are released to the blood. In this way, the rate and timing of liver enzyme and plasmaprotein synthesis, as well as the effects of hormones, chemicals, and various forms of stress, have been determined (Putnam, 1975). However, our understanding of liver gene expression has progressed by utilizing permanent cell lines derived from natural hepatomas, such as the human HepG2 and Hep3B (Knowles et al., 1980) and the rat H4II cell lines (Deschatrette et al., 1985), that express differentiated liver functions. By introducing appropriately constructed plasmids into cultured cell lines of different origin, it has been possible to identify the *cis*-regulatory elements responsible for tissue-specific expression of many liver genes (reviewed by De Simone and Cortese, 1988). In some cases the boundaries of the *cis*-acting DNA segments, important in developmental control of transcription, have also been defined *in vivo* in transgenic mice (Ciliberto et al., 1987; Dente et al., 1988; Hammer et al., 1987; Pinkert et al., 1987; Rüther et al., 1987; Tripodi et al., 1991).

The identification and characterization of *trans*-acting factors were made possible by developing cell-free systems that reproduce *in vitro* the transcription selectivity observed *in vivo* (Gorski et al., 1986). The interaction of regulatory protein with liver-specific promoter and enhancer elements has been studied in detail both *in vitro* and *in vivo* by direct footprinting in which methylation interference is performed on the isolated nuclei or on the whole cell (Becker et al., 1986; Church et al., 1985).

General Architecture of Liver-Specific Promoters

Promoter Modularity

In this chapter, we limit our attention to those promoters for which a precise identification of *cis*-acting sequences has been reported. A summary of the information available in the literature is presented in Figure 1. The *cis*-regulatory signals governing liver-specific gene expression are interspersed throughout several kilobases of 5′- and 3′- noncoding sequences (De Simone and Cortese, 1988). However, in most cases the proximal promoter region contains sufficient information for efficient and tissue-specific transcription. It has thus been possible to study the 250–300 bp upstream of the cap site in greater detail. Binding sites for tissue-specific and ubiquitous factors are concomitantly present in the same promoter, suggesting that both kinds of transactivators are required for maximal

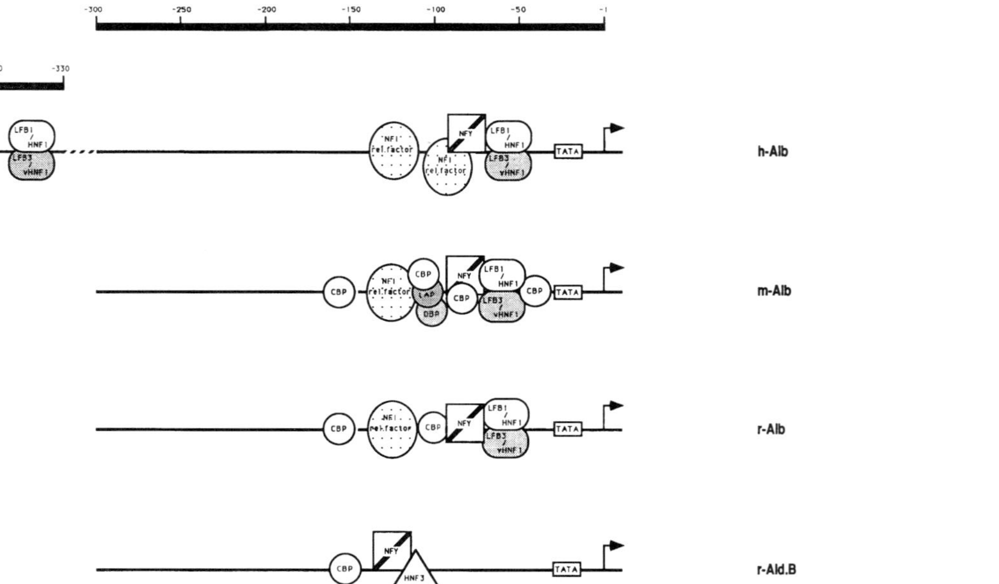

Figure 1   Schematic representation of factors interacting with promoters of liver-specific genes. Numbers at top of figure refer to distance in base pairs from transcription start site. The factor, its abbreviation (in parentheses), and the source of data follow. Human albumin (h-Alb): Nicosia and Monaci, unpublished observations; Hardon et al. (1988); Paonessa et al. (1988); Frain et al. (1990); mouse albumin (m-Alb): Gorski et al. (1986); Lichtsteiner et al. (1987); Baumheuter et al. (1988); Cereghini et al. (1988); Lichtsteiner and Schibler (1989); Izban and Papaconstantinou (1989); Maire et al. (1989); Descombes et al. (1990); Wuarin et al. (1990); rat albumin (r-Alb): Babiss et al. (1987); Heard et al. (1987); Raymondjean et al. (1988); Herbomel et al. (1989); Jose-Estanyol et al. (1989); Tronche et al. (1990); rat aldolase B (r-Ald B): Tsutsumi et al. (1989); Ito K et al. (1990); Raymondjean et al. (1991).

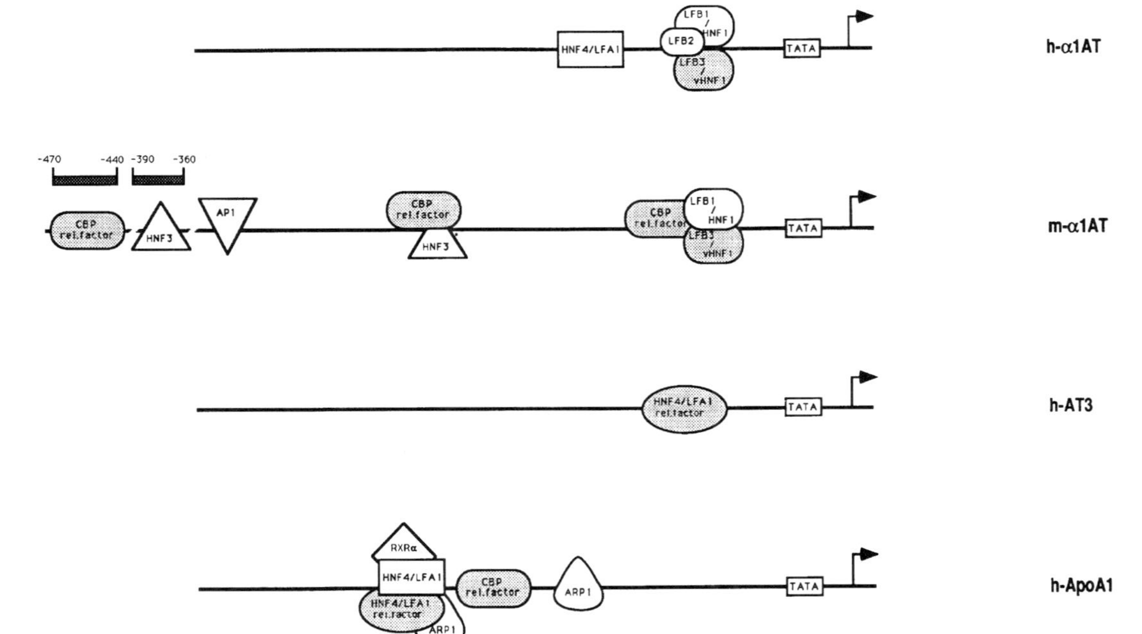

Figure 1 (Cont.) Human α_1-antitrypsin (h-α_1AT): De Simone et al. (1987); Shen et al. (1987); Monaci et al. (1988); Baumhueter et al. (1988); Hardon et al. (1988); Kugler et al. (1988); Cereghini et al. (1988); Courtois et al. (1988); Li et al. (1988); Ramji et al. (1991); Rangan and Das (1990); mouse α_1-antitrypsin (m-α_1AT): Costa et al. (1988a); Grayson et al. (1988a, 1988b); Costa et al. (1989); human antithrombin (h-AT3): Lucero et al. (1989); Ochoa et al. (1989); human apolipoprotein A1 (h-ApoA1): Hardon et al. (1988); Papazafiri et al. (1991); Rottman et al. (1991); Widom et al. (1991).

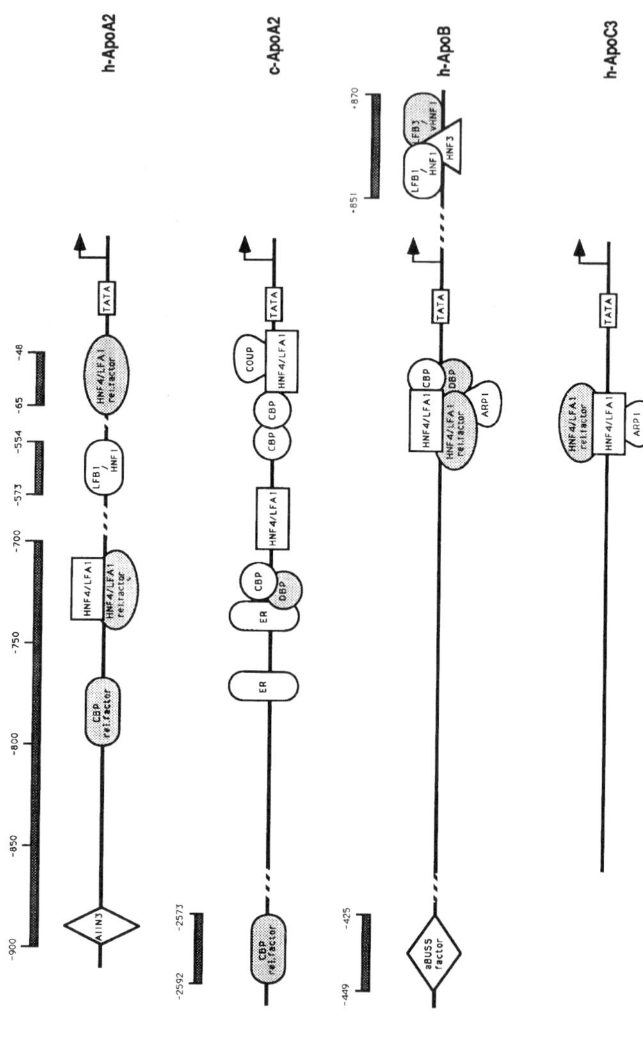

Figure 1 (Cont.) Human apolipoprotein A2 (h-ApoA2): Lucero et al. (1989); chicken apolipoprotein A2 (c-ApoA2): Wijnholds et al. (1988, 1991); Hoodless et al. (1990); Beekman et al. (1991); human apolipoprotein B (h-ApoB): Kardassis et al. (1990a, 1990b); Metzger et al. (1989); Ross et al. (1991); Brooks et al. (1991); human apolipoprotein C3 (h-ApoC3): Reue et al. (1988); Leff et al. (1989); Ogami et al. (1990, 1991).

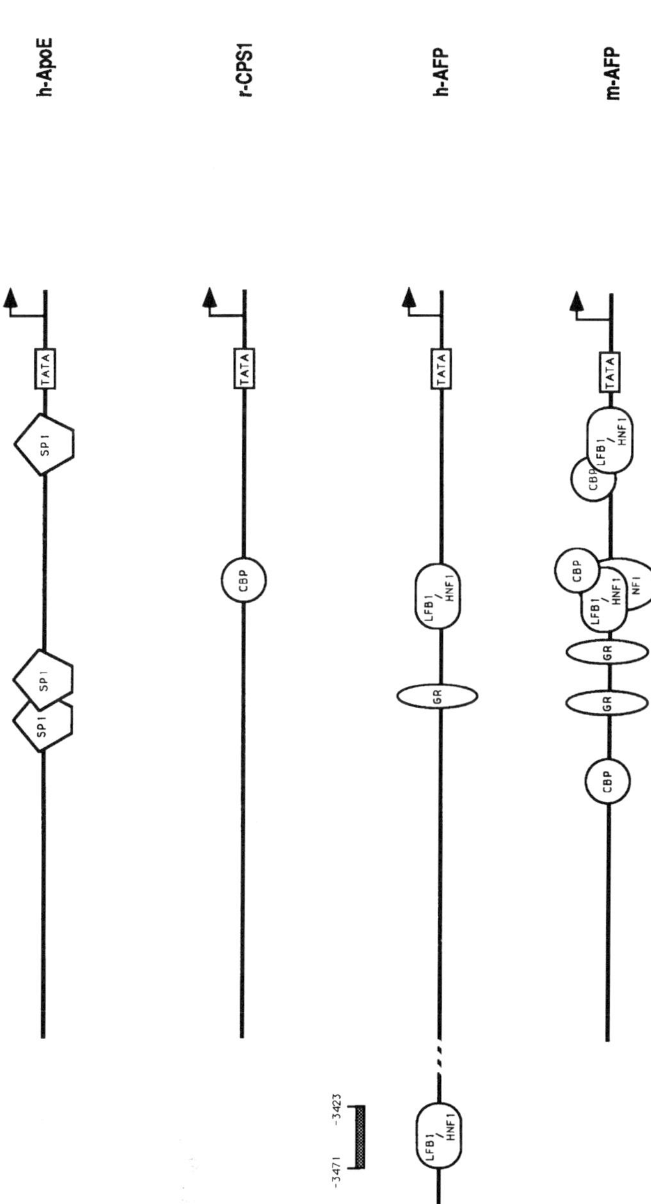

Figure 1   (Cont.) Human apolipoprotein E (h-ApoE): Pajk et al. (1988); Chang DJ et al. (1990); rat carbamyl phosphate synthetase 1 (r-CPS1): Howell et al. (1989); human α-fetoprotein (h-AFP): Sawadaishi et al. (1988); Nakabayashi et al. (1989); Nakao et al. (1990); mouse α-fetoprotein (m-AFP): Feuerman et al. (1989); Zhang DE et al. (1990).

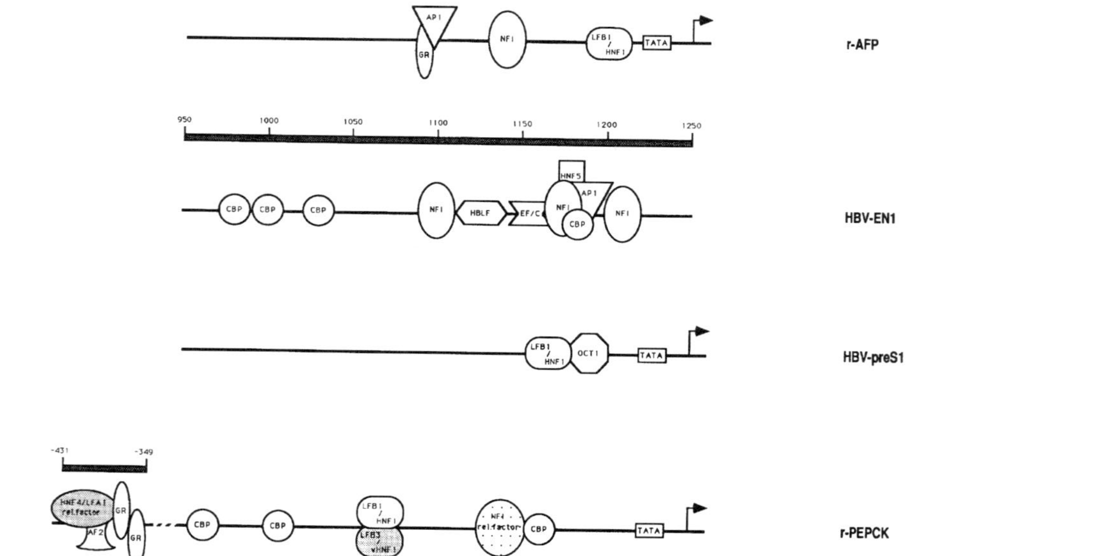

Figure 1   (Cont.) Rat α-fetoprotein (r-AFP): Guertin et al. (1988); Jose-Estanyol et al. (1989); Zhang et al. (1991); hepatitis B virus enhancer 1 (HBV-EN1): Honigwachs et al. (1989); Lòpez-Cabrera et al. (1991); Faktor et al. (1990); Trujillo et al. (1991); Dikstein et al. (1990b); Guo et al. (1991); Hu and Siddiqui (1991); hepatitis B virus presequence 1 surface antigen (HBV-preS1): Raney et al. (1990); Chang H-K and Ting (1989); Zhou and Yen (1991); Chang H-K et al. (1989); rat phosphoenolpyruvate carboxykinase (r-PEPCK): Petersen et al. (1988); Quinn et al. (1988); Benvenisty et al. (1989); Roesler et al. (1989); Park et al. (1990); Forest et al. (1990); Klemm et al. (1990); Quinn and Granner (1990); Trus et al. (1990); Imai et al. (1990); Shoshani et al. (1991); Lucas et al. (1991); Faber et al. (1991).

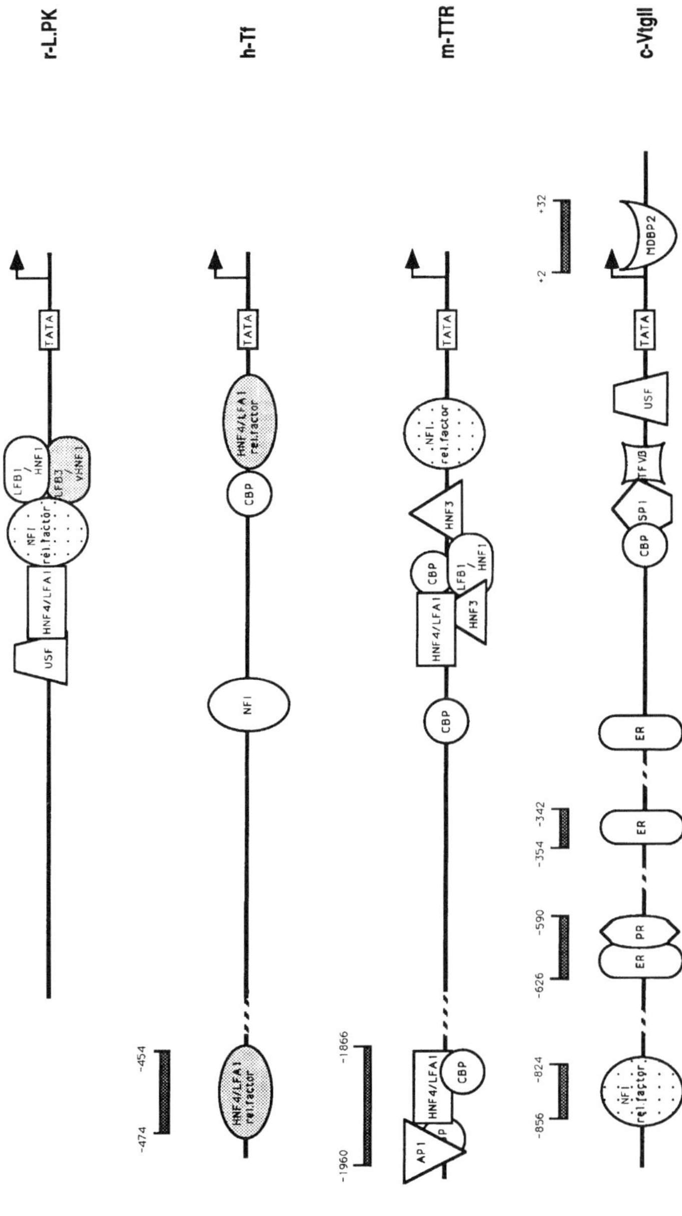

Figure 1   (Cont.) rat-L. pyruvate kinase (r-L.PK): Vaulont et al. (1989a, 1989b); Ramji et al. (1991); Yamada et al. (1990); Cognet et al. (1991); human transferrin (h-Tf): Brunel et al. (1988); Ochoa et al. (1989); Mendelzon et al. (1990); mouse transthyretin (m-TTR): Costa et al. (1988a, 1988b); Grayson et al. (1988a); Costa et al. (1990); chicken vitellogenin II (c-VTgII): Burch et al. (1988); Cato et al. (1988); Philipsen et al. (1988); Pawlak et al. (1990); Vaccaro et al. (1990); Seal et al. (1991).

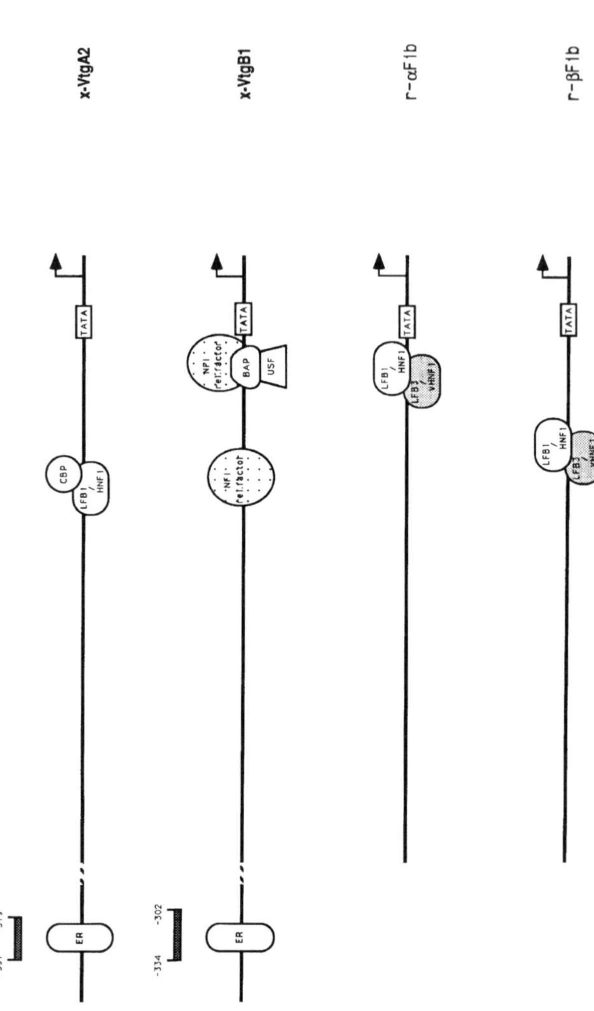

Figure 1 (Cont.) *Xenopus* vitellogenin A2 (x-VtgA2): Klein-Hitpass et al. (1988); Kaling et al. (1991); *Xenopus* vitellogenin B1 (x-VtgB1): Chang and Shapiro (1989); Martinez and Wahli (1989); Kugler et al. (1990); Corthésy et al. (1989, 1990, 1991); rat α-fibrinogen (r-αFib): Courtois et al. (1987, 1988); Baumhueter et al. (1988); Cereghini et al. (1988); rat β-fibrinogen (r-βFib): Courtois et al. (1987, 1988); Baumhueter et al. (1988); Cereghini et al. (1988); Kugler et al. (1988).

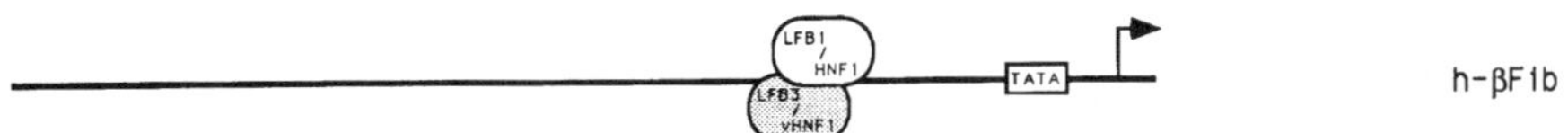

Figure 1   (Cont.) human β-fibrinogen (h-βFib): Baumhueter et al. (1988); Cereghini et al. (1988); Huber et al. (1990).

transcription rates in the hepatocyte. Although the position of *cis* elements in liver-enriched factors is variable, binding sites for C/EBP, LF-B1/HNF-1α, HNF-3, or HNF-4/LF-A1 are always present in the proximal promoter or in the distal enhancer sequences. However, their relative orientation can vary, as in the case of LF-B1/HNF-1α binding sites: these are oriented in one direction in the promoters of α_1-antitrypsin (α_1-AT), albumin, and α-fetoprotein (AFP), and in the opposite direction in the promoters of α- and β-fibrinogen. Similarly, the HNF-4/LF-A1 binding sites present in the promoters of α_1-antitrypsin, apoliopoprotein (Apo) A1, A2, B, C3, and pyruvate kinase are in the opposite orientation with respect to those found in the regulatory regions of transthyretin, Apo A4, and haptoglobin (Hp) genes. Therefore, it is possible to conclude that a unique architecture of *cis*-acting elements is not essential to their function.

In all hepatic promoters, at least one binding site for a liver-specific or liver-enriched transcription factor is located within the first 100–120 bp upstream of the cap site, and frequently this is the protein–DNA site of interaction that maps nearest to the TATA box. Whenever it has been possible to establish a direct correlation between functional and binding data, a clear hierarchy has emerged of the importance of DNA elements, indicating that these tissue-specific proximal promoter elements act as rate-limiting signals for promoter activity. For instance, in the mouse albumin promoter the two elements centered at position -60 and -100 are responsible for most transcriptional activity in liver cells or in nuclear extracts from rat liver. These two elements bind LF-B1/HNF-1α and members of the C/EBP family, respectively (Gorski et al., 1986; Maire et al., 1989; Tronche et al., 1989). In the case of the human α_1-AT promoter, the activity in transiently transfected hepatoma cells or in liver extracts is dependent on the binding sites for HNF-4/LF-A1 (-110 bp from the cap site) and LF-B1/HNF-1α (-70 bp from the cap site). Both elements can stimulate strong and tissue-specific transcription (Monaci et al., 1988). The relative importance of the two elements emerges, however, when their contribution to the promoter activity is analyzed in the context of the entire gene in transgenic mice (Tripodi et al., 1991). In this case, disruption of the HNF-4/LF-A1 site has a detrimental effect on promoter activity only in embryos, while mutation of the more proximal LF-B1/HNF-1α binding site drastically reduces α_1-AT transcription, both throughout development and in the adult animal. It is therefore likely that the various *cis*-acting elements present in liver-specific promoters might play different roles during development and differentiation.

The Interplay of Different DNA-Binding Proteins on the Same Promoter: Cooperative Interaction and Mutually Exclusive Binding

The modular structure of hepatic promoters suggests that each discrete element activates transcription independently. However, in some cases the transcriptional output of a promoter containing multiple binding sites is higher than the sum of the activities displayed by its individual elements. The common interpretation of these findings is that cooperative interactions between neighboring *trans*-acting factors reciprocally augment their transcriptional potency. For instance, DNA methylation in the LF-B1/HNF-1α binding site of the mouse albumin promoter has little effect on transcription efficiency, only if the surrounding *cis*-acting elements are present. Binding of LF-B1/HNF-1α to its cognate sequence apparently is greatly enhanced by cooperative interaction with the factors binding to nearby sequences (Lichtsteiner and Schibler, 1989; Tronche et al., 1989). Likewise, two LF-B1/HNF-1α molecules that bind simultaneously to sites distant from each other in the C-reactive protein (CRP) promoter function in synergy to activate gene expression (Toniatti et al., 1990b).

The high number of closely spaced, or sometimes overlapping, DNA-binding sites in the promoters of many liver genes also implies that only one subset of these elements is occupied *in vivo*. For instance, by *in vitro* footprinting of the mouse transthyretin gene with rat liver nuclear extracts, it has been shown that the proximal promoter region can bind to NF1, HNF-3, LF-B1/HNF-1α, C/EBP, and HNF-4/LF-A1 and that the distal enhancer binds C/EBP, HNF-4/LF-A1, and AP1. However, *in vivo* only the recognition sequences for NF1 and HNF-3 in the promoter and C/EBP in the enhancer are occupied in actively transcribing rat liver nuclei (Mirkovitch and Darnell, 1991).

Increasing evidence has accumulated on binding different factors to the same DNA sequence. Families of transcription factors characterized by functionally identical DNA-binding properties include members specifically enriched in hepatocytes such as the leucine zipper C/EBP family, the extra large homeodomain-containing factors LF-B1/HNF-1α and LF-B3/HNF-1β, and the HNF-4/LF-A1 class of factors that belong to the steroid receptor superfamily (see pp. 181–189). The question is which of these proteins is responsible for individual promoter activity *in vivo*. It is becoming increasingly clear that related factors have distinct transcriptional properties: although C/EBP and IL-6DBP recognize a common binding site, only IL-6DBP can transduce the stimulative signal of the Interleukin-6 (IL-6) cytokine to the transcriptional apparatus (Poli et al., 1990). In the HNF-4/LF-A1 and ARP-1 pair, although HNF-4/LF-A1 is a positive regulator of apolipoprotein A1 (ApoA1) transcription in HepG2 cells the ubiquitous ARP-1 protein can repress ApoA1 promoter activity in the same cells by interacting with the same *cis* element (Ladias and Karathanasis, 1991). Interestingly, transcription inhibition by ARP-1 is observed only when its cognate sequence is present in the context of the native ApoA1 promoter, as if ARP-1 function depended on the particular array of surrounding DNA-binding sites.

The confirmation that some of the *cis*-acting elements characterized *in vitro* or in transfected cell lines do indeed play a role *in vivo* is dramatically provided by analyzing inherited human diseases. In the Leyden variant of hemophilia B, the disease (which is chromosome X-linked) is characterized by severely reduced plasma levels (less than 10% the normal level) of the liver-specific blood clotting factor IX, levels that may increase to 40%–80% after puberty (Briët et al., 1982). Mutations have been found at position +13 after the cap site that result in the loss of binding of transcription factor C/EBP (Crossley and Brownlee, 1990). The mutant promoter cannot be transactivated by a C/EBP expression vector in cotransfection experiments. More interestingly, a different factor IX promoter alteration was identified in a patient whose clinical conditions failed to improve after puberty. In this case, a double mutation was observed: the first mutation maps at position −20 and disrupts the binding site for HNF-4/LF-A1; the second mutation, at position −26, destroys a functional androgen-responsive element. This is a possible explanation for the lack of clinical recovery after puberty (Crossley et al., 1992).

Hormonal Regulation of Liver Gene Expression: The TAT Gene

The expression of a large group of liver genes is under the control of a complex network of hormonal stimuli. The liver gene modulation that takes place during inflammation is well known and goes under the name of "acute-phase response"; its features are extensively described on pp. 200–210. Other hormones that play a major role in influencing liver gene expression include glucocorticoids, growth hormone, insulin, glucagon, and somatostatin. Summarizing all data concerning liver gene regulation by hormones is beyond the scope of this review, but it is important to consider the case of the tyrosine aminotransferase gene (TAT). This gene has been studied in great detail because it is a typical example of the role that multiple hormonal control plays in developmental and tissue-specific regulation.

The features of TAT gene regulation are (a) TAT expression is restricted to hepatocytes (Hargrove and Granner, 1985); (b) expression is undetectable before birth and increases rapidly within a few hours after birth (Greengard, 1970); (c) gene transcription is increased by glucocorticoids and glucagon, the latter acting through an increase in intracellular cAMP levels (Granner and Beale, 1985; Hashimoto et al., 1984; Schmid et al., 1987); (d) the activity of these two hormones is somehow connected with the developmental program of TAT expression, as shown by the fact that administration of glucagon *in utero* leads to prenatal TAT gene expression, further augmented by dexamethasone (Ruiz-Bravo and Ernest, 1982); (e) expression of this gene is negatively regulated in non-hepatic cells. This is demonstrated by the fact that the TAT gene is part of the group of genes controlled by the tissue-specific extinguisher 1 (TSE-1) in hepatoma × fibroblast intertypic hybrids, together with phosphoenol pyruvate carboxykinase (PEPCK) and argininosuccinate synthetase (AS) (Lem et al., 1988; Ruppert et al., 1990; Thayer and Fournier, 1989) (see also pp. 196–200); (f) finally, cell-specific expression and hormonal regulation are strictly linked to each other, as demonstrated by the finding that TSE-1 repression in monochromosomal hepatoma × fibroblast hybrids can be easily reversed by treatment with cAMP (Ruppert et al., 1990; Thayer and Fournier, 1989).

Analyzing the chromatin structure of the TAT gene has led to the discovery of several various DNase I-hypersensitive site(s) (Grange et al., 1989; Nitsch et al., 1990). The −2.5 and −5.5 kb sites correspond to glucocorticoid-responsive units (GRUs), which interact cooperatively to promote glucocorticoid regulation (Grange et al., 1989). While the downstream site is absolutely necessary for hormonal induction, the upstream site only functions to potentiate the activity of the other. Both GRUs, however, have a complex structure characterized by not only multiple glucocorticoid-responsive elements (GREs) but also recognition sites for the two liver-enriched transcription factors C/EBP and HNF-5 (Grange et al.,1991).

The chromatin structure of the downstream GRU has been analyzed in detail by genomic footprinting (Reik et al., 1991). A change in chromatin structure occurs very rapidly after addition of dexamethasone, and transcriptional activation of the TAT promoter is strictly related to this change; a local disruption of the nucleosomal pattern occurs, the C/EBP and HNF-5 sites are occupied by the cognate factors, but this is rapidly reversed on withdrawal of hormone. In contrast to this finding, the direct binding *in vivo* of the glucocorticoid receptor to GREs has never been detected, and in those cases where GRE and HNF-5 sites overlap, only the latter interacts stably with DNA during induction. These results suggest that the glucocorticoid receptor operates by a hit-and-run mode of interaction with the TAT GRU, probably by giving access to the transcription factor HNF-5 which would be otherwise antagonized by the presence of nucleosomes (Rigaud et al., 1991).

The third element, located 3.6 kb upstream of the cap site, contains the cAMP-responsive element (CRE). This region has also been analyzed by genomic footprinting, which has revealed profound changes in chromatin structure and binding of the transcription factor CREB following a rise in cAMP levels (Weih et al., 1990). The CRE is the inhibition target of the product of the TSE-1 locus (Boshart et al., 1990), and the mechanism by which this gene product represses the function of the enhancer has been clarified (Boshart et al., 1991; Jones et al., 1991) (see pp. 197–199).

The Interplay Between Promoters and Enhancers Regulates Transcription of the HBV Genome

The proximal promoter elements of several liver-specific genes and some of their enhancer elements have been characterized in detail. However, the distance between enhancers and promoters is such that their interaction and interdependence can be studied better in the small genome of viruses such as that of hepatitis B virus (HBV).

The genome of the hepatotropic virus HBV is efficiently transcribed and replicated only in hepatic cells. The information for this hepatospecificity is very compactly stored within the sequence of four promoters and two enhancers (Figure 2). The three major transcripts present in HBV-infected cells are 3.6, 2.1, and 2.4 kb long. The 3.6-kb RNA originates from the C gene promoter, while the 2.1- and 2.4-kb RNAs are generated by transcription, starting from the SP2 and SP1 promoter, respectively (Ganem and Varmus, 1987). An RNA 0.7–0.9 kb long originates from the

X gene promoter (Kaneko and Miller, 1988; Siddiqui et al., 1987). The activity of these four promoters is strongly influenced by the presence of at least two enhancer elements: Enh1, which maps approximately at nt 966–1308 (Guo et al., 1991; Shaul et al., 1985; Tognoni et al., 1985), and Enh2, which has been mapped between nt 1630 and 1750 (Yee, 1989). Finally, a glucocorticoid-responsive element (GRE) has been mapped between positions 351 and 366 (Tur-Kaspa et al., 1986) that is capable of confering glucocorticoid inducibility to Enh1 (Tur-Kaspa et al., 1988).

Each individual enhancer can potentiate transcription only from one subset of promoters. Enh1 is active mainly in the X and core promoters (Bulla and Siddiqui, 1988; Hu and Siddiqui, 1991; Lòpez-Cabrera et al., 1990, 1991). Intriguingly, Enh2 has only a minor effect on the neighboring genes but strongly stimulates transcription starting from the remote SP1 and SP2 promoters (Yuh and Ting, 1990). The reasons for this interesting effect are as yet unclear but probably involve productive interactions between particular subsets of *trans*-acting factors occupying specific DNA sites. Zhou and Yen (1990) suggested that Enh2 could be more effective on the SP2 than on the SP1 promoter. If confirmed, this finding might explain an interesting feature of chronic HBV infections; because the downstream portion of the X gene including Enh2 is frequently deleted on HBV DNA integration into the host chromosome during chronic infection, loss of Enh2 could lead to an increase in the ratio of large SP1 versus middle and small SP2 Ags and therefore might contribute to the formation of "ground-glass" cells frequently seen in liver biopsies of patients with chronic hepatitis B (Hoofnagle et al., 1987).

Enh2 has a modular structure resembling the SV40 72-bp enhancer, with two small boxes named a and b (23-bp and 12-bp long, respectively) that are both required for full liver-specific enhancer activity (Wang et al., 1990; Yee, 1989; Yuh and Ting, 1990). Several overlapping binding activities have been detected in this region, but only a few of them have been identified as C/EBP like (Yuh and Ting, 1991). However, the most extensively studied transcription element is Enh1, where different functional regions have been identified, each interacting with one or several putative *trans*-acting factors (Ben-Levy et al., 1989; Patel et al., 1989). The basal enhancer element has been localized between nt 1116 and 1170 (see Figure 1), which contains the binding site for the ubiquitous transcription factor EF-C (Dikstein et al., 1990a) and for a novel liver-specific transcription factor that has not as yet been fully characterized (Guo et al., 1991; Trujillo et al., 1991). Accessory Enh1 elements have been

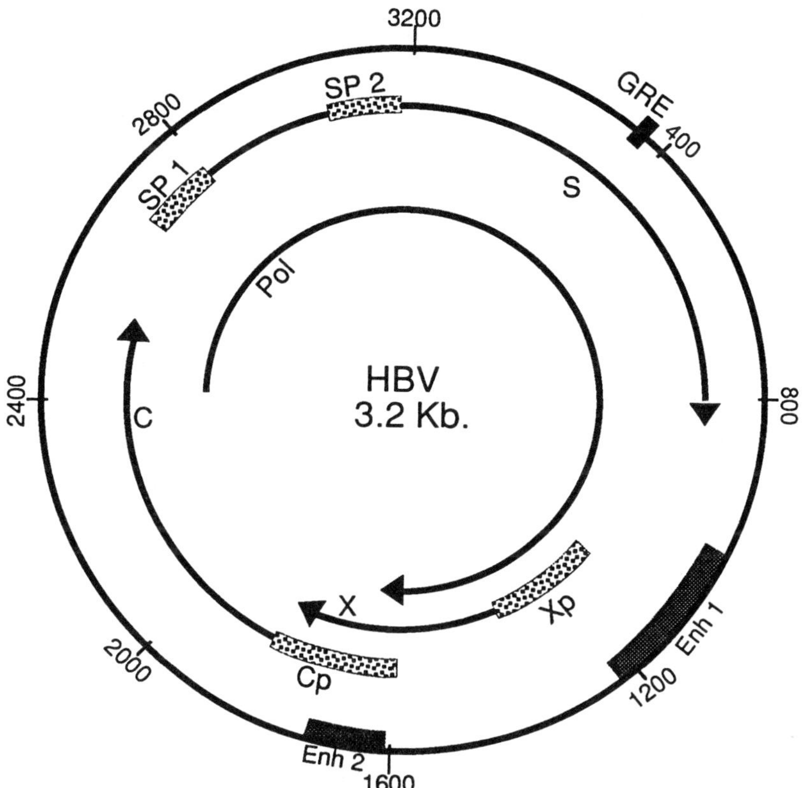

Figure 2   Genetic organization of HBV genome. Open reading frames (ORFs) Pol, S, C, and X are represented by *arrows* (Blum et al., 1989). *Dotted boxes* refer to promoters SP1 (Raney et al., 1990), SP2 (Raney et al., 1989), Cp (Lòpez-Cabrera et al., 1990), and Xp (Trujillo et al., 1991). Locations of (GRE) (Tur-Kaspa et al., 1988) and two enhancers (Enh1: Hu and Siddiqui, 1991; Enh2: Yuh and Ting, 1990) are also indicated.

identified in the region from nt 1170 to 1240 (Guo et al., 1991; Trujillo et al., 1991). Interestingly, the same DNA segment has proven to be a crucial component of the X minimal promoter element mapped between nt 1168 and 1323 (see Figure 2).

Functional analysis of the accessory Enh1 region has allowed the identification of partially overlapping binding sites for NF-1-like, AP-1, and C/EBP-related proteins (Faktor et al., 1990; Dikstein et al., 1990b). In particular, the presence of C/EBP-binding sites with different affinities seems to be the basis of an interesting observation: cotransfection of a reporter gene under the control of the whole HBV Enh1 with increasing amounts of a C/EBP expression vector gives transactivation at low doses, while at high doses C/EPB acts as a repressor (Dikstein et al., 1990b; Trujillo et al., 1991). At low C/EBP concentrations, which correspond to potentiation of transcription, only high-affinity sites are occupied. At high concentrations, protection of low-affinity sites leads to displacement of other factors (EF-C, HNF-5, etc.), and this is probably responsible for shutting down Enh1 activity. Transcriptional inhibition by C/EBP is also observed when the core promoter (which contains several C/EBP-binding sites) or the full-length HBV genome are cotransfected with elevated amounts of C/EBP expression vector (Lòpez-Cabrera et al., 1991; Pei and Shih, 1990). These findings are particularly intriguing because members of the C/EBP family are modulated during inflammation (Poli et al., 1990), thus suggesting an elaborate regulation of HBV transcription during both acute and chronic hepatitis. In addition, because C/EBP levels are higher in resting versus proliferating hepatocytes (Birkenmeier et al., 1989; Friedman et al., 1989), HBV gene expression might be potentiated in infected proliferating cells.

Architecture of Liver-Specific Transcription Factors

In describing the molecular architecture of liver-specific transcription factors, we refer to the structural taxonomy based on DNA-binding domains that has been published by Harrison (1991).

The LF-B1/HNF-1α Family: A Novel Class of Non-Helix-Turn-Helix Homeodomain Proteins

LF-B1/HNF-1α. The LF-B1/HNF-1α group of proteins has been shown to bind to *cis*-acting elements that match the consensus sequence g/aGTT AAT N ATTAACc/a. As anticipated from the dyad symmetry of the consensus binding site, LF-B1/HNF-1α binds to DNA as a dimer (Frain et al., 1989). The functional domains of LF-B1/HNF-1α, as those of several other transcription factors, have been dissected and characterized in detail (Chouard et al., 1990; Nicosia et al., 1990). The sequences required for transcriptional activation are located in the C-terminal half (amino acids 282 to 628), whereas the DNA-binding activity maps in the first 281 amino acids.

In the DNA-binding domain of LF-B1/HNF-1α regions A, B, and C have been identified (Nicosia et al., 1990). Of these three regions, region A (amino acids 1–32) has been shown to mediate dimerization of the protein. Region B (amino acids 100–184) and region C (amino acids 198–281) show limited homology to the POU-specific (POU$_S$) and to the POU-homeo (POU$_{HD}$) motifs, respectively (see Figure 1). The putative homeodomain has an extra insertion of 20 amino acids compared to the canonical homeobox (Finney, 1990); in fact, this region of LF-B1/HNF-1α has been also referred to as the XL (extra-large) homeobox. A recombinant peptide corresponding to the XL homeobox of LF-B1/HNF-1α behaves in solution as an independently folded unit whose tertiary structure, as determined by multidimensional nuclear magnetic resonance (NMR) spectroscopy and x-ray christallography, shows important differences from the structure of canonical homeodomains in that a novel loop structure connects helix II with helix III in place of the usual tight turn found in the helix-turn-helix DNA-binding proteins (Leitung et al., 1993; Ceska et al., 1993).

The isolated XL homeodomain binds as a monomer to a half-palindromic site, but shows diminished sequence specificity. In addition to the XL homeodomain, the B region is required for proper recognition of LF-B1/HNF-1α sites (Tomei et al., 1992). Interestingly, a recombinant protein consisting of only the B and C domains is a monomer in solution, but forms dimers on DNA binding. It can thus be concluded that the B region of LF-B1/HNF-1α is the domain responsible for restricting the specificity of DNA binding, possibly through the formation of DNA-stabilized protein–protein interactions. It is worth pointing out that the POU$_S$ regions of two other transcription factors, namely GHF-1/Pit1 and Oct1 (Ingraham et al., 1990; Verrijzer et al., 1990) have also been implicated in establishing site specificity. Based on the sequence similarity to the POU-A subregion and because of the analogy of function with the POU$_S$ module of Pit1 and Oct1, the B-region of LF-B1/HNF-1α has also been termed Ψ-POU.

Region A has been shown to drive DNA-independent dimerization (Nicosia et al., 1990). The thermal and chemical equilibrium of unfolding in this peptide was studied by circular dichroism (CD) spectroscopy (De Francesco et al., 1991), which led to the conclusion that the dimerization domain of LF-B1/HNF-1α can fold into a stable domain structure independently of the rest of the protein. The secondary structure of a synthetic peptide corresponding to the A region has been revealed by multidimensional NMR (Pastore et al., 1991). Three structurally distinct regions can be distinguished: the N-terminal region from residues 1 to 6 is extended and may be involved in stabilization of the dimer through the formation of a short β sheet. Two α-helical regions span residues 7 to 18 and from 23 to 32. The three-dimensional structure closest to the experimental NMR data, and in agreement with the extraordinary conformational stability of this peptide, is a four-helix bundle generated by the juxtaposition of two helical hairpins (A. Pastore, personal communication).

The C-terminal moiety of LF-B1/HNF-1α (amino acids 281–628) can function as an independent activation domain because fusion of this protein fragment to the heterologous DNA-binding domain of nuclear factor 1 (NFI) or of GAL4 is sufficient to confer transcriptional stimulative activity to the resulting chimeric protein in the yeast *Saccharomyces cerevisiae* (P. Monaci and A. Nicosia, unpublished observations) or in transfected HeLa cells (Toniatti et al., 1993), respectively. When using recombinant LF-B1/HNF-1α, two distinct protein elements ADII (residues 281 to 318) and ADI (residues 547 to 628), located in the C-terminal half of LF-B1/HNF-1α, have been shown to be required for maximum induction of liver-specific promoter activity *in vitro* (Nicosia et al., 1990). ADI is rich in serine and threonine residues, a feature that occurs also in the transcriptional activation domain of the pituitary-specific activator Pit1/GHF-1 (Theill et al., 1989). ADII is characterized by a high proline content, as was originally found in the activation region of CTF-1 (Mermod et al., 1989).

Functional dissection of the LF-B1/HNF-1α transcriptional activation domain has also been studied *in vivo* by cotransfection into cell lines of hepatic or nonhepatic origin (Raney et al., 1991; Toniatti et al., 1993). The crucial role played by the extreme C-terminal serine-/threonine-rich region has been confirmed in this *in vivo* system: a mutant LF-B1/HNF-1α protein lacking residues 506 to 628 retains only about one-half of the transcriptional activity of the wild-type molecule. Further, region 506 to 628 drives GAL4-dependent transcription when linked to the DNA-binding domain of yeast transcription factor GAL4 (Toniatti et al., 1993). A second important activation domain is located between amino acids 440 and 506, and its corresponding peptide is an even more powerful activator of transcription when fused to the binding domain of GAL4. This region is richer overall in glutamine residues as compared to the rest of the protein. Finally, in contrast to the results obtained by *in vitro* transcription, the region of the protein immediately adjacent to the DNA-binding domain (i.e., ADII–amino acids 281–318) does not seem to have any function in transfected cells.

Recombinant LF-B1/HNF-1α produced in *E. coli* stimulates transcription *in vitro* (P. Monaci and A. Nicosia, unpublished data). However, in this case the level of activation is 100 fold lower than that of the *vaccinia*-expressed LF-B1/HNF-1α and of the native protein from rat liver. This finding suggests that posttranslational modifications are essential elements of the LF-B1/HNF-1α transcriptional activating surface. In fact, LF-B1/HNF-1α purified from liver tissue is found to be posttranscriptionally modified by the addition of O-linked *N*-acetylglucosamine (Lichtsteiner and Schibler, 1989). The precise extent of glycosylation and the localization of the modified residues have not yet been studied.

Other members of the family. LF-B3/HNF-1β recognizes the same target sequence as LF-B1/HNF-1α. This protein was originally identified as a DNA-binding activity in the dedifferentiated H5 and C2 rat hepatoma cell lines, or in hepatoma/fibroblasts cell hybrids (Baumhueter et al., 1988; Cereghini et al., 1988).

The cDNA coding for LF-B3/HNF-1β has been cloned independently in several laboratories (De Simone et al., 1991; Mendel et al., 1991a; Rey-Campos et al., 1991). Cotransfection experiments have shown that LF-B3/HNF-1β is a transcriptional activator as powerful as LF-B1/HNF-1α (De Simone et al., 1991). The most conserved regions between LF-B3/HNF-1β and LF-B1/HNF-1α are the XL homeobox (91% homologous), the ψ-POU (82%), and the N-terminal dimerization (79%) domains. On the whole, the C-terminal portions do not show overall such an extensive similarity. However, the glutamine-enriched activation domain of LF-B1/HNF-1α is quite well conserved in LF-B3/HNF-1β, and a deletion mutant that lacks the last 87 amino acids (i.e., the glutamine-enriched stretch) fails to transactivate (V. De Simone, unpublished data).

It has been shown that LF-B3/HNF-1β can form heterodimers with LF-B1/HNF-1α both *in vitro* and *in vivo* (De Simone et al., 1991; Rey-Campos et al., 1991) through the homologous dimerization domains. However, the functional relevance of the heterodimer is not yet clear.

DCoH: A cofactor that regulates the dimerization of LF-B1/HNF-1α. The cDNA coding for a protein factor that copurifies with LF-B1/HNF-1α has been cloned (Mendel et al., 1991b). The 11-kDa protein encoded by this cDNA has been termed DCoH and appears to selectively stabilize LF-B1/HNF-1α dimers. DCoH does not bind to DNA directly, but forms stable heterotetrameric complexes with LF-B1/HNF-1α. Although DCoH has no transcriptional activity when fused to the GAL4 DNA-binding domain, its engagement in the heterotetrameric complex eventually results in a strong enhancement of the transcriptional activity of LF-B1/HNF-1α. It has been suggested that DCoH might determine the developmental specificity of LF-B1/HNF-1α and related proteins by guiding and stabilizing the formation of transcriptionally active complexes (Mendel et al., 1991b).

The bZIP Proteins in Liver

C/EBP: The prototype bZIP protein. C/EBP was initially identified as a heat-stable DNA-binding protein present in rat liver nuclei that is capable of binding to the CCAAT motif present in several viral promoters (Graves et al., 1986) as well as to the core homology common to many viral enhancers (Johnson et al., 1987). It was subsequently shown that C/EBP binds to the *cis*-acting regulatory elements of several liver-specific genes, playing an important role in their transcriptional regulation (De Simone and Cortese, 1988).

C/EBP binds to the symmetry-related sequence as a dimer. The dimerization is mediated by a conserved domain called the "leucine zipper," and DNA is contacted by an adjacent region particularly rich in positively charged amino acids (Lamb and McKnight, 1991). McKnight and his coworkers have also shown that two subregions of C/EBP contribute to activating transcription (Friedman and McKnight, 1990). Both these elements retain their activation potential when fused to the DNA-binding domain of a nonrelated protein. It has been suggested that the first domain, located near the N-terminus of the protein (amino acids 62 to 89), folds as an α-helix, stabilized by intrahelical salt bridges and by a canonical N-cap structure. The extreme N terminus of C/EBP (amino acids 1 to 61) seems to play a role in conferring further stability to this activating domain and modulates the lability of the whole protein. The second activating domain is located between amino acids 150 and 196 and appears to be composed of three elements that show functional redundancy. When the elements are removed one at a time, the activating potential is not affected; conversely, deleting all three elements at once leads to a sharp decrease in transactivation. More recently, another laboratory confirmed that region 171 to 245 behaves as a relatively weak transactivating domain. Within this domain, a 45-amino-acid region that contains 12 prolines and 6 histidines (amino acids 171 to 215) is sufficient for transactivation. Analysis of mutants in a highly conserved proline-octamer region within this domain indicates that a specific proline content is not crucial to transactivation (Pei and Shih, 1991).

C/EBP-related proteins. The cloning of cDNAs and genes coding for a number of proteins structurally related to C/EBP has been reported by several laboratories, thus defining a family of C/EBP-related transcription factors. The various proteins were termed differently, as summarized in Table 1 (and references therein). All possess DNA-binding domains highly homologous to that of C/EBP, and as a consequence all the members of this family share identical DNA-binding specificity. In addition, the proteins belonging to this group can cross-heterodimerize via the conserved leucine-zipper domain. Despite the high degree of similarity among all the members of the C/EBP family in the carboxy-terminal regions, the amino-terminal regions, which carry the information for transcriptional activation, are very divergent. This difference might be responsible for different roles and functions played by homodimers and heterodimers in cell physiology.

Table 1. Terminology for C/EBP related transcription factors

Designation	Species	References
C/EBPα	Rat	Landschulz et al., 1988
NF-IL6	Human	Akira et al., 1990
IL-6DBP	Rat	Poli et al., 1990
rNF-IL6	Rat	Metz and Ziff, 1991
CRP2	Rat	Williams SC et al., 1991
C/EBPβ	Mouse	Cao et al., 1991
AGP/EBP	Mouse	Chang CJ et al., 1990
Ig/EBP-1	Mouse	Roman et al., 1990
NF-IL6β	Human	Kinoshita et al., 1992
CELF	Rat	Kageyama et al., 1991
C/EBPδ	Mouse	Cao et al., 1991
CRP3	Mouse	Williams SC et al., 1991
CRP1	Rat	Williams SC et al., 1991

DBP and VBP. DBP and VBP are two related bZIP proteins that bind to a subset of the C/EBP sites (Iyer et al., 1991; Mueller et al., 1990). DBP was in fact originally isolated as a transcription factor that binds to the D site of the rat serum albumin promoter (Mueller et al., 1990), whereas VBP was cloned as a protein binding to chicken vitellogenin II promoter

(Iyer et al., 1991). It is not yet clear whether the differences between the primary sequence of DBP and that of VBP are species specific or represent two different members of a novel family of transcription factors. This second hypothesis is more probable especially in the light of functional studies that have been made (see pp. 192–193).

The basic region in both DBP and VBP is similar to that of the C/EBP family members, but their leucine-zipper domain presents significant differences from the canonical C/EBP zipper. The most apparent feature of the leucine zippers of DBP and VBP is that neither has a perfect leucine heptade repeat; DBP has only two leucines of four (I-L-L-V), and VBP only three (I-L-L-L). Further, in contrast to that of C/EBP, the leucine zipper of DBP has no potential to form intrahelical ion pairs, which are known to confer extrastability to canonical coiled-coil structures (Iyer et al., 1991).

It is interesting to note, in the DBP and VBP proteins, that sequences that lie outside the bZIP domain contribute significantly to the binding affinity for DNA (Iyer et al., 1991). This behavior was unexpected, as all the other bZIP factors that have been thoroughly characterized can be truncated very near the bZIP domain and still retain full DNA-binding activity, or sometimes bind the DNA with an affinity higher than that of the full-length protein (Descombes and Schibler, 1991).

The HNF-3 Family: The Forkhead Domain

HNF-3 was described as a DNA-binding activity interacting with the mouse transthyretin (TTR) gene promoter, the α_1-AT proximal enhancer region, and the far upstream region of the rat phosphoenolpyruvate carboxykinase (PEPCK) gene (Costa et al., 1989; Ip et al., 1990). It was subsequently shown that this activity consists of three major DNA-binding proteins, called HNF-3α, HNF-3β, and HNF-3γ. The first protein of this set to be purified and cloned was HNF-3α (Lai et al., 1990). HNF-3β and HNF-3γ were cloned soon after by exploiting their sequence homology (Lai et al., 1991). The three proteins bind to the same DNA sequences with different affinities. Sequence comparison of the three proteins reveals a highly conserved 110-amino-acid region within the portion of the proteins required for DNA binding, but the remaining parts are highly divergent: HNF-3γ is about 140 residues shorter than HNF-3α and HNF-3β, and there are only two other very short stretches of conserved sequence toward the C-terminal end.

The HNF-3 proteins are highly homologous in the very same three regions in the *Drosophila* homeotic gene *forkhead* product (Weigel and Jäckle, 1990). Because the DNA-binding motif found in the HNF-3 family and in the *forkhead* gene product is not found in any other protein described previously, they define a new class of DNA-binding proteins. However, the mode in which the proteins bearing this novel motif bind to their DNA cognate sequence is not yet known. A cluster of basic amino acids (RRQKREK) is always present near the C-terminus end of the conserved region within the DNA-binding domain, and similar sequences are known to mediate the DNA contacts of bZIP proteins, but the binding modality must be different because the forkhead proteins interact with DNA as monomers.

The HNF-4 and ARP-1 Pair

HNF-4 was first identified as a factor that binds to the TTR promoter (Costa et al., 1989). Subsequently, analysis of the deduced protein primary sequence, revealed that this protein in fact belongs to the superfamily of nuclear steroid-thyroid hormone receptors (Sladek et al., 1990). However, no ligand for HNF-4 has been identified so far.

Between amino acids 50 and 116, HNF-4 contains a region with two potential four-cysteine zinc fingers that is 40%–60% homologous to the DNA-binding domain of other members of the steroid hormone receptor family. This family can be subclassified according to the amino acid sequence found in the "knuckle" region of the first finger (Harrison, 1991), which has been identified as critical to determining the binding specificity of different receptors (Forman and Samuels, 1990). In HNF-4, the sequence of this region (DGCKG) shares some similarity with that of the thyroid hormone receptor (EGCKA). On the C-terminus side of the zinc finger DNA-binding domain of HNF-4, after a stretch of sequence that has no similarity to any known protein, there is a large hydrophobic region (amino acids 133 to 373) that is reminiscent of the ligand-binding domain of other receptors of the same superfamily (20%–37% identity).

Like the other steroid receptors, HNF-4 binds DNA as a homodimer. The receptor dimerization function of the steroid-thyroid hormone receptors is restricted to a series of heptad repeats found within the ligand-binding domain (Forman and Samuels, 1990). In the corresponding region of HNF-4, at least 12 heptad repeats can be found. Because the regions responsible for ligand-binding and dimerization overlap in many nuclear receptors, it may very well be that the amino acid conservation observed is caused solely by a similar dimerization surface and not by the existence of a ligand-binding pocket in HNF-4.

HNF-4 bears a proline-rich region (23%) at the C-terminus (amino acids 400–447) and three serine-threonine-rich regions scattered throughout the molecule (amino acids 15–44, 129–161, and 398–426). Similar regions have previously been shown to correlate with the activation of transcription in LF-B1/HNF-1α (Nicosia et al., 1990) and NF1 (Mermod et al., 1989).

The isolation and molecular cloning of a protein that binds to the same sequence in the ApoA1 gene promoter and which is also recognized by HNF-4 has been recently reported (Ladias and Karathanasis, 1991). ARP-1 also binds to a thyroid hormone-responsive element and to regulatory regions of the ApoB, apolipoprotein CIII (ApoCIII), insulin, and ovalbumin genes. This protein, designated ARP-1 (ApoAI regulatory protein-1), also belongs to the steroid-thyroid hormone receptor superfamily. HNF-4 and ARP-1 are clearly different, yet they are much more alike than the other members of the family. Also, ARP-1 binds to DNA as a dimer, and its dimerization domain was localized in the carboxyl-terminal region. In transfection experiments, ARP-1 acts as a negative modulator of transcription; in fact, it downregulates both ApoAI and ApoCIII gene expressions (Ladias and Karathanasis, 1991). Moreover, a three- to fourfold excess of ARP-1 prevents the activating effect of HNF-4. This effect probably results from competition for the same target sequence, because ARP-1 and HNF-4 bind to the ApoAI recognition site with the same affinity. However, they markedly differ in their affinity to other binding sites: for example, the α_1-AT promoter site is only recognized by HNF-4. Although both bind DNA as homodimers, no heterodimer formation occurs between the two *in vitro* translated proteins (J. Darnell, personal communication). An important recent finding is that HNF-4 has a *Drosophila* homolog with which it shares more than 90% sequence identity in the DNA-binding domain (Lai and Darnell, 1991; Zhong et al., 1993).

Tissue Distribution and Regulation of Liver-Specific Transcription Factors

The HNF-1 Family

LF-B1/HNF-1α and LF-Be/HNF-1β are not restricted to liver cells. In the adult rat and mouse, the corresponding mRNAs and proteins accumulate in organs such as kidney, intestine, stomach, and pancreas (Baumhueter et al., 1990; De Simone et al., 1991; Rey-Campos et al., 1991). LF-B3/HNF-1β mRNA is also expressed in the lung (De Simone et al., 1991). LF-B1/HNF-1α and LF-B3/HNF-1β are differentially expressed in rat hepatoma cells lines. LF-B1/HNF-1α mRNA is not present in dedifferentiated hepatoma variants (C2 and H5), whereas LF-B3/HNF-1β transcripts are detectable both in differentiated (H4 and C2rev7) and dedifferentiated (C2 and H5) cell lines (Baumhueter et al., 1990; Cereghini et al., 1990; De Simone et al., 1991).

The tissues where LF-B1/HNF-1α and LF-B3/HNF-1β mRNA are expressed are not embryologically related; their only common feature is that they become specialized epithelia. The early and sequential expression of LF-B3/HNF-1β and LF-B1/HNF-1α in epithelial cells suggests that these factors play a role in implementing an "epithelial" genetic program (De Simone et al., 1991). In fact, LF-B3/HNF-1β is expressed earlier than HNF-1/HNF-1α during mouse, rat, and *Xenopus* development. This sequential expression of the two genes has been monitoried in detail by *in situ* hybridization during rat kidney organogenesis. LF-B3/HNF-1β transcripts can be detected in mesoderm-derived cells as soon as they are induced to differentiate into a polarized epithelium, while LF-B1/HNF-1α transcripts appear only at a later stage, when the three different segments of the nephron become apparent (Lazzaro et al., 1992). The sequential activation of the two genes is also observed in organ cultures of nephrogenic mesenchyme (Lazzaro et al., 1992).

In F9 cells, as in kidney development, LF-B3/HNF-1β and LF-B1/HNF-1α are also sequentially activated on induction of differentiation by retinoic acid (De Simone et al., 1991; Kuo C-J et al., 1991). Taken together, these data suggest that LF-B3/HNF-1β expression is an early event in the differentiation of polarized epitelia and probably "primes" cells to the expression of LF-B1/HNF-1α.

Although LF-B1/HNF-1α and LF-B3/HNF-1β mRNAs are restricted to a limited number of tissues, steady-state mRNA levels in different tissues are not only determined by different rates of transcription. In fact it has been shown, by nuclear run-on transcription assay, that an extraordinarily high rate of LF-B1/HNF-1α transcription in spleen is not accompanied by significant accumulation of the mRNA (Xanthopoulos et al., 1991). Paradoxically, liver and lung, which show the highest levels of LF-B1/HNF-1α mRNA and protein, transcribe the LF-B1/HNF-1α gene at a very low rate. These experiments indicate that posttranscriptional control mechanisms play an important role.

The promoter region of the LF-B1/HNF-1α gene was isolated and functionally characterized by Tian and Schibler (1991). *In vivo* and *in vitro* analysis identified a *cis*-acting element located upstream of the TATA box, which is necessary and sufficient for the transcription of LF-B1/HNF-1α (Kuo C-J et al., 1992; Tian and Schibler, 1991). This element, located between position -82 and -40, is a high-affinity binding site for HNF-4. It has also been observed that overexpression of the LF-B1/HNF-1α protein downregulates the transcription of a reporter gene under the control of the LF-B1/HNF-1α promoter (Tomei et al., 1993). It is thus possible that LF-B1/HNF-1α participates in regulating its own expression through a mechanism that is yet to be elucidated.

Regulation of bZIP Liver Proteins

C/EBP. High levels of C/EBP expression have been observed in all tissues that metabolize lipids and cholesterol-related compounds at a high rate, such as liver, gut, lung, white and brown adipose tissue, adrenal gland, and placenta (Birkenmeier et al., 1989). In fact, it has been shown that C/EBP activates a variety of adipose-specific genes (Christy et al., 1989; Herrera et al., 1989; Ro and Roncari, 1991).

C/EBP expression seems to be limited to terminally differentiated, growth-arrested cells. In the adult liver, C/EBP is detectable only in the nuclei of mature hepatocytes (Birkenmeier et al., 1989). On the contrary, rapidly dividing rat or human hepatoma cells express C/EBP at a much lower level than do adult liver cells (Friedman et al., 1989). In addition, disruption of cell–cell contacts following liver cell dispersion results in a dramatic reduction of the transcription rate of the C/EBP gene (Xanthopolous et al., 1989). C/EBP is restricted to the terminally differentiated cells also in adipose tissues: cultured adipoblasts that do not express this gene during proliferative growth show a strong induction of C/EBP when stimulated to differentiate into adipocytes (Christy et al., 1989). All these observations led to the speculation that C/EBP might function to affect or maintain the differentiated, nonproliferative state. It was further observed that the levels of C/EBP mRNA in the liver and intestine of rodents display a coordinate induction just before birth (Birkenmeier et al., 1989), and that C/EBP transcripts are present in several regions of adult mouse brain but not in fetal brain (Kuo CF et al., 1990).

Interestingly, overexpression of C/EBP in cultured adipoblasts leads to a direct arrest of mitotic cell growth, and promotes terminal cell differentiation in conjunction with adipogenic hormones (Umek et al., 1991). However, ectopic expression of C/EBP in other cell types and overexpression in hepatoma cell lines does not affect *per se* the differentiation pattern of the cells or expression of the endogenous target genes.

The exact mechanism that regulates the cell-type-specific expression of C/EBP has not yet been well described. Nuclear run-on transcription measurements have revealed that the basis of the restricted cellular distribution of the C/EBP transcript is transcriptional regulation of the gene (Xanthopoulos et al., 1989, 1991). As with LF-B1/HNF-1α, the occurrence of a C/EBP binding site on the gene promoter (between nt -203 and -176), which gives a different DNase I footprinting pattern with differentiated versus nondifferentiated preadipocytes nuclear extracts, suggests the C/EBP might regulate the rate of transcription of its own gene (Christy et al., 1991).

C/EBP-related proteins. Each member of the C/EBP-like superfamily of transcription factors has a unique pattern of expression in the various tissues. Ig/EBP mRNA is essentially present in all the organs tested, although the relative levels vary almost 20 fold, the highest levels being in early B cells (Roman et al., 1990). C/EBPβ/IL-6DBP is most abundant in liver, intestine, lung, and adipose tissues, whereas C/EBPδ mRNA seems to be expressed primarily in lung, intestinal, and adipose tissues (Cao et al., 1991). Transcripts corresponding to CRP1 are present only at extremely low levels in all the tissues originating from adult animals (Williams SC et al., 1991).

DBP. Although transcripts of the DBP gene are present in all tissues tested with the exception of testis, high levels of DBP protein have been detected only in liver (Mueller et al., 1990). For this reason DBP can be defined as a "liver-specific" transcription factor. These findings strongly suggest that the tissue specificity of DBP is determined posttranscriptionally. Because of a unique structure found in the 5′ untranslated region of DBP mRNA, it has been suggested that the tissue-specificity of DBP might be determined by a translational mechanism.

Both DBP and its mRNA accumulate to significant amounts in rat liver only during adult life. In addition, DBP and its transcript are rapidly downregulated upon chemically induced liver regeneration, to reappear only in the reconstituted liver (Mueller et al., 1990). These observations suggest that DBP might be involved in the proliferation control of hepatocytes.

The expression of this transcription factor has a unique feature: DBP accumulates according to a stringent, free-running circadian rhythm (Wuarin and Schibler, 1990). In adult rat liver, the level of DBP is below detection in the morning, rises sharply in the afternoon, and peaks around 8 P.M. Noticeable, transcription of the albumin gene, the putative target of DBP, is subject to the same circadian fluctuations. DBP mRNA also oscillates in the same fashion in spleen and brain. In the latter two tissues the protein does not accumulate in significant amounts, which rules out an autocatalytical role for DBP in controlling its own transcription.

VBP transcripts have been found in all tissues examined, including testis (Iyer et al., 1991). In contrast to DBP, VBP gene expression in the liver occurs early in embryo development and is not controlled by a circadian rhythm.

HNF-3: Tissue Distribution and Regulation

HNF-3α and β mRNAs are transcribed in liver, lung, intestine, and HNF-3γ mRNA is transcribed in liver, testis, and intestine (Lai et al., 1991). It thus appears that expression of the HNF-3 family transcription factors is restricted to derivatives from the lining of the primitive gut. It is worth pointing out that *forkhead*, the *Drosophila* homolog of HNF-3, is expressed in the terminal regions of the embryo, where cells invaginate to generate the anterior and posterior gut structures. This finding has led to the hypothesis that the HNF-3 gene family may contribute to the differentiation of organs derived from the endodermal tissues of the primitive intestine. Experiments based on steady-state mRNA measurements, nuclear run-on transcription assays, and determination of DNA-binding activity (Lai et al., 1990; Xanthopulos et al., 1991) indicate that the tissue-restricted expression pattern is primarily regulated at transcriptional level, at least in the case of HNF-3α.

HNF-4: Tissue Distribution and Regulation

HNF-4 mRNA is present in liver, lung, and intestine, but not in brain, spleen, fat, lung, or heart (Sladek et al., 1990). In addition, posttranscriptional events control the final tissue distribution of the HNF-4 protein (Xanthopulos et al., 1991). The *Drosophila* homolog of HNF-4 has been cloned (Zhong et al., 1993): the *Drosophila* gene is expressed in developing gut tissue, but not in other embryonic tissues.

Transactivators Involved in Liver Oncogenesis

There is strong epidemiologic evidence for a linkage between chronic HBV infection and hepatocellular carcinoma (HCC) (Blum et al., 1989), and integrated HBV DNA is found in more than 90% of HCCs in patients with chronic HBV infection (Tiollais et al., 1985). It has been suggested that transformation might be a consequence of the *cis*-activation of proto-oncogene transcription at an integration site (Dejean et al., 1986; Möröy et al., 1986; zur Hausen, 1991). Other evidence, however, suggests that HBV might encode for an oncogene.

The X open reading frame (ORF) of hepatitis B virus encodes a protein expressed during HBV infection and in HCC (Haruna et al., 1991; Wollerscheim et al., 1988). Although the role of X in the viral life cycle is still unclear, several reports in the literature show that it is able to transactivate a large number of viral and cellular RNA polymerase II (pol II) promoters in different cell types (Colgrove et al., 1989; Faktor et al., 1990; Seto et al., 1990; Siddiqui et al., 1989; Spandau and Lee, 1988; Twu and Robinson, 1989; Twu and Schloemer, 1987; Twu et al., 1989; Zhou et al., 1990), and also RNA polymerase III (pol III) promoters (Aufiero and Schneider, 1990). This activity does not appear to occur through direct DNA-binding but is mediated through *cis* elements that interact with transcription factors activated by protein kinase C (PKC) (NFkB, AP1, Ap2) (Kekulé et al., 1993; Seto et al., 1990; Siddiqui et al., 1989; Twu et al., 1989). Virus-derived as well as recombinant X protein has been shown to display serine-threonine kinase activity *in vitro*, both on itself or on histone H1 as substrates (Wu et al., 1990). This suggests that posttranslational modification could be the mechanism by which X exerts its transactivating activity. However, the recent finding that HBV-X interacts with CREB and with ATF2 *in vitro*, increasing their affinity for CRE sequence in the HBV enhancer (Maguire et al., 1991), suggests an alternative mechanism involving protein–protein interactions.

The central role played by X in the pathogenesis of hepatocarcinoma is underlined by two indications: (1) NIH3T3 cells and fetal mouse hepatocytes with stably integrated X ORF acquire higher saturation density and form tumors in nude mice (Shirakata et al., 1989); (2) transgenic mice expressing HBV-X develop hepatocarcinoma (Kim et al., 1991). However, these latter results have not been confirmed in a very similar transgenic mice model in which X mRNA (0.8 kb) is abundantly produced in the liver but no X protein can be detected and no HCC develops (M. Tripodi, personal communication). Even transgenic mice expressing the X Ag under the control of the α_1-AT liver-specific promoter do not develop HCC (Lee TH et al., 1990). Moreover, transgenic mice carrying the entire HBV genome, expressing all HBV mRNAs and showing DNA replicative intermediates in liver and kidney cells, never develop carcinomas (Farza et al., 1988). It is not possible, on the basis of these contradictory reports, to conclude that X is actually the liver oncogene of HCC. If this is the case, it is also not clear whether the molecular mechanisms of X Ag-induced carcinogenesis are linked to the transactivation potential or to other still unknown properties of this protein, and further investigation is needed to clarify this tissue.

Kekulé et al. (1990) reported that a 3′ truncation of the preS2/S *reading frame* which leaves the S2 region intact, gives rise to another protein with transactivating properties. Interestingly, a 3′ truncation preS2/S region is found integrated in many HBV-related HCCs, and the same region, directly cloned from a HBsAg-positive human hepatoma tissue is capable of transactivating heterologous promoters (Caselmann et al., 1990).

In this context, report of a specific and high-frequency mutational hot spot in the p53 gene in human HCCs is of considerable interest. When patients with HCCs from China and South Africa were analyzed, in nearly all mutations in the tumor-suppressor gene p53 were localized in codon 249 (Bressac et al., 1991; Hsu et al., 1991). This situation differs from that in carcinomas derived from different tissues in which p53 mutations have been found clustered in exons 5–8 but scattered over different codons (Hollstein et al., 1991). It is worth observing that China and South Africa are areas where HBV infection and aflatoxins have been epidemiologically linked to HCC (Popper et al., 1987; Slagle etal., 1991). These findings raise the intriguing possibility that mutations in the p53 gene, in the context of a chronically HBV-infected liver, may be contributing to hepatocyte transformation. The availability of the mutant p53/249 will open up the possibility of testing its activity as a liver-specific "oncogene," particularly as p53 is reported to act both as a relatively sequence-specific DNA-binding protein and as a transcriptional activator (Fields and Jang, 1990; Kern et al., 1991a; O'Rourke et al., 1990; Raycroft et al., 1990). In fact, the possibility cannot be excluded that the liver-specific p53 mutation may affect the transcription of genes specifically controlling hepatocyte proliferation (such as growth factors or their receptors), through the modified DNA-binding or transactivating properties, as it has already been reported for other p53 mutant proteins (Kern et al., 1991b; Raycroft et al., 1991).

Identification of Liver-Specific Regulatory Pathways Through the Use of Somatic Cell Genetics

The Extinction Phenomenon

Fusing cells of different types generates hybrids in which either silent genes are activated or active ones are repressed (Davidson, 1974; Davidson et al., 1968; Gourdeau and Fournier, 1990). It was initially observed that hybrids obtained by crossing hepatoma cells that express several differentiated functions of the adult hepatocyte with a variety of other cell types failed to express liver-specific traits (Bortolotti and Weiss, 1972a, 1972b; Schneider and Weiss, 1971). The propagation in culture of isolated hybrid clones results in the reexpression of previously extinguished functions, an event correlated to the loss of chromosomes selectively derived from one parent. These data provided the first clues that extinction has a genetic basis, the result of specific genetic loci mapping on different chromosomes (reviewed by Gordeau and Fournier, 1990).

Crosses between cells derived from the same embryonic lineage but with diverging differentiation patterns have also been studied to elucidate the mechanisms underlying the spatial and temporal activation or repression of specific genes during development. Fusions between primary hepatocytes with dedifferentiated hepatoma cells, or fusions of differentiated hepatoma cells with dedifferentiated variants of hepatocytes, have been used to investigate the differential fate of several genes during liver development. This type of cross results in a variety of phenotypes (Gourdeau and Fournier, 1990). These results, as well as data obtained in other systems, suggest that the phenotypic diversity resulting in intratypic crosses may result from limiting amounts of regulatory factors.

Although the extinction phenomenon has been known for more than 25 years, the molecular mechanisms underlying this event have been partially elucidated only recently.

Tissue-Specific Extinguisher 1 (Tse-1)

Killary and Fournier (1984) showed that in microcell hybrids retaining a single chromosome from the nonexpressing partner, that is, the mouse fibroblast chromosome 11 or its human homolog 17, the TAT gene was repressed although not completely. This effect was related to the presence of a genetic locus defined tissue-specific extinguisher 1 (Tse-1), not syntenic with the structural genes encoding mouse or rat TAT, mapping on a different chromosome, and derived from cells of any histotype. In addition to TAT, Tse-1 coordinately represses the activity of a subset of liver genes, such as PEPCK, AS, carbamyl phosphate synthetase I (CPSI), serine dehydratase (SDT), and two anonymous cDNA clones termed X1 and X2 in a very specific and selective manner (Lem et al., 1988; Ruppert et al., 1990; Thayer and Fournier, 1989).

All the Tse-1 target genes are, in fact, regulated by glucagon via the cAMP signal transduction pathway, and all are activated at birth (Boshart et al., 1990; Ruppert et al., 1990). The repression by Tse-1 in microcell hybrids can be reversed by activating the cAMP pathway, indicating a functional antagonism between Tse-1 and the cAMP signaling (Weih et al., 1990). Further, the cAMP-responsive element (CRE) located upstream to the TAT transcription start site is the target for gene repression by Tse-1.

Two groups using independent approaches have identified the gene product of the Tse-1 locus as the regulatory subunit $R1\alpha$ of the cAMP-dependent protein kinase (Boshart et al., 1991; Jones et al., 1991). High levels of expression of this subunit can sequester the catalytic subunits and abolish their activity. This block, in turn, determines inhibition of the protein kinase A (PKA-) dependent phosphorylation of the serine 133 of transcription factor CREB. The dephosphorylated protein shows a reduced binding of the TAT CRE, which then results in lack of TAT gene transcription. The $R1\alpha$ subunit has been found to be particularly abundant in nonliver cells or in dedifferentiated hepatoma cell lines; that is, in all cells that do not express the TAT gene. In differentiated hepatoma cells, on the contrary, expression of this subunit is undetectable, and when the cells are stably transfected with the $R1\alpha$ cDNA under the control of an inducible promoter, they acquire a TSE-1 phenotype, thus reproducing what occurs in extinguished hybrids. cAMP binds and inactivates the regulatory subunit and this results in activation of the catalytic subunit. The foregoing explains why CREB phosphorylation increases when microcell hybrids or hepatoma cells transfected with the $R1\alpha$ cDNA are treated with cAMP, resulting in an enhanced binding on the CREB recognition sequence and reversion of the extinction phenotype (Boshart et al., 1991; Jones et al., 1991). These findings suggest that activation at birth of the neoglucogenic enzymes results from the concomitant occurrence of two events: (1) a peak of glucagon and somatostatin in the blood, which increases cAMP levels, and (2) a liver-specific downregulation of $R1\alpha$ gene expression.

A set of genes, some of which overlap those repressed by Tse-1, are positively regulated by another important locus, termed *alf*, as the factor indicated by the albino letal mutation. This is the gene product of a DNA segment located on mouse chromosome 7, whose deletion in homozygous animals fails to activate a group of liver-specific genes (Gluecksohn-Waelsch, 1979). This suggested that a positive *trans*-acting factor encoded by this region is required for the proper expression of genes to achieve the full repertoire of adult hepatocyte functions. The activity of this factor is antagonized by Tse-1, which operates as a negative modulator on the expression of these genes. Moreover, albino mutant mice do not respond to stimulation with glucocorticoids or cAMP, suggesting that these inducers may cooperate with the *trans*-acting factor *alf* to activate the expression of liver genes. Clues about the nature of this gene product and the role it plays in this complex metabolic pathway have been disclosed by (Ruppert et al., 1992). It was already known that the *alf* locus maps near the pigmentation locus, which codes for the enzyme tyrosinase. Using probes derived from the tyrosinase locus and making use of chromosome walking and jumping techniques, Ruppert isolated the bona fide *alf* locus at a distance of about four to five megabases. This DNA region contains the coding region for the enzyme fumarylacetoacetate hydrolase (FAH), an enzyme involved in the metabolic pathway for tyrosine degradation. FAH mRNA is not synthesized in *alf* mice, and the gene shows gross rearrangements. Interestingly, FAH deficiency in humans is the primary cause of a disease called "human hereditary tyrosinemia type I," which is characterized by the accumulation of toxic intermediate metabolites, impairment of liver function, and death at 3 to 5 months. The close relationship between the *alf* status and this human disease is clear, but the final proof that FAH is the only gene responsible for the *alf* phenotype lies in a rescue experiment with transgenic *alf* mice carrying the wild-type FAH gene.

Tissue-Specific Extinguisher 2

Another genetic locus responsible for the extinction of albumin, alcohol dehydrogenase and Liv.10 genes has been identified (Chin and Fournier, 1989). Transfection in somatic cell hybrids of chimeric constructs having regulatory sequences from the albumin 5′ flanking region fused to a reporter gene allowed the target sequences of extinction to be located (Petit et al., 1986) in the same promoter elements that mediate liver-specific expression of the gene (Cereghini et al., 1987; Lichtsteiner et al., 1987; Maire et al., 1989). In this case, therefore, the regulatory sequences able to confer liver specificity also mediate the extinction of the gene. Moreover, by using microcell hybrids, a fibroblast locus assigned to mouse chromosome 1 has been identified as responsible for the repression of the genes (Chin and Fournier, 1989). This locus has been designated tissue-specific extinguisher 2 (Tse-2).

The mechanism of Tse-2 action is at present unknown. Analysis of the *cis*-regulatory sequences of the albumin gene revealed the crucial role played by transcription factor LFB1/HNF-1α (see also pp. 174–175). Surprisingly, the LF-B1 gene is expressed in microcell hybrids in which albumin is extinguished and Tse-2 is retained (Bulla et al., 1992).

α_1-*Antytripsin Gene Extinction*

α_1-Antytripsin is extinguished in hepatoma x fibroblast hybrids, and repression is reproduced on a transfected chimeric gene (Bulla et al., 1992). By transfecting α_1-AT-CAT constructs, the target sequences for gene extinction were localized to the same sequence elements that mediate tissue-specific expression of the gene: the binding sites for transcription factors LF-B1/HNF-1α and HNF-4/LF-A1 located in the proximity of the TATA box (see pp. 174–175). When α_1-AT-extinguished hybrids were tested for the presence of LF-B1/HNF-1α mRNA and protein, neither transcriptional activity nor protein-binding activity was found. Thus, extinction of α_1-AT expression is linked to lack of LF-B1/HNF-1α activity, but this cannot be the only cause of α_1-AT extinction. In fact, in somatic hybrids where the LF-B1/HNF-1α cDNA was transiently cotransfected with an α_1-AT promoter chloramphenicol acetyltransferase (CAT) fusion, the latter was efficiently expressed but there was no reactivation of expression of the extinguished α_1-AT chromosomal genes. The same result was obtained when interspecies hybrids were generated by fusing hepatoma cells with cells of different origin, stably transfected with the LF-B1/HNF-1α cDNA. These results indicate that a hierarchy of regulatory genes is involved in the extinction of α_1-AT, which is probably also linked to stable changes in the chromatin configuration of the extinguished genes.

The Acute-Phase Response

Transcriptional Regulation by Cytokines

The acute-phase (AP) response is characterized by dramatic changes in the blood levels of several plasma proteins, which are predominantly syn-

thesized in the liver in response to infection, inflammation, or other pathological conditions (for reviews, see Baumann, 1989; Ciliberto, 1989; Fey
and Gauldie, 1990). Two classes of AP proteins can be distinguished:
the so-called positive AP reactants (e.g., hemopexin (Hpx), CRP, haptoglobin (Hp), α_1-acid glycoprotein (α_1-AGP), whose plasma concentrations increase during inflammation; and negative AP reactants, including
albumin and transferrin, whose levels decrease. The changes in the levels
of both positive and negative AP reactants can be correlated with variations in the level of mRNA accumulation or, more precisely, changes
in the transcription rate of the corresponding genes. In particular, the
upregulation of positive AP genes is mainly at transcriptional level. The
study of AP has, therefore, attracted substantial interest not only because
of clinical importance, but also because it provides an excellent system to
elucidate the molecular mechanisms involved in the modulation of gene
expression.

Several hormones and cytokines have been shown to play an important role in the regulation of AP response. These include interleukin-1
(IL-1) (Ramadori et al., 1985; Woo et al., 1987), tumor necrosis factor
(TNF) (Darlington et al., 1986; Perlmutter et al., 1986), Il-6 (Castell et
al., 1988; Gauldie et al., 1987; Morrone et al., 1988), leukemia inhibitory
factor (LIF) (Baumann and Wong, 1989), glucocorticoids (Baumann et
al., 1987a; Brasier et al., 1990a; Marinkovic' and Baumann, 1990; Ron
et al., 1990), interleukin-11 (IL-11) (Baumann and Schendel, 1991), and
transforming growth factor-β (TGF-β) (Morrone et al., 1989). Among
these, IL-1 and IL-6 are by far the principal mediators. Overall, AP genes
can be divided into two major categories: class I and II (Baumann et al.,
1987b). Maximal induction of genes in class I requires IL-1 plus IL-6
and a combination of these two plus glucocorticoids, and includes CRP,
complement C3, Hp, HpX, and SAA. Class II genes, including α_1-AT,
fibrinogen, and α_2-macroglobulin (α_2-M) serum amyloid A (SAA), are
regulated only by IL-6 or combinations of IL-6 plus glucocorticoids.

It is beyond the scope of this chapter to summarize what is known
about functional activities, receptor interactions, and signal transduction
pathways of the cytokines involved in the acute-phase response; several
excellent reviews have been published (Hirano, 1991; Van Snick, 1990).
Here we report what is known about the molecular mechanisms of activation of liver-specific promoters induced by the two main acute-phase
mediators IL-1 and IL-6.

Structure of Acute-Phase Responsive Promoters

During the past few years substantial progress has been made in identifying the *cis*-acting, acute-phase-responsive elements (APREs) required for the induction of several positive AP genes. The general strategy has been to analyse the activity of deletion or substitution mutants in the 5′ flanking regions of the gene, linked to the coding sequence of a reporter gene such as bacterial CAT, in transient transfections using established hepatoma cell lines (e.g., HepG2, Hep3B). For induction experiments, the cells are treated with either conditioned medium from peripheral monocytes stimulated with lipopolysaccaride (LPS) or with recombinant hormone or cytokines. In several cases, individual APREs have also been tested for their ability to stimulate inducible transcription of heterologous promoters, such as the early SV40 promoter and TATA box. In some cases, these studies have been extended to transgenic mice (Ciliberto et al., 1987; Dente et al., 1988; Dewey et al., 1990).

A schematic representation of the location of APREs identified in the 5′ flanking regions of several AP genes (Figure 3) reveals a significant degree of heterogeneity in the structure and location of APREs in the 5′ flanking region of the AP-responsive genes that have been extensively characterized. In some cases, such as in Hpx, SAA2β, complement C3, and angiotensinogen, a single element has been found. For other genes, such as Hp, CRP, and α_2M, two distinct and functionally independent elements have been mapped. The presence of redundant APREs is probably responsible for a higher degree of *in vivo* gene induction by the corresponding cytokine. With the exception of the rat α_1-AGP gene, all APREs are located in the proximity of the cap site, and in most cases right upstream of the TATA box. As discussed for constitutive liver-specific promoters, it is also possible here that the *trans*-acting factor(s) involved in cytokine response might be performing a dual role of facilitating chromatin decondensation in the proximity of the cap site and directly promoting assembly of the preinitiation complex.

Regulation of CRP gene expression shows a peculiar feature with respect to the other AP genes. Maximal expression of the CRP gene in some cultured cells requires both IL-1 and IL-6 (Castell et al., 1988; Ganapathi et al., 1988). IL-6 acts at transcriptional level via two distinct and redundant IL-6REs (see Figure 3) (Arcone et al., 1988; Ganter et al., 1989), which function independently from each other, as removal of the downstream IL-6RE does not affect inducibility (Toniatti et al., 1990a). In contrast, a specific IL-1 RE has not been found in the CRP promoter, and

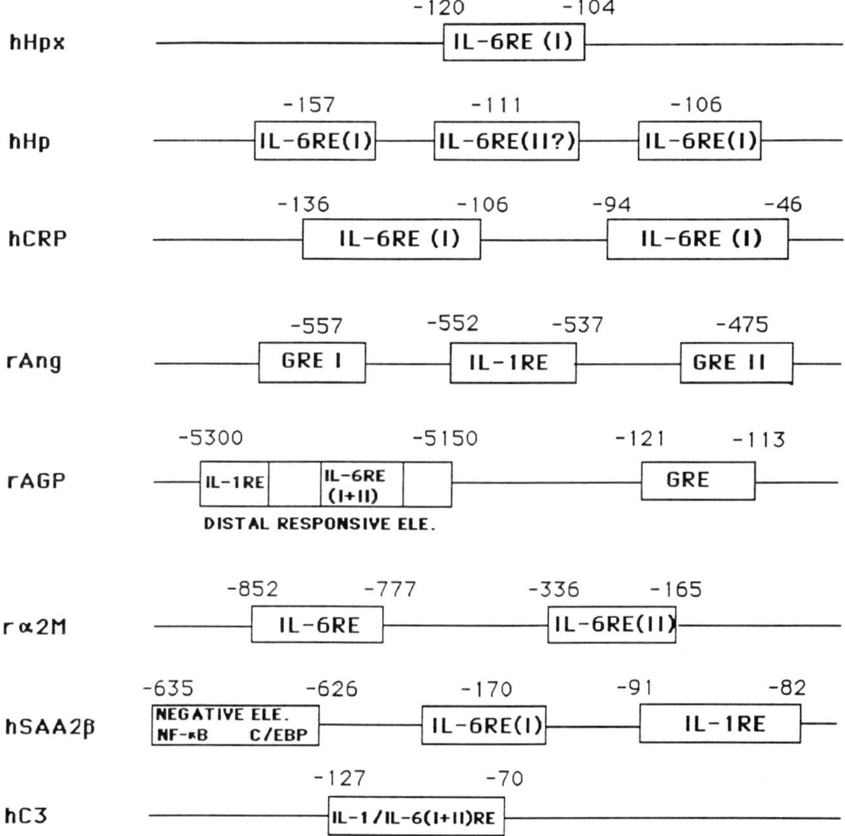

Figure 3   Schematic representation of APREs in 5′ flanking regions of positive AP genes.   Nature of various APREs and interaction between them are described in detail in the text. Abbreviations: h, human; r, rat; Hpx, hemopexin; Hp, haptoglobin; CRP, C-reactive protein; Ang, angiotensinogen; AGP, α_1-acid glycoprotein; α_2M, α_2-macroglobulin; SAA2β, serum amyloid A β gene; C3, complement C3; IL-6RE (I), type-I IL-6 responsive element; IL-6RE (II), type-II IL-6 responsive element. For the Hp gene, precise positions of three IL-6REs are not known; 3′ boundaries of 6-bp mutations, which affect IL-6 inducibility, are thus shown. Similarly, central position of two GREs in rAng is shown. The 142-bp distal responsive element of the rAGP gene contains both IL-1 (position 1 to 36) and IL-6 (position 63 to 117) responsive elements. Figure is based on these sources: hHpx (Poli and Cortese, 1989); rHp (Oliviero and Cortese, 1989); hCRP (Arcone et al., 1988; Majello et al., 1990; Toniatti et al., 1990a, 1990b); rAng (Brasier et al., 1990a, 1990b; Ron et al., 1990); rAGP (Prowse and Baumann, 1988; Won and Baumann, 1991); rα_2M (Kunz et al., 1989; Hattori et al., 1990); hSAA2β (Edbrooke et al., 1991); hC3 (Wilson DR et al., 1990).

IL-1 acts only as a translational modulator of CRP expression. Response to IL-1 is abolished when the first 15 bp of the CRP transcript are replaced by the leader sequence from a different mRNA (Ganter et al., 1989).

Transcription Factors Induced by IL-1 and IL-6

Analysis of the promoter regions of IL-6-induced genes has resulted in the identification of two classes of *cis*-acting sequences, type I and II, that mediate IL-6 induction. Type I elements have been initially identified in the promoter of the Hpx, Hp, and CRP genes (Arcone et al., 1988; Oliviero et al., 1987; Poli et al., 1989). They interact with a series of heat-resistant proteins present constitutively in the nuclei of hepatoma cells, all of which share the same DNA-binding specificity. Besides these proteins, a specific DNA-binding activity is reproducibly induced by IL-6 via a posttranslational mechanism (Majello et al., 1990; Oliviero and Cortese, 1989; Poli and Cortese, 1989). The alignment of all these sequences allows a consensus sequence, agTgNNGYAA, to be defined. It has recently been shown that these sequences interact with the C/EBP class of transcription factors. Several members of this class have been isolated (see pp. 185–187), but only IL-6DBP has been shown to be modulated by IL-6. This factor has been independently isolated by several laboratories from rat, mouse, and human, and designated IL-6DBP (Poli et al., 1990), LAP (Descombes et al., 1990), CRP2 (Williams SC et al., 1991), and C/EBPβ (Cao et al., 1991) for the rat, AGP/EBP (Chang CJ et al., 1990) for the mouse, and NF-IL-6 (Akira et al., 1990) for the human (see Table 1). This protein, but not C/EBP, is able to transactivate synthetic promoters formed by multimeric type I IL-6 REs in an IL-6-dependent manner (Poli et al., 1990). In addition, immunoprecipitation experiments performed with antibodies against IL-6DBP demonstrated that the DNA-binding activity of this protein is positively modulated by IL-6 (Poli et al., 1990). C/EBPδ/NFIL-6β has also been shown to be involved in the IL-6 signal transduction pathway of gene activation (Ramji et al., 1993; Kinoshita et al., 1992).

The activation mechanism of IL-6DBP is not yet fully understood. IL-6DBP is constitutively expressed in Hep3B cells, and its levels are not increased by IL-6 (Ramji et al., 1992, 1993). Although Metz and Ziff (1991) have recently shown that in PC12 cells, cAMP stimulates IL-6DBP to *trans*-locate to the nucleus and induce *c-fos* transcription, in human hepatoma cells no change in the nuclear translocation of IL-6DBP can be observed in the presence or absence of IL-6 (Ramji et al., 1992,

1993). It is thus likely that posttranslational modification of pre-existing nuclear IL-6DBP is the principal mechanism of its activation in hepatic cells. This modification must be a subtle one, because no significant alterations in the electrophoretic mobility of IL-6DBP can be observed either by one- or two-dimensional gel electrophoresis (unpublished data). The amino terminus of IL-6DBP contains the information for both constitutive and IL-6-induced transcriptional activation. Within this region, two distinct domains have been identified, one that is required for basal transactivation and the other which is the target of IL-6 activation (G. Ciliberto, D. Ramji, and R. Cortese, unpublished).

All members of the C/EBP family share the property of forming heterodimers *in vivo* and *in vitro*. Thus, heterodimeric interactions between the different members may play an important role in the modulation of AP gene transcription, as has already been shown for C/EBP and IL-6DBP (Poli et al., 1990). C/EBP activates transcription in an IL-6-independent manner, and its DNA-binding activity is not induced by IL-6. However, when both IL-6DBP and C/EBP are transfected in equimolar amounts into Hep3B cells, the activity of the reporter plasmid in unstimulated cells is reproducibly reduced in comparison to transfection of C/EBP alone. In IL-6-treated cells, however, the activation observed is substantially higher than the sum of values obtained when C/EBP and IL-6DBP are separately transfected. The simplest explanation of these results is that heterodimers between C/EBP and IL-6DBP are less active in transactivation in the absence of IL-6 and more active than both C/EBP and IL-6DBP homodimers in the presence of IL-6.

Interestingly, IL-6DBP, CRP1, C/EBP, and C/EBP-δ all contain a conserved cysteine residue at or near the C terminus immediately following the leucine zipper (Williams SC et al., 1991), which allows the formation of covalently linked dimers *in vitro*. Although no gross difference between covalent and noncovalent dimers in binding to DNA was detected, it is conceivable that linked dimers stabilize subunit interactions *in vivo* and thus maintain a homeostatic population of dimers within a cell. In this framework, changes in the redox potential of the cell could provide a regulatory mechanism for modulating transcription.

These observations suggest that modulation of gene transcription by IL-6 could occur through AP-mediated changes in expression levels of different C/EBP members and their ability to form heterodimers with IL-6DBP. In support of this view, Akira et al. (1990) have shown that NF-IL-6 mRNA is present in trace amounts in mouse liver and is induced severalfold on injection of LPS or IL-6. In contrast, C/EBP protein concentration decreases considerably after IL-6 stimulation (Isshiki et al., 1991). It is interesting to note that a similar regulation takes place in cells of nonhepatic origin that express the same set of transcription factors. For example, Ron and colleagues have recently observed that in 3T3-L1 fibroblasts, which can be differentiated *in vitro* into adipocytes (Samuelsson et al., 1991; Umek et al., 1991), TNF induces a decrease in C/EBP protein levels and a reciprocal increase of IL-6DBP and most likely also in other members of the C/EBP family (Ron and Habener, 1992).

IL-6DBP/LAP contains three in-frame AUGs that are differentially recognized by the translational machinery because of a leaky ribosome scanning mechanism (Descombes and Schibler, 1991). The first two are close to each other, and their utilization leads to the formation of transcriptionally active proteins LAP* (the longer) and LAP (the shorter). The third AUG occurs upstream of the DNA-binding domain, and its use generates the transcriptional inhibitor LIP, which lacks the activation domain and has a higher DNA-binding affinity. Transcriptional activation does not reflect the absolute amount of LAP but rather the ratio of LAP to LIP. For example, a modest increase in the LAP/LIP ratio results in a significantly higher activation of the target genes. The LAP/LIP ratio increases about fivefold during terminal rat liver differentiation and is thus likely to play a role in modulating liver-specific gene expression in intact animals. However, this mechanism is not involved in IL-6 induction because removal of the third AUG does not affect IL-6 inducibility of reporter genes in cotransfection experiments. Further, no alteration of LAP/LIP ratio has been observed on IL-6 stimulation (D.P. Ramji, unpublished data).

It must be emphasized that the presence of a binding site for IL-6DBP does not necessarily imply that the gene is induced by IL-6. For example, IL-6DBP interacts with the promoter of the albumin gene, which is a negative AP reactant. This suggests that sequences adjacent to the IL-6 recognition site or its location with respect to other transcription factors might determine whether a particular gene is regulated by IL-6. In addition, as all the C/EBP-related factors interact with identical recognition sequences, the partners used in heterodimeric interaction may also play a role in determining IL-6 inducibility. Binding sites for IL-6DBP/NF-IL-6 have also been identified in several viral enhancers and the promoter regions of cytokines IL-6, IL-8, TNFα, and G-CSF (Akira et al., 1990).

The NF-IL-6 site in the IL-6 promoter occurs at positions -158 to -145, and mediates induction by IL-1, IL-6, TNF, and LPS (Akira et al., 1990). In addition, NF-IL-6 also interacts with its own promoter (T. Kishimoto, personal communication). This suggests that an autocrine amplification loop exists in which IL-6-dependent modification of IL-6DBP/NF-IL-6 induces its own expression, which then stimulates the synthesis and secretion of IL-6.

Type II IL-6 REs have been initially identified in the promoter of the rat α_2M gene. This promoter has two IL-6 REs (from -852 to -777 and -336 to -165) (see Figure 3) (Kunz et al., 1989). The proximal element has been studied in greater detail. It contains the core motif CTGGGAA (Hattori et al., 1990), which is also present in the promoter regions of rat α-, β- and γ-fibrinogen (Fowlkes et al., 1984), rat T kininogen (Fung and Schreiber, 1987), the Hp B site (Oliviero and Cortese, 1989), rat transthyretin (Fung et al., 1988), murine SAA A (Lowell et al., 1986), and rat α_1-AGP (Prowse and Baumann, 1988). This core motif, which has been defined as type II IL-6RE, specifically interacts with an IL-6-inducible transcription factor differing from IL-6DBP and designated IL-6RE-BP (Hattori et al., 1990). This protein has a higher molecular weight than IL-6DBP, is heat sensitive, and requires *de novo* protein synthesis for its activation (Hocke et al., 1992).

IL-1-induced promoters are all characterized by the presence of binding sites for transcription factor NF-κB. This factor was originally identified as a DNA-binding protein, present in the nucleus of mature B cells, that interacts with the κB site in the immunoglobulin enhancer (Sen and Baltimore, 1986). However, it has since become apparent that NF-κB is involved in the inducible expression of a large number of genes in different cell types (see review by Lenardo and Baltimore, 1989). These include cytokines (IL-6 and β-interferon), cytokine receptors (α-chain of IL-2 receptor), major histocompatibility antigens, AP genes (SAA and angiotensinogen), and a variety of viruses such as HIV, SV40, and cytomegalovirus.

A characteristic feature of NF-κB is that it is posttranslationally regulated by nucleocytoplasmic transport in several nonlymphocyte cells. The NF-κB complex appears to be a heterodimer of 50- and 65-kd subunits that interact with the κB motifs (Baeuerle and Baltimore, 1989). This complex is maintained in an inactive form in the cytoplasm by virtue of the association between the 65-kDa subunit and a family of inhibitor proteins referred to as I-κB (Baeuerle and Baltimore, 1988a,

1988b). On induction by a variety of agents [e.g., LPS, phorbol myristate acetate (PMA), TNF, double-stranded RNA, IL-1], I-κB becomes phosphorylated and releases NF-κB, which is then translocated to the nucleus (Ghosh and Baltimore, 1990). The cDNA clones for both the 50- and 65-kDa subunits have recently been isolated (Ghosh et al., 1990; Kieran et al., 1990; Nolan et al., 1991); p50 is synthesized as a larger protein, and then processed to its functional size. The sequence in the amino terminus of the protein is homologous to the proto-oncogene c-*rel*, its viral counterpart v-*rel*, and the *Drosophila* maternal effect gene *dorsal*, all of which display cytoplasmic-nuclear translocation (Gilmore and Temin, 1986; Roth et al., 1989; Rushlow et al., 1989). Interestingly, also the amino terminus of p65 contains extensive similarities to c-*rel* (Nolan et al., 1991). The homology of both p50 and p65 with *rel* has recently generated enough interest to search for related factors. DNA–protein cross-linking and proteolytic mapping studies have revealed that NF-κB binding activity consists of at least four proteins (p50, p55, p75, and p85) (Molitor et al., 1990). While p55 and p50 are closely related to the 50-kDa subunit of NF-κB, p75 and p85 exhibit different properties and p85 in particular apears to be identical to c-*rel* (Ballard et al., 1990). Similarly, immunoprecipitation of extracts from a variety of avian cells with anti-*rel* antibodies coprecipitate a wide variety of proteins with molecular weight ranging from 40 to 125 kDa (Davis et al., 1990; Lim et al., 1990; Morrison et al., 1989; Simek and Rice, 1988; Tung et al., 1988). Binding to NF-κB site could, therefore be achieved through the generation of homo- and heterodimers between different members of the family of *rel*-related proteins. Of course, as previously discussed for the transcriptional activators of the C/EBP family, heterodimerization between these different members may potentially play an important role in modulating gene transcription.

In addition to several *rel*-related transcription factors, at least two types of inhibitory I-κB proteins have been isolated, I-κB-α, and I-κB-β (Zabel and Baeuerle, 1990). Phosphorylation of I-κB-α by the PKC pathway activates the DNA-binding activity of NF-κB *in vitro* (Ghosh and Baltimore, 1990). In contrast, I-κB-β can be regulated by phosphorylation with either PKA or PKC (Kerr et al., 1991). This suggests that different I-κB species may be utilized for induction of transcription by various agents that act *via* distinct intracellular messenger systems.

Several cDNAs coding for very large proteins containing zinc finger motifs interacting with the κB site have also been described (Adams et al., 1991; Baldwin et al., 1990). Thus, two completely different binding motifs (zinc fingers and *rel* homology region) can interact with the same DNA sequence. Although these two classes of transcription factors are unlikely to interact through heterodimerization, they may play a role in regulating transcription by competing for binding to the κB recognition sequences. Use of cloned cDNAs should in the future allow us to understand the role of each component in regulation of gene transcription. It is also important to remember, in the regulation of the IL-1-induced rat α_1-AGP gene, that its IL-1RE interacts with a heat-labile, basic nuclear protein designated NF-AB that is distinct from NF-κB (Won and Baumann, 1991). The human NF-AB binding activity is detectable only after 13 h of cytokine treatment (Won and Baumann, 1991).

In conclusion, IL-1 gene regulation through the κB recognition sequence could occur in several different manners: (a) heterodimerization between different *rel* family members, (b) differential utilization of the I-κB species, (c) modulation of the activity of I-κB by cytokine-induced phosphorylation, (d) involvement of novel DNA-binding proteins such as NF-AB, and (e) competition with zinc finger proteins.

The Interplay of Different Transcription Factors on Acute-Phase Genes Promoters

Several interesting examples of interplay between transcription factors on the promoter surface of AP genes have been described. In the case of CRP, each type I IL-6 RE has a complex structure characterized by the presence of adjacent binding sites for transcription factor LF-B1/HNF-1 and a factor of the C/EBP family (Toniatti et al., 1990b). Both sites have a weak affinity for their corresponding factor. Mutations in either of the two sites lead to loss of response to IL-6, which is probably caused by a cooperative interaction between adjacently bound transcription factors.

Glucocorticoid induction of the rat α_1-AGP gene requires a functional GRE (-120 to -107) together with sequences located between -106 and -42 (Klein et al., 1988). IL-6DBP or its mouse homolog, AGP/EBP, interacts with two sites within the -120 to -42 region (Williams PM et al., 1991). The upstream site (HA; -113 to -104) overlaps the GRE, while the downstream binding site (HB) occurs between positions -81 to -72. Mutational analysis has shown that occupation of the HB site is necessary for maximal glucocorticoid induction, while occupation of the HA site interferes with maximal induction (Williams PM et al., 1991). Also, the 5' flanking region from -121 to -70 of the mouse

α_1-AGP gene contains binding sites for C/EBP-related factors, although their exact role in inducible transcription has not yet been determined (Chang C-J et al., 1990).

The 5′ flanking region of the angiotensinogen gene contains a central APRE, which mediates induction by IL-1 and TNF, flanked on either side by two classical glucocorticoid responsive elements (GRE). These elements interact in a cooperative manner, because mutation of the IL-1/TNF RE abolishes glucocorticoid induction and point mutations in the GREs result in no IL-1 induction (Brasier et al., 1990a). The same APRE shows a complex structure because it interacts, in a mutually exclusive manner, with members of the C/EBP-family and with a cytokine-inducible protein that is indistinguishable from NF-κB (Ron et al., 1990). According to a model proposed by Brasier et al. (1990b), basal transcription in unstimulated cells is caused by the occupancy of APRE by b-Zip proteins. On IL-1 induction, NF-kB translocates to the nucleus, and, by virtue of its fivefold higher DNA-binding affinity, displaces the other factor(s) from interacting with the APRE and this results in increased transcription.

NF-κB and C/EBP members also play a key role in the expression of the serum amyloid A (SAA) genes. The human SAA2β gene contains an IL-1 and phorbol ester-inducible element (position −82 to −91), which interacts with inducible NF-κB-like proteins (Edbrooke et al., 1991). Full IL-6 induction requires an additional IL-6RE located 170 bp upstream of the CAP site, which contains a binding site for factors of the C/EBP family (P. Woo; personal communication). Another negatively acting element is also present between positions −626 and −635. This region contains a NF-κB recognition sequence, but footprint analysis shows that a constitutive nuclear factor interacts with a C/EBP recognition site 3′ to the NF-κB motif. This protein is displaced by binding of NF-κB-like protein on IL-1 stimulation (Edbrooke et al., 1991). Also, the mouse SAA3 promoter contains two adjacent C/EBP-binding sites in the proximal region that enhance basal expression in a hepatocyte-specific manner, and a distal 68-bp region confers responsiveness to both conditioned medium from mixed lymphocytes and recombinant IL-1 (Huang et al., 1990). For the rat SAA1 gene, an NF-κB recognition sequence has also been identified at a similar position as the proximal site in the human SAA2β gene. In addition, four regions interating with nuclear factors were identified upstream of the NF-kB sequence. Two of these regions interact with C/EBP or related factors, with the distal site having an affinity about 10 fold higher (Li and Liao, 1991).

Forcing Expression of Exogenous Genes in Liver Cells

Transgenic Mice

The isolation of liver-specific promoters and enhancers has allowed the generation of transgenic mice in which the expression of several genes has been targeted to hepatocytes. The research in this field was directed mainly at two different aspects: studies of liver carcinogenesis and generation of mouse models of human diseases (Adams and Cory, 1991; Hanahan, 1989).

Four different liver-specific promoters/enhancers have been used to drive expression of oncogenes in the liver: the mouse albumin promoter/enhancer (Lee GH et al., 1990; Pinkert et al., 1987; Sandgren et al., 1989), the mouse major urinary promoter (Held et al., 1989), the human α_1-AT promoter (Sepulveda et al., 1989), and the human C-reactive protein (CRP) promoter (Rüther et al., 1993). In the first three agents, the oncogene is expressed constitutively at very high levels and, at least when the α_1-AT promoter is used, transcription is detectable before birth in 12.5-day-old embryos (Butel et al., 1990; Sepulveda et al., 1989). On the contrary, CRP promoter expression is detectable only after birth on injection of irritants (Ciliberto et al., 1987). In this last case, therefore, oncogene expression is switched on only in particular experimental conditions (Rüther et al., 1993).

The oncogene whose pathological effects on liver cells have been studied in greater detail is the SV40 large T (Adams and Cory, 1991). All liver-specific promoter/SV40 fusions tested so far give rise to transgenic mice with a 100% incidence of HCC. SV40 gene expression in general temporally precedes the development of HCC, and when tumors appear they have the characteristic of multifocal nodules. Histologically, the progression from the initial liver damage (hyperplasia/dysplasia) to the appearance of hepatocellular carcinoma (HCC) is remarkably similar (Butel et al., 1990; Held et al., 1989; Sandgren et al., 1989).

The other two oncogenes tested in liver have not shown to be as powerful as SV40 T-Ag. In fact, only 60% of alb-*myc* mice develop detectable HCC after more than 1 year of life (Sandgren et al., 1989). On the contrary, alb-*ras* mice show severe and diffuse hyperplasia and mild dysplasia with loss of the normal hepatic architecture and perinatal death.

Targeting of different oncogenes in liver can also be considered a valid system to study oncogene cooperation (Hunter, 1991) because the appearance of HCC in transgenic mice carrying a combination of two oncogenes expressed in the liver is much faster (Sandgren et al., 1989). This is further supported by the finding that alb-SV40 transgenic mice show a high mutation rate of the endogenous c-H-*ras* oncogene at the second position of codon 61 in transformed versus nontransformed tissues (Lee GH et al., 1990). Finally, association of the oncoprotein with the p53 tumor suppressor gene has been observed in α_1-AT/SV40 transgenic mice (Sepulveda et al., 1989). This result acquires particular relevance in view of the finding that high-frequency p53 mutations occur spontaneous human HCCs (see pp. 194–196). Hepatic targeting of oncogenes in transgenic mice represents an excellent model for the study of liver-specific gene expression because cell lines can be derived from these hepatocarcinomas in which the influence of specific oncogenes on the expression of markers of the differentiated state can be directly assessed (Jallat et al., 1990; Paul et al., 1988).

Liver-specific expression of exogenous genes in transgenic mice has proved important for elucidating mechanisms responsible for disease development. One clear example is the transgenic approach to hepatitis B virus- (HBV-induced) liver damage. Transgenic mice have been extensively used to address this issue. Mice have been developed that carry the pre-S1, pre-S2, S, and X coding region under the control of the albumin enhancer/promoter complex (Chisari et al., 1987). Interestingly, liver damage is observed that is proportional to the HbSAg expression level. Where expression is very high, intracellular accumulation of large-envelope protein is observed together with cell injury, the appearance of characteristic "ground-glass" hepatocytes, and accompanying inflammatory reactions (Chisari et al., 1989); the appearance of ground-glass hepatocytes is a common feature of chronically infected human liver (Hoofnagle et al., 1987). These alterations are followed by the appearance of regenerative nodules, by the expression of the liver tumor marker α-fetoprotein (AFP), and finally by the inexorable development of a clearly visible HCC at about 12 months of age (Dunsford et al., 1990). It is not yet clear if this model of hepatocellular carcinogenesis is precisely mirroring what happens in humans nor is it known what role is played in hepatocarcinogenesis by the extensive regeneration that follows cell damage and necrosis. In fact, even though X Ag expression has never been observed, it cannot be excluded that small and undetectable levels of X Ag expression might be responsible for HCC development (see pp. 194–196).

Another example is the analysis of the effect of the α_1-AT mutant Z allele. The Z mutant contains a lysine instead of a glutamic acid at position 342 (Perlmutter, 1991). In ZZ homozygous patients, there is a high incidence of development of pulmonary emphysema and liver cirrhosis (Sifers et al., 1989). The molecular basis of the cirrhosis is constituted by impaired secretion of the Z protein from hepatocytes, leading to accumulation of the Z protein in the hepatic endoplasmic reticulum that predisposes to liver damage (Perlmutter, 1991). Two laboratories have reported generation of transgenic mice harboring the mutant Z allele under the control of its own promoter (Carlson et al., 1989; Dycaico et al., 1988). Interestingly, in both cases when high levels of the transgene are expressed, a liver pathology develops that is characterized by initial tissue damage and necrosis (from accumulation of Z protein in the cells), followed by inflammatory response, neonatal hepatitis, and growth failure (Geller et al., 1990). This progression is in part similar to that observed in HBsAg transgenic mice, thus supporting the idea that accumulation of proteins in liver cells may by itself be responsible for the development of hepatocellular injury and accompanying inflammatory response. The absence of cirrhosis in these mice might simply reflect genetic resistance of mice to the development of this kind of pathology. This intrinsic genetic difference might therefore be responsible for the evolution of different pathologies in the two species following the same biochemical alterations.

In the case of lipid metabolism disorders, liver-targeted expression of transgenes also has begun to furnish valid indications about ethiology and therapy of these diseases (Chajek-Shaul et al., 1991; Rubin et al., 1991; Walsh et al., 1989). For instance, overexpression of the human Apo-CIII driven by its own liver-specific promoter results in a severe hypertriglyceridemia (HTG), thus suggesting an intriguing and plausible ethiopathogenetic mechanism for both primary and associated HTG (Ito Y et al., 1990). Moreover, the promoter-mediated, liver-specific overexpression of human low-density lipoprotein (LDL) receptors protects mice from the increase in plasma LDL cholesterol that results from the ingestion of a high-fat diet (Yokode et al., 1990). These last two studies open new and unexpected possibilities for the study of lipid metabolic disorders, their relationship with the occurrence of atherosclerosis and miocardial infarction, and, eventually, their therapy. In connection to this, is the reported biochemical and phenotypical correction of ornithine-transcarbamylase (OTC) deficiency in Spf-ash mice by virtue of the targeted liver expression of the rat OTC gene under the control of its own promoter (Shimada et al., 1991).

Liver-Targeted Somatic Gene Therapy

Retroviral vectors have been used to introduce foreign genes into hepatocytes. Differentiated, quiescent hepatocytes cannot be infected by retroviruses, but it has been demonstrated that primary hepatocytes cultured *in vitro* under particular growth and media conditions are susceptible to retroviral infection, presumably during a period of transient and partial dedifferentiation *in vitro* (Ledley et al., 1987; Wilson JM et al., 1988; Wolff et al., 1987). Hepatocyte primary cultures can thus be infected *in vitro* with retroviral vectors potentially expressing any gene desired and then transplanted back to the living organism. This technology has been recently improved by the use of strong liver-specific promoters previously defined in both *in vitro* (cultured hepatoma cell) and *in vivo* (transgenic mice) studies. So far, various liver-specific promoters have been successfully used: the human α_1-AT promoter (Parker Ponder et al., 1991; Peng et al., 1988), or the rat PEPCK promoter (Hatzoglou et al., 1990). In this last case, the linked bovine growth hormone gene has shown to be subjected to dietary regulation like the natural PEPCK enzyme.

By injecting retrovirally infected hepatocytes directly into the blood stream (portal vein), only transient expression (1 week or less) has been reported (Wilson JM et al., 1990). Much more stable expression has been obtained when genetically altered hepatocytes are transplanted into the spleen, from where they migrate to the liver parenchyma and here persist indefinitely, more than 1 year (Parker Ponder et al., 1991). Another promising approach is based on the possibility of infecting *in vivo* hepatocytes with recombinant retrovirus; this approach has been reported to be feasible if partial hepatectomy (two-thirds of total liver mass) is performed immediately or 24 h before liver perfusion with virus (Ferry et al., 1991; Hatzoglou et al., 1990). In the last case, genetically modified hepatocytes actively producing a marker gene (β-galactosidase) were detected as long as 3 months after the infection (Ferry et al., 1991). Very recently the retrovirus gene transfer technology has been successfully applied in targeting genes to dog liver (Kay et al., 1992).

Several other systems have been used to transfer genetic material (DNA) to liver cells, including direct injection (Dubensky et al., 1984), bombardment of the organ with DNA-coated gold particles (Yang et al., 1990), and injection in the portal vein of vesicle complexes containing

DNA and nuclear proteins (Kaneda et al., 1989). One of the most promising methods uses a DNA carrier system consisting of asialoorosomucoid covalently linked to poly-L-lysine: this last polycationic molecule can bind DNA in a noncovalent and nondamaging manner (Wu and Wu, 1988). Also, in this case, the liver specificity of the system has been improved by linking the selected gene to a liver-specific promoter (the mouse albumin promoter/enhancer sequence). When this complex is injected via the tail vein in rats, DNA is delivered specifically to hepatocytes through their unique asialoglycoprotein receptors. Partial hepatectomies showed that expression of the targeted gene is maintained up to several weeks (Wu et al., 1989). Using this technique, genetically analbuminemic rats show partial correction of their genetic defect *in vivo* on delivery of the human albumin cDNA (Wu et al., 1991).

Conclusions

The study of promoters and enhancers that are active in hepatocytes has led to the discovery of several liver-enriched transcription factors during the past years. This was probably to be expected, owing to the complexity of hepatic cell functions. However, none of these transcription factors are found exclusively in hepatocytes; rather, they are all expressed, although to a lesser degree, in other tissues. Also, none of these genes is able to trigger the liver phenotype in nonliver cells the way *myoD* triggers the muscle phenotype in nonmuscle cells. There is thus a possibility that either the master liver regulatory gene has not yet been isolated, or that it simply does not exist and instead the differentiated functions of hepatocytes depend on the concerted action of several factors working together. Recent studies have focused on trying to establish the hierarchial importance of the various liver-enriched factors and their relative role along the genetic program leading to hepatocyte differentiation. So far, the data are only preliminary and incomplete. For instance, it appears that efficient transcription of the LF-B1/HNF1α gene might depend on HNF-4 (Kuo C-J et al., 1992). Also the expression pattern of a *Drosophila* homolog of HNF-4 suggests a determination role in formation (Zhong et al., 1993). However, this solitary observation does not yet allow us to conclude if HNF-4 "comes" before HNF-1, and more studies are needed to clarify this issue.

The enormous progress that has been made in the functional characterization of several liver-enriched transcription factors and in the study of their interplay in regulating hepatocyte gene expression will soon lead to a full understanding of the processes underlying the development of the liver. In this context, the use of newly developed differentiated cell lines and tissue culture conditions will be of great help. Such important advances are of practical relevance when one considers that this knowledge could be utilized in the near future in the therapy of metabolic disorders and of infectious and neoplastic diseases of the liver.

Notes in Proof. DCoH, the cofactor that stabilizes dimerization of XL-homeobox proteins has been recently shown to be identical to a 4a-carbinolamine dehydratase, an enzyme of phenylalanine metabolism (Citron et al., 1992). Recently, a new member of C/EBP related factors (named CHOP) has been cloned (Ron and Habener, 1992). CHOP does not possess a DNA binding domain but has a perfectly Leucin zipper which allows its heterodimerization with all other C.EBP related proteins. By so doing CHOP functions as dominant-negative inhibitor transcription.

Acknowledgments. Many thanks to K. Caswell, J. Clench, and T. Gobbi for their assistance in editing and revising this review. This study was also supported by the Consiglio Nazionale delle Ricerche target projects "Biotecnology and Bioinstrumentation" and "Genetic Engineering" and by NATO grant 860816 to G. Ciliberto.

References

Adams BS, Leung K, Hanley EW, Nabel GJ (1991): Cloning of RκB, a novel DNA-binding protein that recognizes the interleukin-2 receptor α chain κB site. *New Biol* 3:1063–1073.

Adams JM, Cory S (1991): Transgenic models of tumor development. *Science* 254:1161–1167.

Akira S, Isshiki H, Sugita T, Tanabe O, Kinoshita S, Nishio Y, Nakajima T, Hirano T, Kishimoto T (1990): A nuclear factor for IL-6 expression (NF-IL6) is a member of a C/EBP family. *EMBO J* 9:1897–1906.

Arcone R, Guaiandi G, Ciliberto G (1988): Identification of sequences responsible for acute-phase induction of human C-reactive protein. *Nucleic Acids Res* 16:3195–3207.

Aufiero B, Schneider RJ (1990): The hepatitis B virus X-gene prcduct trans-activates both RNA polymerase II and III promoters. *EMBO J* 9:497–504.

Babiss LE, Herbst RS, Bennett AL, Darnell JE Jr (1987): Factors that interact with the rat albumin promoter are present both in hepatocytes and othei cell types. *Genes & Dev* 1:256–267.

Baeuerle PA, Baltimore D (1988a): Activation of DNA-binding activity in an apparently cytoplasmic precursor of the NF-κB transcription factor. *Cell* 53:211–217.

Baeuerle PA, Baltimore D (1988b): I-κB: A specific inhibitor of the NF-κB transcription factor. *Science* 242:540–546.

Baeuerle PA, Baltimore D (1989): A 65-kD subunit of active NF-κB is required for inhibition of NF-κB by I-κB. *Genes & Dev* 3:1689–1698.

Baldwin AS Jr, LeClair KP, Singh H, Sharp PA (1990): A large protein containing zinc finger domains binds to related sequence elements in the enhancers of the class I major histocompatibility complex and kappa immunoglobulin genes. *Mol Cell Biol* 10:1406–1414.

Ballard DW, Walker WH, Doerre S, Sista P, Molitor JA, Dixon EP, Peffer NJ, Hannik M, Greene WC (1990): The V-*rel* oncogenes encodes a κB enhancer binding protein that inhibits NF-κB function. *Cell* 63:803–814.

Baumann H (1989): Hepatic acute phase reaction *in vivo* and *in vitro*. In *Vitro Cell Dev Biol* 25:115–126.

Baumann H, Schendel P (1991): Interleukin-11 regulates the hepatic expression of the same plasma protein genes as interleukin-6. *J Biol Chem* 266:20424–20427.

Baumann H, Wong GG (1989): Hepatocyte-stimulating factor III shares structural and functional identity with leukemia inhibitory factor. *J Immunol* 143:1163–1167.

Baumann H, Richards C, Gauldie J (1987a): Interaction among hepatocyte-stimulating factors, interleukin-1 and glucocorticoids for the regulation of acute phase plasma proteins in human hepatoma (HEPG2) cells. *J Immunol* 139:2144–4128.

Baumann H, Onorato V, Gauldie J, Jahreis GP (1987b): Distinct sets of acute phase plasma proteins are stimulated by separate human hepatocyte-stimulating factors and monokines in rat hepatoma cells. *J Biol Chem* 262:9756–9768.

Baumhueter S, Courtois G, Crabtree GR (1988): A variant nuclear protein in dedifferentiated hepatoma cells binds to the same functional sequences in the β fibrinogen gene promoter as HNF-1. *EMBO J* 7:2485–2493.

Baumhueter S, Mendel DB, Conley PB, Kuo CJ, Turk C, Graves MK, Edwards CA, Courtois G, Crabtree GR (1990): HNF-1 shares three sequence motifs with the POU domain proteins and is identical to LF-B1 and APF. *Genes & Dev* 4:372–379.

Becker PB, Gloss B, Schmid W, Strähle U, Schütz G (1986): *In vivo* protein-DNA interactions in a glucocorticoid response element require the presence of the hormone. *Nature* (Lond) 324:686–688.

Beekman JM, Wijnholds J, Schippers IJ, Pot W, Gruber M, Ab G (1991): Regulatory elements and DNA-binding proteins mediating transcription from the chicken very-low-density apolipoprotein II gene. *Nucleic Acids Res* 19:5371–5377.

Ben Ze'ev A, Robinson GS, Bucher NLR, Farmer SR (1988): Cell-cell and cell-matrix interactions differentially regulate the expression of hepatic and cytoskeletal genes in primary cultures of rat hepatocytes. *Proc Natl Acad Sci USA* 85:2161–2165.

Ben-Levy R, Faktor O, Berger I, Shaul Y (1989): Cellular factors that interact with the hepatitis B virus enhancer. *Mol Cell Biol* 9:1804–1809.

Benvenisty N, Nechushtan H, Cohen H, Reshef L (1989): Separate *cis*-regulatory elements confer expression of phosphoenolpyruvate carboxykinase (GTP) gene in different cell lines. *Proc Natl Acad Sci USA* 86:1118–1122.

Birkenmeier EH, Gwynn B, Howard S, Jerry J, Gordon JI, Landschultz WH, McKnight SL (1989): Tissue-specific expression, developmental regulation, and genetic mapping of the gene encoding CCAAT/enhancer binding protein. *Genes & Dev* 3:1146–1156.

Blum HE, Gerok W, Vyas GN (1989): The molecular biology of hepatitis B virus. *Trends Genet* 5:15–158.

Bortolotti R, Weiss MC (1972a): Expression of differentiated functions in hepatoma cell hybrids. II. Aldolase. *J Cell Physiol* 79:211–224.

Bortolotti R, Weiss MC (1972b): Expression of differentiated functions in hepatoma cell hybrids. VI. Extinction and reexpression of liver alcohol dehydrogenase. *Biochimie* 54:195–201.

Boshart M, Weih F, Nichols M, Schuetz G (1991): The tissue-specific extinguisher locus Tse-1 encodes a regulatory subunit of cAMP-dependent protein kinase. *Cell* 66:849–859.

Boshart M, Weih F, Schmidt A, Fournier REK, Schuetz G (1990): A cyclic AMP response element mediates repression of tyrosine aminotransferase gene transcription by the tissue-specific extinguisher locus Tse-1. *Cell* 61:905–916.

Brasier AR, Ron D, Tate JE, Habener JF (1990a): Synergistic enhancers located within an acute phase responsive enhancer modulate glucocorticoid induction of angiotensinogen gene transcription. *Mol Endocrinol* 4:1921–1933.

Brasier AR, Ron D, Tate JE, Habener JF (1990b): A family of constitutive C/EBP-like DNA binding proteins attenuate the IL-1α induced, NFκB mediated *trans*-activation of the angiotensinogen gene acute-phase response element. *EMBO J* 9:3933–3944.

Bressac B, Kew M, Wands J, Ozturk M (1991): Selective G to T mutations of p53 gene in hepatocellular carcinoma from southern Africa. *Nature* (Lond) 350:429–431.

Briët E, Bertina RM, van Tilburg NH, Veltkamp JJ (1982): Hemophilia B Leyden. A sex-linked hereditary disorder that improves after puberty. *N Eng J Med* 306:788–790.

Brooks AR, Blackhart BD, Haubold K, Levy-Wilson B (1991): Characterization of tissue-specific enhancer elements in the second intron of the human apolipoprotein B gene. *J Biol Chem* 266:7848–7859.

Brunel F, Ochoa A, Schaeffer E, Boissier F, Guillou Y, Cereghini S, Cohen GN, Zakin MM (1988): Interactions of DNA-binding proteins with the 5′ region of the human transferrin gene. *J Biol Chem* 263:10180–10185.

Bulla GA, De Simone V, Cortese R, Fournier REK (1992): Extinction of α_1-antitrypsin gene expression in somatic cell hybrids: Evidence for multiple controls. *Genes & Dev* 6:316–327.

Bulla GA, Siddiqui A (1988): The hepatitis B virus enhancer modulates transcription of the hepatitis B virus surface antigen gene from an internal location. *J Virol* 62:1437–1441.

Burch JBE, Evans M, Friedman TM, O'Malley PJ (1988): Two functional estrogen response elements are located upstream of the major chicken vitellogenin gene. *Mol Cell Biol* 8:1123–1131.

Butel JS, Sepulveda AR, Finegold MJ, Woo SLC (1990): SV40 large T antigen directed by regulatory elements of the human alpha-1-antitrypsin gene. *Intervirology* 31:85–100.

Cao Z, Umek RM, McKnight SL (1991): Regulated expression of three C/EBP isoforms during adipose conversion of 3T3-L1 cells. *Genes & Dev* 5:1538–1552.

Carlson JA, Rogers BB, Sifers RN, Finegold MJ, Clift SM, DeMayo FJ, Bullock DW, Woo SLC (1989): Accumulation of PiZ alpha$_1$-antitrypsin causes liver damage in transgenic mice. *J Clin Invest* 83:1183–1190.

Cascio S, Zaret KS (1991): Hepatocyte differentiation initiates during endodermal-mesenchymal interactions prior to liver formation. *Development* 113:217–225.

Castell JV, Gomez-Lechon MJ, David M, Hirano T, Kishimoto T, Heinrich PC (1988): Recombinant human interleukin-6 (IL-6, BSF-2, HSF) regulates the synthesis of acute phase proteins in human hepatocytes. *FEBS Lett* 232:347–350.

Cato ACB, Heitlinger E, Ponta H, Hitpass LK, Ryffel GU, Bailly A, Rauch C, Milgrom E (1988): Estrogen and progesterone receptor-binding sites on the chicken vitellogenin II gene: synergism of steroid hormone action. *Mol Cell Biol* 8:5323–5330.

Caselmann WH, Meyer M, Kekulé AS, Lauer V, Hofschneider PH, Koshy R (1990): A trans-activator function is generated by integration of hepatitis B virus pre S/S sequences in human hepatocellular carcinoma DNA. *Proc Natl Adac Sci USA* 87:2970–2974.

Cereghini S, Blumenfeld M, Yaniv M (1988): A liver-specific factor essential for albumin transcription differs between differentiated and dedifferentiated rat hepatoma cells. *Genes & Dev* 2:957–974.

Cereghini S, Yaniv M, Cortese R (1990): Hepatocyte dedifferentiation and extinction is accompanied by a block in the synthesis of mRNA coding for the transcription factor HNF1/LFB1. *EMBO J* 9:2257–2263.

Cereghini S, Raymondjean M, Garcia-Carranca A, Herbomel P, Yaniv M (1987): Factors involved in control of tissue-specific expression of albumin gene. *Cell* 50:627–638.

Ceska TA, Lamers M, Monaci P, Nicosia A, Cortese R, Suck D (1993): The structure of a variant homeodomian present in the rat liver transcription factor LFB1/HNF1 and implications for DNA binding. *EMBO J*, in press.

Chajek-Shaul T, Hayer T, Walsh A, Breslow JL (1991): Expression of the human apolipoprotein A-I gene in transgenic mice alters high density lipoprotein (HDL) particle size distribution and diminishes selective uptake of HDL cholesteryl esters. *Proc Natl Acad Sci USA* 88:6731–6735.

Chambaz J, Cardot P, Pastier D, Zannis VI, Cladaras C (1991): Promoter elements and factors required for hepatic transcription of the human ApoA-II gene. *J Biol Chem* 266:11676–11685.

Chang C-J, Chen T-T, Xei H-Y, Chen D-S, Lee S-C (1990): Molecular cloning of a transcription factor, AGP/EBP, that belongs to members of the C/EBP family. *Mol Cell Biol* 10:6642–6653.

Chang DJ, Paik Y-K, Leren TP, Walker DW, Howlett GJ, Taylor JM (1990): Characterization of a human apolipoprotein E gene enhancer element and its associated protein factors. *J Biol Chem* 265:9496–9504.

Chang H-K, Ting L-P (1989): The surface gene promoter of the human hepatitis B virus displays a preference for differentiated hepatocytes. *Virology* 170:176–183.

Chang H-K, Wang B-Y, Yuh C-H, Wei C-L, Ting L-P (1989): A liver-specific nuclear factor interacts with the promoter region of the large surface protein gene of human hepatitis B virus. *Mol Cell Biol* 9:5189–5197.

Chang T-C, Shapiro DJ (1989): An NF1-related vitellogenin activator element mediates transcription from the estrogen-regulated *Xenopus laevis* vitellogenin promoter. *J Biol Chem* 265:8176–8182.

Chin AC, Fournier REK (1989): Tse-2: A trans-dominant extinguisher of albumin gene expression in hepatoma hybrid cells. *Mol Cell Biol* 9:3736–3743.

Chisari FV, Filippi P, Buras J, McLachlan A, Popper H, Pinkert CA, Palmiter RD, Brinster RL (1987): Structural and pathological effects of synthesis of hepatitis B virus large envelope polypeptide in transgenic mice. *Proc Natl Acad Sci USA* 84:6909–6913.

Chisari FV, Klopchin K, Moriyama T, Pasquinelli C, Dunsford HA, Sell S, Pinkert CA, Brinster RL, Palmiter RD (1989): Molecular pathogenesis of hepatocellular carcinoma in hepatitis B virus transgenic mice. *Cell* 59:1145–1156.

Chouard T, Blumenfeld M, Bach I, Vandekerckhove J, Cereghini S, Yaniv M (1990): A distal dimerization domain is essential for DNA-binding by the atypical HNF1 homeodomain. *Nucleic Acids Res* 18:5853–5863.

Christy RJ, Kaestner KH, Geiman DE, Lane MD (1991): CCAAT/enhancer binding protein gene promoter: Binding of nuclear factors during differentiation of 3T3-L1 preadipocytes. *Proc Natl Acad Sci USA* 88:2593–2597.

Christy RJ, Yang VW, Ntambi JM, Geiman DE, Landschulz WH, Friedman AD, Nakabeppu Y, Kelly TJ, Lane MD (1989): Differentiation-induced gene expression in 3T3-L1 preadipocytes: CCAAT/enhancer binding protein interacts with and activates the promoters of two adipocyte-specific genes. *Genes & Dev* 3:1323–1335.

Church GM, Ephrussi A, Gilbert W, Tonegawa S (1985): Cell-type-specific contacts to immunoglobulin enhancers in nuclei. *Nature* (Lond) 313:798–801.

Ciliberto G (1989): Transcriptional regulation of acute phase response genes with emphasis on the human C-reactive protein gene. In: *Acute Phase Proteins and the Acute Phase Response*, Pepys M, ed., pp. 29–46. New York: Springer-Verlag.

Ciliberto G, Arcone R, Wagner EF, Ruether U (1987): Inducible and tissue-specific expression of human C-reactive protein in transgenic mice. *EMBO J* 6:4017–4022.

Citron BA, Davis MD, Milstein S, Gutierrez J, Mendel DB, Crabtree GR, Kaufman S (1992): Identity of 4a-carbinolamine dehydratase, a component of the phenylalanine hydroxylation system, and DCoH, a transregulator of homeodomain proteins. *Proc Natl Acad Sci USA* 89:11891–11894.

Clayton DF, Darnell JE Jr (1983): Changes in liver-specific compared to common gene transcription during primary culture of mouse hepatocyte. *Mol Cell Biol* 3:1552–1561.

Clayton DF, Harrelson AL, Darnell JE Jr (1985a): Dependence of liver-specific transcription on tissue organization. *Mol Cell Biol* 5:2623–2632.

Clayton DF, Weiss M, Darnell JE Jr (1985b): Liver-specific RNA metabolism in hepatoma cells: variations in transcription rates and mRNA levels. *Mol Cell Biol* 5:2633–2641.

Cognet M, Bergot M-O, Kahn A (1991): *cis*-Acting DNA elements regulating expression of the liver pyruvate kinase gene in hepatocytes and hepatoma cells. Evidence for tissue specific activators and extinguisher. *J Biol Chem* 266:7368–7375.

Colgrove R, Simon G, Ganem D (1989): Transcriptional activation of homologous and heterologous genes by the hepatitis B virus X gene product in cells permissive for viral replication. *J Virol* 63(9):4019–4026.

Corthésy B, Claret F-X, Wahli W (1990): Estrogen receptor level determines sex-specific *in vitro* transcription from the *Xenopus* vitellogenin promoter. *Proc Natl Acad Sci USA* 87:7878–7882.

Corthésy B, Corthésy-Theulaz I, Cardinaux J-R, Wahli W (1991): A liver protein fraction regulating hormone-dependent *in vitro* transcription from the vitellogenin genes induces their expression in *Xenopus* oocytes. *Mol Endocrinol* 5:159–169.

Corthésy B, Hipskind R, Theulaz I, Wahli W (1988): Estrogen-dependent *in vitro* transcription from the vitellogenin promoter in liver nuclear extracts. *Science* 239:1137–1139.

Costa RH, Grayson DR, Darnell JE Jr (1989): Multiple hepatocyte-enriched nuclear factors function in the regulation of transthyretin and α_1-antitrypsin genes. *Mol Cell Biol* 9:1415–1425.

Costa RH, Grayson DR, Xanthopoulos KG, Darnell JE Jr (1988a): A liver-specific DNA-binding protein recognizes multiple nucleotide sites in regulatory regions of transthyretin, alpha-1-antitrypsin, albumin, and simian virus

40 genes. *Proc Natl Acad Sci USA* 85:3840–3844.

Costa RH, Lai E, Grayson DR, Darnell JE Jr (1988b): The cell-specific enhancer of the mouse transthyretin (prealbumin) gene binds a common factor at one site and a liver-specific factor(s) at two other sites. *Mol Cell Biol* 8:81–90.

Costa RH, Van Dyke TA, Yan C, Kuo F, Darnell JE Jr (1990): Similarities in transthyretin gene expression and differences in transcription factors: Liver and yolk sac compared to choroid plexus. *Proc Natl Acad Sci USA* 87:6589–6593.

Courtois G, Baumhueter S, Crabtree GR (1988): Purified hepatocyte nuclear factor 1 interacts with a family of hepatocyte-specific promoters. *Proc Natl Acad Sci USA* 85:7937–7941.

Courtois G, Morgan JG, Campbell LA, Fourel G, Crabtree GR (1987): Interaction of a liver-specific nuclear factor with the fibrinogen and α_1-antitrypsin promoters. *Science* 238:688–692.

Crossley M, Ludwig M, Starwell KM, De Vos P, Olek K, Brownlee GG (1992): Recovery from hemophilia B Leyden: an androgen-responsive element in the factor IV promoter. *Science* 257:377–379.

Crossley M, Brownlee GG (1990): Disruption of a C/EBP binding site in the factor IX promoter is associated with haemophilia B. *Nature* (Lond) 345:444–446.

Crossley M, Ludwig M, Stowell K, Olek K, Brownlee GG: A mutation in an LF-A1/HNF4 site unmasks an overlapping androgen responsive element and explains the clinical recovery of haemophilia B Leyden patients (in manuscript).

Darlington GH, Wilson DR, Lachman LB (1986): Monocyte-conditioned medium, interleukin-1 and tumor necrosis factor stimulate the acute phase response in human hepatoma cells *in vitro*. *J Cell Biol* 103:787–793.

Davidson RL (1974): Gene expression in somatic cell hybrids. *Annu Rev Genet* 8:195–218.

Davidson RL, Ephrussi B, Yamamoto K (1968): Regulation of melanin synthesis in mammalian cells as studied by somatic hybridization. I. Evidence for negative control. *J Cell Physiol* 72:115–127.

Davis JN, Bargmann W, Bose HR Jr (1990): Identification of protein complexes containing the c-*rel* proto-oncogene product in avian hematopoietic cells. *Oncogene* 5:1109–1115.

De Francesco R, Pastore A, Vecchio G, Cortese R (1991): Circular dichroism study on the conformational stability of the dimerization domain of transcription factor LFB-1. *Biochemistry* 30(1):143–147.

De Simone V, Cortese R (1988): The transcriptional regulation of liver-specific gene expression. In: *Oxford Surveys on Eukaryotic Genes*, McLean N, ed, pp. 51–90. New York: Oxford University Press.

De Simone V, Ciliberto G, Hardon E, Paonessa G, Palla F, Lundberg L, Cortese R (1987): *Cis-* and *trans*-acting elements responsible for the cell-specific expression of the human α_1-antitrypsin gene. *EMBO J* 6:2759–2766.

De Simone V, De Magistris L, Lazzaro D, Gerstner J, Monaci P, Nicosia A,

Cortese R (1991): LFB3, a heterodimer-forming homeoprotein of the LFB1 family, is epxressed in specialized epithelia. *EMBO J* 10(6):1435–1443.

Dejean A, Bougueleret L, Grzeschik K-H, Tiollais P (1986): Hepatitis B virus DNA integration in a sequence homologous to v-*erb*-A and steroid receptor genes in a hepatocellular carcinoma. *Nature* (Lond) 322:70–72.

Dente L, Rüther U, Tripodi M, Wagner EF, Cortese R (1988): Expression of human α_1-acid glycoprotein genes in cultured cells and in transgenic mice. *Genes & Dev* 2:259–266.

Deschatrette J, Fougere-Deschatrette C, Corcos L, Schimke RT (1985): Expression of the mouse serum albumin gene introduced into differentiated and dedifferentiated rat hepatoma cells. *Proc Natl Acad Sci USA* 82:765–769.

Descombes P, Schibler U (1991): A liver-enriched transcriptional activator protein, LAP, and a transcriptional inhibitory protein, LIP, are translated from the same mRNA. *Cell* 67:569–579.

Descombes P, Chojkier M, Lichtsteiner S, Falvey E, Schibler U (1990): LAP, a novel member of the C/EBP gene family, encodes a liver-enriched transcriptional activator protein. *Genes & Dev* 4:1541–1551.

Dewey MJ, Rheaume C, Berger FG, Baumann H (1990): Inducible and tissue-specific expression of rat α-acid glycoprotein in transgenic mice. *J Immunol* 144:4392–4398.

Dikstein R, Faktor O, Ben-Levy R, Shaul Y (1990a): Functional organization of the hepatitis B virus enhancer. *Mol Cell Biol* 10:3683–3689.

Dikstein R, Faktor O, Shaul Y (1990b): Hierarchic and cooperative binding of the rat liver nuclear protein C/EBP at the hepatitis B virus enhancer. *Mol Cell Biol* 10:4427–4430.

Di Persio CM, Jackson DA, Zaret KS (1991): The extracellular matrix coordinately modulates liver transcription factors and hepatocyte morphology. *Mol Cell Biol* 11:4405–4414.

Dubensky TW, Campbell BA, Villareal LP (1984): Direct transfection of viral and plasmid DNA into the liver or spleen of mice. *Proc Natl Acad Sci USA* 81:7529–7533.

Dunsford HA, Sell S, Chisari FV (1990): Hepatocarcinogenesis due to chronic liver cell injury in hepatitis B virus transgenic mice. *Cancer Res* 50:3400–3407.

Dycaico MJ, Grant SGN, Felts K, Nichols WS, Geller SA, Hager JH, Pollard AJ, Kohler SW, Short HP, Jirik FR, Hanahan D, Sorge JA (1988): Neonatal hepatitis induced by α_1-antitrypsin: A transgenic mouse model. *Science* 242:1409–1412.

Edbrooke MR, Foldi J, Cheshire JK, Li F, Faulkes DJ, Woo P (1991): Constitutive and NF-κB-like proteins in the regulation of the serum amyloid A gene by interleukin 1. *Cytokines* 3:380–388.

Faber S, Ip T, Granner D, Chalkley R (1991): The interplay of ubiquitous DNA-binding factors, availability of binding sites in the chromatin, and DNA methylation in the differential regulation of phosphoenolpyruvate carboxykinase gene expression. *Nucleic Acids Res* 19:4681–4688.

Faktor O, Budlovsky S, Ben-Levy R, Shaul Y (1990): A single element within the hepatitis B virus enhancer binds multiple proteins and responds to multiple stimuli. *J Virol* 64:1861–1863.

Farza H, Hadchouel M, Scotto J, Tiollais P, Babinet C, Pourcel C (1988): Replication and gene expression of hepatitis B virus in a transgenic mouse that contains the complete viral genome. *J Virol* 62:4144–4152.

Ferry N, Duplessis O, Houssin D, Danos O, Heard J-M (1991): Retroviral-mediated gene transfer into hepatocytes *in vivo*. *Proc Natl Acad Sci USA* 88:8377–8381.

Feuerman MH, Godbout R, Ingram RS, Tilghman SM (1989): Tissue-specific transcription of the mouse α-fetoprotein gene promoter is dependent on HNF-1. *Mol Cell Biol* 9:4204–4212.

Fey GH, Gauldie J (1990): The acute phase response of the liver in inflammation. In: *Progress in Liver Diseases*, Popper H, Schaffner F, ed. Philadelphia: WB Saunders.

Fields S, Jang SK (1990): Presence of a potent transcription activating sequence in the p53 protein. *Science* 24:1046–1049.

Finney M (1990): The homeodomain of the transcription factor LF-B1 has a 21 amino acid loop between helix 2 and helix 3. *Cell* 60:5–6.

Forest CD, O'Brien RM, Lucas PC, Magnuson MA, Granner DK (1990): Regulation of phosphoenolpyruvate carboxykinase gene expression by insulin. Use of the stable transfection approach to locate an insulin responsive sequence. *Mol Endocrinol* 4:1302–1310.

Forman BM, Samuels HH (1990): Interactions among a subfamily of nuclear hormone receptors. The regulatory zipper model. *Mol Endocrinol* 4:1293–1301.

Fowlkes DM, Mullis NT, Comeau CM, Crabtree GR (1984): Potential basis for regulation of the coordinately expressed fibrinogen genes: Homology in the 5′ flanking regions. *Proc Natl Acad Sci USA* 81:2313–2316.

Frain M, Hardon EM, Ciliberto G, Sala Trepat JM (1990): Binding of a liver-specific factor to the human albumin gene promoter and enhancer. *Mol Cell Biol* 10:991–999.

Frain M, Swart G, Monaci P, Nicosia A, Stampfli S, Frank R, Cortese R (1989): The liver-specific transcription factor LF-B1 contains a highly diverged homeobox DNA binding domain. *Cell* 59:145–157.

Fraslin JM, Kneip B, Vaulont S, Glaise D, Munnich A, Guguen-Guilloze C (1985): Dependence of hepatocyte-specific gene expression on cell-cell interaction in primary culture. *EMBO J* 4:2487–2491.

Friedman AD, McKnight SL (1990): Identification of two polypeptide segments of CCAAT/enhancer-binding protein required for transcriptional activation of the serum albumin gene. *Genes & Dev* 4:1416–1426.

Friedman AD, Landschultz WH, McKnight SL (1989): CCAAT/enhancer binding protein activates the promoter of the serum albumin gene in cultured hepatoma cells. *Genes & Dev* 3:1314–1322.

Fung WP, Thomas T, Dickson PW, Aldred AR, Milland J, Dziadek M, Power B, Hudson P, Schreiber (1988): Structure and expression of the rat transthyretin (prealbumin) gene. *J Biol Chem* 263:480–488.

Fung W-P, Schreiber G (1987): Structure and expression of the genes for major acute phase α_1-protein (thiostatin) and kininogene in the rat. *J Biol Chem* 262(19):2208–2308.

Ganapathi MK, Schultz D, Mackiewicz A, Samols O, Ity S-I, Brabenec A, Macintyre SS, Kushner I (1988): Differential regulation of human serum amyloid A, C-reactive protein, and other acute phase proteins by cytokines in Hep 3B cells. *J Immunol* 141:564–569.

Ganem D, Varmus HE (1987): The molecular biology of the hepatitis B viruses. *Annu Rev Biochem* 56:651–693.

Ganter U, Arcone R, Toniatti C, Morrone G, Ciliberto G (1989): Dual control of C-reactive protein gene expression by interleukin-1 and interleukin-6. *EMBO J* 8:3773–3779.

Gauldie J, Richards C, Harnish D, Lansdorf P, Baumann H (1987): Interferon $\beta2$/B-cell stimulatory factor type 2 shares identity with monocyte-derived hepatocyte-stimulating factor and regulates the major acute phase protein response in liver cells. *Proc Natl Acad Sci USA* 84:7251–7255.

Gebhardt R (1990): Altered acinar distribution of glutamine synthetase and different growth response of cultured enzyme-positive and -negative hepatocytes after partial hepatectomy. *Cancer Res* 50:4407–4410.

Geller SA, Nichols WS, Dycaico MJ, Felts KA, Sorge JA (1990): Histopathology of α_1-antitrypsin liver disease transgenic mouse model. *Hepatology* 12(I):40–47.

Ghosh S, Baltimore D (1990): Activation *in vitro* of NF-κB by phosphorylation of its inhibitor IκB. *Nature* (Lond) 344:678–682.

Ghosh S, Gifford AM, Riviere LR, Tempst P, Nolan GP, Baltimore D (1990): Cloning of the p50 DNA binding subunit of NF-κB: Homology to *rel* and *dorsal*. *Cell* 62:1021–1029.

Gilmore TD, Temin HM (1986): Different localization of the product of the V-*rel* oncogene in chicken fibroblasts and spleen cells correlates with transformation by REV-T. *Cell* 44:791–800.

Gleiberman AS, Abelev GI (1985): Cell position and cell interactions: Expression of fetal phenotype of hepatocytes. *Int Rev Cytol* 95:229–266.

Gluecksohn-Waelsch S (1979): Genetic control of morphogenetic and biochemical differentiation: Lethal albino deletions in the mouse. *Cell* 16:225–237.

Gorski K, Carneiro M, Schibler U (1986): Tissue-specific *in vitro* transcription from the mouse albumin promoter. *Cell* 47:767–776.

Gourdeau H, Fournier REK (1990): Genetic analysis of mammalian cell differentiation. *Annu Rev Cell Biol* 6:69–94.

Grange T, Roux J, Rigaud G, Pictet R (1989): Two remote glucocorticoid responsive units interact cooperatively to promote glucocorticoid induction of rat tyrosine minotransferase gene expression. *Nucleic Acids Res* 17:8695–8709.

Grange T, Roux J, Rigaud G, Pictet R (1991): Cell-type-specific activity of two glucocorticoid responsive units of rat tyrosine aminotransferase gene is associated with multiple binding sites for C/EBP and a novel liver-specific nuclear factor. *Nucleic Acids Res* 19:131–139.

Granner DK, Beale EG (1985): Regulation of the synthesis of tyrosine aminotransferase and phosphoenolpyruvate carboxykinase by glucocorticoid hormones. In: *Biochemical Actions of Hormones XII*, Litwack G, ed., pp. 89–138. New York: Academic Press.

Graves B, Johnson PF, McKnight SL (1986): Homologous recognition of a promoter domain common to the MSV LTR and the HSV tk gene. *Cell* 44:565–576.

Grayson DR, Costa RH, Xanthopoulos KG, Darnell JE Jr (1988a): A cell-specific enhancer of the mouse α_1-antitrypsin gene has multiple functional regions and corresponding protein-binding sites. *Mol Cell Biol* 8:1055–1066.

Grayson DR, Costa RH, Xanthopoulos KG, Darnell JE Jr (1988b): One factor recognizes the liver-specific enhancers in alpha-1-antitrypsin and transthyretin genes. *Science* 239:786–788.

Greengard O (1970): The developmental formation of enzymes in rat liver. In: *Mechanism of Hormone Action I*, Litwarck G ed., pp. 53–85. New York: Academic Press.

Guertin M, Larue H, Bernier D, Wrange O, Chevrette M, Gingras M-C, Bélanger L (1988): Enhancer and promoter elements directing activation and glucocorticoid repression of the α-fetoprotein gene in hepatocytes. *Mol Cell Biol* 8:1398–1407.

Guo W, Bell KD, Ou J-H (1991): Characterization of the hepatitis B virus EnhI enhancer and X promoter complex. *J Virol* 65:6686–6692.

Hammer RE, Krumlauf R, Camper SA, Brinster RL, Tilghman S (1987): Diversity of α-fetoprotein gene expression in mice is generated by a combination of separate enhancer elements. *Science* 235:53–58.

Hanahan D (1989): Transgenic mice as probes into complex systems. *Science* 246:1265–1275.

Hardon EM, Frain M, Paonessa G, Cortese R (1988): Two distinct factors interact with the promoter regions of several liver-specific genes. *EMBO J* 7:1711–1719.

Hargrove JL, Granner DK (1985): Biosynthesis and intracellular processing of tyrosine aminotransferase. In: *Transaminases*, Christen P, Metzler PE, eds., pp. 511–532. New York: Wiley.

Harrison SC (1991): A structural taxonomy of DNA-binding domains, Review article. *Nature* (Lond) 353:715–719.

Haruna Y, Hayashi N, Katayama K, Yuki N, Kasahara A, Sasaki Y, Fusamoto H, Kamada T (1991): Expression of X protein and hepatitis B virus replication in chronic hepatitis. *Hepatology* 13(3):417–421.

Hashimoto S, Schmid W, Schuetz G (1984): Transcriptional activation of the rat liver tyrosine aminotransferase gene by cAMP. *Proc Natl Acad Sci USA* 81:6637–6641.

Hattori M, Abraham LJ, Northemann W, Fey GH (1990): Acute-phase reaction induces a specific complex between hepatic nuclear proteins and the interleukin 6 response element of the rat α_2-macroglobulin gene. *Proc Natl Acad Sci* 87:2364–2368.

Hatzoglou M, Lamers W, Bosch F, Wynshaw-Boris A, Clapp DW, Hanson RW (1990): Hepatic gene transfer in animals using retroviruses containing the promoter from the gene for phosphoenolpyruvate carboxykinase. *J Biol Chem* 265:17285–17293.

Heard J-M, Herbomel P, Ott M-O, Mottura-Rollier A, Weiss M, Yaniv M (1987): Determinants of rat albumin promoter tissue specificity analyzed by an improved transient expression system. *Mol Cell Biol* 7:2425–2434.

Held WA, Mullins JJ, Kuhn NJ, Gallagher JF, Gu GD, Gross KW (1989): T antigen expression and tumorigenesis in transgenic mice containing a mouse major urinary protein/SV40 T antigen hybrid gene. *EMBO J* 8:183–191.

Herbomel P, Rollier A, Tronche F, Ott M-O, Yaniv M, Weiss M (1989): The rat albumin promoter is composed of six distinct positive elements within 130 nucleotides. *Mol Cell Biol* 9:4750–4758.

Herrera R, Ro HS, Robinson GS, Xanthopoulos KG, Spiegelman BM (1989): A direct role for C/EBP and the AP-I-binding site in gene expression linked to adipocyte differentiation. *Mol Cell Biol* 9:5331–5339.

Hirano T (1991): Interleukin 6 (IL-6) and its receptor: Their role in plasma cell neoplasias. *Int J Cell Cloning* 9:166–184.

Hocke GM, Barry D, Fey GH (1992): Synergistic action of interleukin-6 and glutocorticoids is mediated by the interleukin-6 response element of the rat $\alpha 2$ macroglobulin gene. *Mol Cell Biol* 12:2282–2294.

Hollstein M, Sidransky D, Vogelstein B, Harris CC (1991): p53 mutations in human cancer. *Science* 253:49–53.

Honigwachs J, Faktor O, Dikstein R, Shaul Y, Laub O (1989): Liver-specific expression of hepatitis B virus is determined by the combined action of the core gene promoter and the enhancer. *J Virol* 63:919–924.

Hoodless PA, Roy RN, Ryan AK, Haché RJG, Vasa MZ, Deeley RG (1990): Developmental regulation of specific protein interactions with an enhancerlike binding site far upstream from the avian very-low-density apolipoprotein II gene. *Mol Cell Biol* 10:154–164.

Hoofnagle JH, Shafritz DA, Popper H (1987): Chronic type B hepatitis and the "healthy" HBsAg carrier state. *Hepatology* 7:758–763.

Houssaint E (1980): Differentiation of the mouse hepatic primordium. I. An analysis of tissue interactions in hepatocyte differentiation. *Cell Differ* 9:269–279.

Howell BW, Lagacé M, Shore GC (1989): Activity of the carbamyl phosphate synthetase I promoter in liver nuclear extracts is dependent on a *cis*-acting

C/EBP recognition element. *Mol Cell Biol* 9:2928–2933.

Hsu IC, Metcalf RA, Sun T, Welsh JA, Wang NJ, Harris CC (1991): Mutational hotspot in the p53 gene in human hepatocellular carcinomas. *Nature* (Lond) 350:427–428.

Hu K-Q, Siddiqui A (1991): Regulation of the hepatitis B virus gene expression by the enhancer element I. *Virology* 181:721–726.

Huang JH, Rienhoff HY Jr, Liao WSL (1990): Regulation of mouse serum amyloid A gene expression in transfected hepatoma cells. *Mol Cell Biol* 10(7):3619–3625.

Huber P, Laurent M, Dalmon J (1990): Human β-fibrinogen gene expression. *J Biol Chem* 265(10):5695–5701.

Hunter T (1991): Cooperation between oncogenes. *Cell* 64:249–270.

Imai E, Stromstedt P-E, Quinn PG, Carlstedt-Duke J, Gustafsson JA, Granner DK (1990): Characterization of a complex glucocorticoid response unit in the phosphoenolpyruvate carboxykinase gene. *Mol Cell Biol* 10:4712–4719.

Ingraham HA, Flynn SE, Voss JW, Albert VR, Kapilof MS, Wilson L, Rosenfeld MG (1990): The POU-specific domain of Pit-1 essential for sequence-specific, high affinity DNA binding and DNA-dependent Pit-1-Pit-1 interactions. *Cell* 61:1021–1033.

Ip YT, Poon D, Stone D, Granner DK, Chalkley R (1990): Interaction of a liver-specific factor with an enhancer 4.8 kilobases upstream of the phospho-enolpyruvate carboxykinase gene. *Mol Cell Biol* 10:3770–3781.

Isshiki H, Akira S, Sugita T, Nishio Y, Hashimoto S, Pawlowski T, Suematsu S, Kishimoto T (1991): Reciprocal expression of NF-IL6 and C/EBP in hepato-cytes: Possible involvement of NF-IL6 in acute phase protein gene expression. *New Biol* 3:63–70.

Ito K, Tanaka T, Tsutsumi R, Ishikawa K, Tsutsumi K-I (1990): Two different HNF1-like transcription activators in the liver bind to the same region of the rat aldolase B promoter. *Biochem Biophys Res Commun* 173:1337–1343.

Ito Y, Azrolan N, O'Connell A, Walsh A, Breslow JL (1990): Hypertriglyc-eridemia as a result of human Apo CIII gene expression in transgenic mice. *Science* 249:790–793.

Iyer SV, Davis DL, Seal SN, Burch JBE (1991): Chicken vitellogenin gene-binding protein, a leucine zipper transcription factor that binds to an important control element in the chicken vitellogenin II promoter, is related to rat DBP. *Mol Cell Biol* 11:4863–4875.

Izban MG, Papaconstantinou J (1989): Cell-specific expression of mouse albu-min promoter. *J Biol Chem* 264:9171–9179. Jallat S, Perraud F, Dalemans W, Ballan DA, Dieterce A, Faure T, Meulien P, Pavirani A (1990): Charac-terization of recombinant human factor IX expressed in transgenic mice and inderived trans-immortalized hepatic cell lines. *EMBO J* 9:3295–3301.

Jefferson DM, Clayton DF, Darnell JE Jr, Reid LM (1984): Post-transcriptional modulation of gene expression in cultured rat hepatocytes. *Mol Cell Biol* 4:1929–1934.

Johnson PF, Landschulz WH, Graves BJ, McKnight SL (1987): Identification of a rat liver nuclear protein that binds to the enhancer core element of three animal viruses. *Genes & Dev* 1:133–146.

Jones KW, Shapero MH, Chevrette M, Fournier REK (1991): Subtractive hybridization cloning of a tissue-specific extinguisher: TSE1 encodes a regulatory subunit of protein kinase A. *Cell* 66:861–872.

Jose-Estanyol M, Pollard A, Foiret D, Dana J-L (1989): A common liver-specific factor binds to the rat albumin and α-fetoprotein promoters *in vitro* and acts as a positive trans-acting factor *in vivo*. *Eur J Biochem* 181:761–766.

Jungermann K (1988): Metabolic zonation of liver parenchyma. *Semin Liver Dis* 8:329–341.

Kaling M, Kugler W, Ross K, Zoidl C, Ryffel GU (1991): Liver-specific gene expression: A-activator-binding site, a promoter module present in vitellogenin and acute-phase genes. *Mol Cell Biol* 11:93–101.

Kageyama R, Sasai Y, Nakansihi S (1991): Molecular characterization of transcription factors that bind to the cAMP responsive region of the substance P precursor gene. *J Biol Chem* 266:15525–15531.

Kaneda Y, Iwai K, Uchida T (1989): Increased expression of DNA controinduced with nuclear protein in adult rat liver. *Science* 243:375–378.

Kaneko S, Miller RH (1988): X-region-specific transcript in mammalian hepatitis B virus-infected liver. *J Virol* 62:3979–3984.

Kardassis D, Hadzopoulou-Cladaras M, Ramji DP, Cortese R, Zannis VI, Cladaras C (1990a): Characterization of the promoter elements required for hepatic and intestinal transcription of the human apoB gene: Definition of the DNA-binding site of a tissue-specific transcriptional factor. *Mol Cell Biol* 10:2653–2659.

Kardassis D, Zannis VI, Cladaras C (1990b): Purification and characterization of the nuclear factor BA1. *J Biol Chem* 265:21733–21740.

Karpen S, Banerjee R, Zelent A, Price P, Acs G (1988): Identification of protein-binding sites in the hepatitis B virus enhancer and core promoter domains. *Mol Cell Biol* 8:5159–5165.

Kay MA, Baley P, Rothenberg S, Leland F, Fleming L, Parker Ponder K, Liu T-J, Finegod M, Darlington G, Pokorny W, Woo SLC (1992): Expression of human a_1-antitrypsin in dogs after autologous transplantation of retroviral transduced hepatocytes. *Proc Natl Acad Sci USA* 89:89–93.

Kekulé AS, Lauer U, Weiss L, Luber B, Hofschneider PH (1993): Hepatitis B virus transactivator HBx uses a tumour promoter signalling pathway. *Nature* 361:742–745.

Kekulé AS, Lauer U, Meyer M, Caselman WH, Hofschneider PH, Koshy R (1990): The preS2/S region of integrated hepatitis B virus DNA encodes a transcriptional transactivator. *Nature* (Lond) 343:457–461.

Kern SE, Kinzler KW, Bruskin A, Jarosz D, Friedman P, Prives C, Vogelstein B (1991a): Identification of p53 as a sequence-specific DNA-binding protein. *Science* 252:1708–1711.

Kern SE, Kinzler KW, Baker SJ, Nigro JM, Rotter V, Levine AJ, Friedman P,

Prives C, Vogelstein B (1991b): Mutant p53 proteins bind DNA abnormally *in vitro*. *Oncogene* 6:131–136.

Kerr LD, Inoue J, Davis N, Link E, Baeuerle PA, Bose HR Jr, Verma IM (1991): The *Rel*-associated pp40 protein prevents DNA binding of Rel and Nf-κB relationship with IκBβ and regulation by phosphorylation. *Genes & Dev* 5:1564–1576.

Kieran M, Blank V, Logeat F, Vandekerckhove J, Lottspeich F, Le Bail O, Urban MB, Kourilsky P, Baeuerle PA, Israële A (1990): The DNA binding subunit of NF-κB and homologous to the rel oncogene product. *Cell* 62:1007–1018.

Killary AM, Fournier REK (1984): A genetic analysis of extinction: *Trans*-dominant loci regulate expression of liver-specific traits in hepatoma hybrid cells. *Cell* 38:523–534.

Kim C-M, Koike K, Saito I, Miyamura T, Jay G (1991): HBx gene of hepatitis B virus induces liver cancer in transgenic mice. *Nature* (Lond) 351:317–320.

Kinoshita S, Akira S, Kishimoto T (1992): A member of the C/EBP family, NF-IL6β, forms a heterodimer and transcriptionally synergizes with NF-IL6. *Proc Natl Acad Sci USA* 89:1473–1476.

Klein ES, Di Lorenzo D, Posseckert G, Beato M, Ringold GM (1988): Sequences downstream of the glucocorticoid regulatory element mediate cyclohexunide inhibitron of steroid induced expression from the rat α_1-acid glycoprotein promoter: Evidence for a labile transcription factor. *Mol Endocrin* 2:1343–1351.

Klein-Hitpass L, Ryffel GU, Heitlinger E, Cato ACB (1988): A 13-bp palindrome is a functional estrogen responsive element and interacts specifically with estrogen receptor. *Nucleic Acids Res* 16:647–663.

Klemm DJ, Roesler WJ, Liu J, Park EA, Hanson RW (1990): *In vitro* analysis of promoter elements regulating transcription of the phosphoenolpyruvate carboxykinase (GTP) gene. *Mol Cell Biol* 10:480–485.

Knowles BB, Howe CC, Aden DP (1980): Human hepatocellular carcinoma cell lines secrete the major plasma proteins and hepatitis B surface antigen. *Science* 209:497–499.

Kugler W, Wagner U, Ryffel GU (1988): Tissue-specificity of gene expression: A common liver-specific promoter element. *Nucleic Acids Res* 16:3165–3174.

Kugler W, Kaling M, Ross K, Wagner U, Ryffel GU (1990): BAP, a rat liver protein that activates transcription through a promoter element with similarity to the USF/MLTF binding site. *Nucleic Acids Res* 18:6943–6951.

Kunz D, Zimmermann R, Heisig M, Heinrich PC (1989): Identification of the promoter sequences involved in the interleukin-6 dependent expression of the rat α_2 macroglobulin gene. *Nucleic Acids Res* 17:1121–1138.

Kuo CF, Xanthopoulos K, Darnell JE Jr (1990): Fetal and adult localization of C/EBP: Evidence for combinatorial action of transcription factors in cell-specific gene expression. *Development* 109:473–481.

Kuo CJ, Mendel DB, Hansen LP, Crabtree GR (1991): Independent regulation of HNF-1 alpha and HNF-1 beta by retinoic acid in F9 teratocarcinoma cells. *EMBO J* 10:2231–2236.

Kuo CJ, Conley PB, Chen L, Sladek FM, Darnell JE Jr, Crabtree GR (1992): A transcriptional hierarchi involved in mammalian cell-type specification. *Nature* (Lond) 355:457–460.

Ladias JAA, Karathanasis SK (1991): Regulation of the apolipoprotein AI gene by ARP-1, a novel member of the steroid receptor superfamily. *Science* 251:561–565.

Lai E, Darnell JE Jr (1991): Transcriptional control in hepatocytes: A window on development. *Trends Biochem Sci* 16:427–430.

Lai E, Prezioso VR, Smith E, Litvin O, Costa RH, Darnell JE Jr (1990): HNF-3A, a hepatocyte-enriched transcription factor of novel structure is regulated transcriptionally. *Genes & Dev* 4:1427–1436.

Lai E, Prezioso VR, Tao WF, Chen WS, Darnell JE Jr (1991): Hepatocyte nuclear factor 3 alpha belongs to a gene family in mammals that is homologous to the *Drosophila* homeotic gene fork head. *Genes & Dev* 5:416–427.

Lamb P, McKnight SL (1991): Diversity and specificity in transcriptional regulation: the benefits of heterotypic dimerization. *Trends Biochem Sci* 16:417–426.

Landschulz WH, Johnson PF, Adashi EY, Grave BJ, McKnight SL (1988): Isolation of a recombinant copy of the gene encoding C/EBP. *Genes & Dev* 2:786–800.

Lazzaro V, De Simone V, De Magistris L, Lehtonen E, Cortese R (1992): LFB-1 and LFB-3 homeoproteins are sequentially expressed during kidney development. *Development* 114:469–479.

Le Douarin NM (1975): An experimental analysis of liver development. *Med Biol* 53:427–455.

Ledley FD, Darlington GJ, Hahn T, Woo SLC (1987): Retroviral gene transfer into primary hepatocytes: Implications for genetic therapy of liver-specific functions. *Proc Natl Acad Sci USA* 84:5335–5339.

Lee G-H, Li H, Ohtake K, Nomura K, Hino O, Furuta Y, Aizawa S, Kitagawa T (1990): Detection of activated c-H-*ras* oncogene in hepatocellular carcinomas developing in transgenic mice harboring albumin promoter-regulated simian virus 40 gene. *Carcinogenesis* 11:1145–1148.

Lee T-H, Finegold MJ, Shen R-F, DeMayo JL, Woo SLC, Butel JS (1990): Hepatitis B virus transactivator X protein is not tumorigenic in transgenic mice. *J Virol* 64:5939–5947.

Leff T, Reue K, Melian A, Culver H, Breslow JL (1989): A regulatory element in the ApoCIII promoter that directs hepatic specific transcription binds to proteins in expressing and nonexpressing cell types. *J Biol Chem* 264:16132–16137.

Leiting B, De Francesco R, Tomei L, Cortese R, Otting G, Wüthrich K (1993): The three-dimensional NMR-solution of the polypeptide fragment 195-268 of the LFB1/HNF1 transcription factor from rat liver comprises a non-classical homeodomain. *EMBO J*, in press.

Lem J, Chin AC, Thayer MJ, Leach RJ, Fournier REK (1988): Coordinate regulation of two genes encoding gluconeogenic enzymes by the trans-dominant locus Tse-1. *Proc Natl Acad Sci USA* 85:7302–7306.

Lenardo MJ, Baltimore D (1989): NF-κB: A pleiotropic mediator of inducible and tissue-specific gene control. *Cell* 58:227–229.

Li X, Liao WS-L (1991): Expression of rat serum amyloid A1 gene involves both C/EBP-like and NF-κB-like transcription factors. *J Biol Chem* 266:15192–15201.

Li Y, Shen RF, Tsai SY, Woo SL (1988): Multiple hepatic *trans*-acting factors are required for *in vitro* transcription of the human alpha-1-antitrypsin gene. *Mol Cell Biol* 8:4362–4369.

Lichtsteiner S, Schibler U (1989): A glycosylated liver-specific transcription factor stimulates transcription of the albumin gene. *Cell* 57:1179–1187.

Lichtsteiner S, Wuarin J, Schibler U (1987): The interplay of DNA-binding proteins on the promoter of the mouse albumin gene. *Cell* 51:963–973.

Lim MY, Davis N, Zhang J, Bose HR Jr (1990): The v-*rel* oncogene product is complexed with cellular proteins including its proto-oncogene product and heat shock protein 70. *Virology* 175:149–160.

Liu JK, Di Persio CM, Zaret KS (1991): Extracellular signals that regulate liver transcription factors during hepatic differentiation *in vitro*. *Mol Cell Biol* 11:773–784.

Lòpez-Cabrera M, Letovsky J, Hu K-Q, Siddiqui A (1990): Multiple liver-specific factors bind to the hepatitis B virus core/pregenomic promoter: Trans-activation and repression by CCAAT/enhancer binding protein. *Proc Natl Acad Sci USA* 87:5069–5073.

Lòpez-Cabrera M, Letovsky J, Hu K-Q, Siddiqui A (1991): Transcriptional factor C/EBP binds to and transactivates the enhancer element II of the hepatitis B virus. *Virology* 183:825–829.

Lowell CA, Potter DA, Stearman RS, Morrow JF (1986): Structure of the murine serum amyloid A gene family. *J Biol Chem* 261:8442–8452.

Lucas PC, O'Brien RM, Mitchell JA, Davis CM, Imai E, Forman BM, Samuels HH, Granner DK (1991): A retinoic acid response element is part of a pleiotropic domain in the phosphoenolpyruvate carboxykinase gene. *Proc Natl Acad Sci USA* 88:2184–2188.

Lucero MA, Sanchez D, Ochoa AR, Brunel F, Cohen GN, Baralle FE, Zakin MM (1989): Interaction of DNA-binding protein with the tissue-specific human apolipoprotein-AII enhancer. *Nucleic Acids Res* 17:2283–2300.

Maguire HE, Hoeffler JP, Siddiqui A (1991): HBV X protein alters the DNA binding specificity of CREB and ATF-2 by protein-promoter interactions. *Science* 252:842–844.

Maire P, Wuarin J, Schibler U (1989): The role of *cis*-acting promoter elements in tissue-specific albumin gene expression. *Science* 244:343–346.

Majello B, Arcone R, Toniatti C, Ciliberto G (1990): Constitutive and IL-6-induced nuclear factors that interact with the human C-reactive protein promoter. *EMBO J* 9:457–465.

Marinkovic' S, Baumann H (1990): Structure, hormonal regulation and identification of the interleukin-6- and dexamethasone-responsive element of the rat haptoglobin gene. *Mol Cell Biol* 10:1573–1583.

Martinez E, Wahli W (1989): Cooperative binding of estrogen receptor to imperfect estrogen-responsive DNA elements correlates with their synergistic hormone-dependent enhancer activity. *EMBO J* 8:3781–3791.

Mendel DB, Hansen LP, Graves MK, Conley PB, Crabtree GR (1991a): HNF-1 alpha and HNF-1 beta (vHNF-1) share dimerization and homeo domains, but not activation domains, and form heterodimers *in vitro*. *Genes & Dev* 5(6):1042–1056.

Mendel DB, Khavari PA, Conley PB, Graves MK, Hansen LP, Admon A, Crabtree GR (1991b): Characterization of a cofactor that regulates dimerization of a mammalian homeodomain protein. *Science* 254:1762–1767.

Mendelzon D, Boissier F, Zakin MM (1990): The binding site for the liver-specific transcription factor Tf-LF1 and the TATA box of the human transferrin gene promoter are the only elements necessary to direct liver-specific transcription *in vitro*. *Nucleic Acids Res* 18:5717–5721.

Mermod N, O'Neill EA, Kelly TJ, Tjian R (1989): The proline-rich transcriptional activator of CTF/NF-1 is distinct from the replication and DNA binding domain. *Cell* 58:741–753.

Metz R, Ziff E (1991): cAMP stimulates the C/EBP-related transcription factor rNFIL-6 to *trans*-locate to the nucleus and induce c-*fos* transcription. *Genes & Dev* 5:1754–1766.

Metzger S, Leff T, Breslow JL (1990): Nuclear factors AF-1 and C/EBP bind to the human ApoB gene promoter and modulate its transcriptional activity in hepatic cells. *J Biol Chem* 265:9978–9983.

Mirkovitch J, Darnell JE Jr (1991): Rapid *in vivo* footprinting technique identifies proteins bound to the TTR gene in the mouse liver. *Genes & Dev* 5:83–93.

Molitor JA, Walker WH, Doerre S, Ballard DW, Greene WC (1990): NF-κB: A family of inducible and differentially expressed enhancer-binding proteins in human T cells. *Proc Natl Acad Sci USA* 87:10028–10032.

Monaci P, Nicosia A, Cortese R (1988): Two different liver-specific factors stimulate *in vitro* transcription from the human α_1-antitrypsin promoter. *EMBO J* 7:2075–2087.

Morrison LE, Kabrun N, Mudri S, Hayman MJ, Enrielto PJ (1989): Viral rel and cellular rel associated with oncogene proteins in transformed and normal cells. *Oncogene Res* 4:677–683.

Morrone G, Cortese R, Sorrentino V (1989): Post-transcriptional control of negative acute phase genes by transforming growth factor beta. *EMBO J* 8:3764–3771.

Morrone G, Ciliberto G, Oliviero S, Arcone R, Dente L, Content J, Cortese R (1988): Recombinant interleukin 6 regulates the transcriptional activation of a set of human acute phase genes. *J Biol Chem* 263:12554–12558.

Möröy T, Marchio A, Etiemble J, Trépo C, Tiollais P, Buendia M-A (1986):

Rearrangement and enhanced expression of c-*myc* in hepatocellular carcinoma of hepatitis virus infected woodchucks. *Nature* (Lond) 324:276–279.

Mueller CR, Maire P, Schibler U (1990): DBP, a liver-enriched transcriptional activator, is expressed late in ontogeny and its tissue specificity is determined post-transcriptionally. *Cell* 61:279–291.

Nakabayashi H, Watanabe K, Saito A, Otsuru A, Sawadaishi K, Tamaoki T (1989): Transcriptional regulation of α-fetoprotein expression by dexamethasone in human hepatoma cells. *J Biol Chem* 261:266–271.

Nakao K, Lawless L, Ohe Y, Miyao Y, Nakabayashi H, Kamiya H, Miura K, Ohtsuka E, Tamaoki T (1990): c-Ha-*ras* down regulates the α-fetoprotein gene but not the albumin gene in human hepatoma cells. *Mol Cell Biol* 10:1461–1469.

Nicosia A, Monaci P, Tomei L, De Francesco R, Nuzzo M, Stunnenberg H, Cortese R (1990): A myosin-like dimerization helix and an extra-large homeodomain are essential elements of the tripartite DNA binding structure of LFB1. *Cell* 61:1225–1236.

Nitsch D, Stewart AF, Boshart M, Mestril R, Weih F, Schuetz G (1990): Chromatin structures of the rat aminotransferase gene relate to the function of its *cis*-acting elements. *Mol Cell Biol* 10:3334–3342.

Nolan GP, Ghosh S, Liou H-C, Tempst P, Baltimore D (1991): DNA binding and IκB inhibition of the cloned p65 subunit of NF-κB, a *rel*-related polypeptide. *Cell* 64:961–969.

Ochoa A, Brunel F, Mendelzon D, Cohen GN, Zakin MM (1989): Different liver nuclear proteins bind to similar DNA sequences in the 5$'$ flanking regions of three hepatic genes. *Nucleic Acids Res* 17:119–133.

Ogami K, Hadzopoulou-Cladaras M, Zannis VI (1990): Promoter elements and factors required for hepatic and intestinal transcription of the human ApoCIII gene. *J Biol Chem* 265:9808–9815.

Ogami K, Kardassis D, Cladaras C, Zannis VI (1991): Purification and characterization of a heat-stable nuclear factor CIIIB1 involved in the regulation of the human ApoC-III gene. *J Biol Chem* 266:9640–9646.

Oliviero S, Cortese R (1989): The human haptoglobin gene promoter: Interleukin-6-responsive elements interact with a DNA-binding protein induced by interleukin-6. *EMBO J* 8:1145–1151.

Oliviero S, Morrone G, Cortese R (1987): The human haptoglobin gene: Transcriptional regulation during development and acute phase induction. *EMBO J* 6:1905–1912.

O'Rourke RW, Miller CW, Kato GJ, Simon KJ, Chen D-L, Dang C-V, Koeffler HP (1990): A potential transcriptional activation element in the p53 protein. *Oncogene* 5:1829–1832.

Pajk Y-K, Chang DJ, Reardon CA, Walker MD, Taxman E, Taylor JM (1988): Identification and characterization of transcriptional regulatory regions associated with expression of the human apolipoprotein E gene. *J Biol Chem* 263:13340–13349.

Paonessa G, Gounari F, Frank R, Cortese R (1988): Purification of a NF1-like DNA-binding protein from rat liver and cloning of the corresponding cDNA. *EMBO J* 7:315–3123.

Papazafiri P, Ogami K, Ramji DP, Nicosia A, Monaci P, Cladaras C, Zannis VI (1991): Promoter elements and factors involved in hepatic transcription of the human ApoA-I gene positive and negative regulators bind to overlapping sites. *J Biol Chem* 266:5790–5797.

Park EA, Roesler WJ, Liu J, Klemm DJ, Gurney AL, Thatcher JD, Schuman J, Friedman A, Hanson RW (1990): The role of the CCAAT/enhancer-binding protein in the transcriptional regulation of the gene for phosphoenolpyruvate carboxykinase (GTP). *Mol Cell Biol* 10:6264–6272.

Parker Ponder K, Gupta S, Leland F, Darlington G, Finegold M, DeMayo J, Ledley FD, Chowdhury JR, Woo SLC (1991): Mouse hepatocytes migrate to liver parenchyma and function indefinitely after intrasplenic transplantation. *Proc Natl Acad Sci USA* 88:1217–1221.

Pastore A, De Francesco R, Barbato G, Castiglione Morelli MA, Motta A, Cortese R (1991): 1H resonance assignment and secondary structure determination of the dimerization domain of transcription factor LFB1. *Biochemistry* 30(1):148–153.

Patel NU, Jameel S, Isom H, Siddiqui A (1989): Interactions between nuclear factors and the hepatitis B virus enhancer. *J Virol* 63:5293–5301.

Paul D, Hohne M, Pinkert C, Piasecki A, Ummelmann E, Brinster RL (1988): Immortalized differentiated hepatocyte lines derived from transgenic mice harboring SV40 T-antigen genes. *Exp Cell Res* 175:354–362.

Pawlak A, Bryans M, Jost J-P (1990): An avian 40-KDa nucleoprotein binds preferentially to a promoter sequence containing one single pair of methylated CpG. *Nucleic Acids Res* 19:1029–1034.

Pei D, Shih C (1991): An "attenuator domain" is sandwiched by two distinct transactivation domains in the transcription factor C/EBP. *Mol Cell Biol* 11:1480–1487.

Pei D, Shih C (1990): Transcriptional activation and repression by cellular DNA-binding protein C/EBP. *J Virol* 64:1517–1522.

Peng H, Armentano D, MacKenzie-Graham L, Shen R-F, Darlington G, Ledley FD, Woo SLC (1988): Retroviral-mediated gene transfer and expression of human phenylalanine hydroxylase in primary mouse hepatocytes. *Proc Natl Acad Sci USA* 85:8146–8150.

Perlmutter DH (1991): The cellular basis for liver injury in a_1-antitrypsin deficiency. *Hepatology* 13(I):172–185.

Perlmutter DH, Dinarello CA, Punsal PJ, Colten HR (1986): Cathetin/tumor necrosis factor regulates hepatic acute-phase gene expression. *J Clin Invest* 78:1349–1354.

Petersen DD, Magnuson MA, Granner DK (1988): Location and characterization of two widely separated glucocorticoid response elements in the phosphoenolpyruvate carboxykinase gene. *Mol Cell Biol* 8:96–104.

Petit C, Levilliers J, Ott MO, Weiss MC (1986): Tissue-specific expression of the rat albumin gene: Genetic control of its extinction in microcell hybrids. *Proc Natl Acad Sci USA* 83:2561–2565.

Philipsen JNJ, Hennis BC, Ab G (1988): *In vivo* footprinting of the estrogen-inducible vitellogenin II gene from chicken. *Nucleic Acids Res* 16:9663–9676.

Pinkert CA, Ornitz DM, Brinster RL, Palmiter RD (1987): An albumin enhancer located 10 kb upstream functions along with its promoter to direct efficient, liver-specific expression in transgenic mice. *Genes & Dev* 1:268–276.

Poli V, Cortese R (1989): Interleukin 6 induces a liver-specific nuclear protein that binds to the promoter of acute-phase genes. *Proc Natl Acad Sci USA* 86:8202–8206.

Poli V, Mancini FP, Cortese R (1990): IL-6DBP, a nuclear protein involved in interleukin-6 signal transduction, defines a new family of leucine zipper proteins related to C/EBP. *Cell* 63:643–653.

Poli V, Silengo L, Altruda F, Cortese R (1989): The analysis of the human hemopexin promoter defines a new class of liver-specific genes. *Nucleic Acids Res* 17:9351–9365.

Popper H, Shafritz DA, Hoofnagle JH (1987): Relation of the hepatitis B virus carrier state to hepatocellular carcinoma. *Hepatology* 7:764–772.

Prowse KR, Baumann H (1988): Hepatocyte-stimulating factor, β_2 interferon and interleukin-1 enhance expression of the rat α_1-acid glycoprotein gene via a distal upstream regulatory region. *Mol Cell Biol* 8:42–51.

Putnam FW (1975): *The Plasma Proteins: Structure, Function and Genetic Control.* New York: Academic Press.

Quinn PG, Granner DK (1990): Cyclic AMP-dependent protein kinase regulates transcription of the phosphoenolpyruvate carboxykinase gene but not binding of nuclear factors to the cyclic AMP regulatory element. *Mol Cell Biol* 10:3347–3364.

Quinn PG, Wong TW, Magnuson MA, Shabb JB, Granner DK (1988): Identification of basal and cyclic AMP regulatory elements in the promoter of the phosphoenolpyruvate carboxykinase gene. *Mol Cell Biol* 8:3467–3475.

Quistorff B (1990): Metabolic heterogeneity of liver parenchymal cells. *Essays Biochem* 25:83–136.

Ramadori G, Sipe JD, Dinarello CA, Mitzel SB, Colten HR (1985): Pretranslational modulation of acute phase hepatic protein synthesis by murine recombinant interleukin 1 (IL-1) and purified human IL-1. I. *Exp Med* 162:930–942.

Ramji DP, Tadros MH, Hardon EM, Cortese R (1991): The transcription factor LF-A1 interacts with a bipartite recognition sequence in the promoter regions of several liver-specific genes. *Nucleic Acids Res* 19:1139–1146.

Ramji DP, Vitelli A, Tronche F, Cortese R, Ciliberto G (1993): The two C/EBP isoforms, IL-6 DBP/NF-IL6 and C/EBPδ/NF-IL6β are induced by IL-6 to promote acute phase gene transcription via different mechanisms. *Nucl Acids Res* 21:289–294.

Ramji DP, Tronche F, Gallinari P, Vitelli A, Ciliberto G, Cortese R (1992): Induction of gene transcription by interleukin-6 (IL-6) requires post-translational modification of IL-6 DBP. In: *IL-6: Physiopathology and Clinical Potentials,* Revel M, ed., pp. 55–62. Raven Press.

Raney AK, Milich DR, McLachlan A (1989): Characterization of hepatitis B virus major surface antigen gene transcriptional regulatory elements in differentiated hepatoma cell lines. *J Virol* 63:3919–3925.

Raney AK, Easton AJ, Milich DR, McLachlan A (1991): Promoter-specific transactivation of hepatitis B virus transcription by a glutamine- and proline-rich domain of hepatocyte nuclear factor 1. *J Virol* 65:5774–5781.

Raney AK, Milich DR, Easton AJ, McLachlan A (1990): Differentiation-specific transcriptional regulation of the hepatitis B virus large surface antigen gene in human hepatoma cell lines. *J Virol* 64:2360–2368.

Rangan VS, Das GC (1990): Purification and biochemical characterization of hepatocyte nuclear factor 2 involved in liver-specific transcription of the human alpha 1-antitrypsin gene. *J Biol Chem* 265:8874–8879.

Raycroft L, Wu H, Lozano G (1990): Transcriptional activation by wild-type but not transforming mutants of the p53 anti-oncogene. *Science* 249:1049–1051.

Raycroft L, Schmidt JR, Yoas K, Hao M, Lozano G (1991): Analysis of p53 mutants for transcriptional activity. *Mol Cell Biol* 11:6067–6074.

Raymondjean M, Cereghini S, Yaniv M (1988): Several distinct "CCAAT" box binding proteins coexist in eukaryotic cells. *Proc Natl Acad Sci USA* 85:757–761.

Raymondjean M, Pichard A-L, Gregori C, Ginot F, Kahn A (1991): Interplay of an original combination of factors: C/EBP, NFY, HNF3, and HNF1 in the rat aldolase B gene promoter. *Nucleic Acids Res* 19:6145–6153.

Reik A, Schuetz G, Stewart AF (1991): Glucocorticoids are required for establishment and maintenance of an alteration in chromatin structure: Induction leads to a reversible disruption of nucleosomes over an enhancer. *EMBO J* 10:2569–2576.

Reue K, Leff T, Breslow JL (1988): Human apolipoprotein CIII gene expression is regulated by positive and negative *cis*-acting elements and tissue-specific protein factors. *J Biol Chem* 263:6857–6864.

Rey-Campos J, Chouard T, Yaniv M, Cereghini S (1991): vHNF1 is a homeoprotein that activates transcription and forms heterodimers with HNF-1. *EMBO J* 10(6):1445–1457.

Rigaud G, Roux J, Pictet R, Grange T (1991): *In vivo* footprinting of rat TAT gene: Dynamic interplay between the glucocorticoid receptor and a liver-specific factor. *Cell* 67:977–986.

Ro HS, Roncari DA (1991): The C/EBP-binding region and adjacent sites regulate expression of the adipose P2 gene in human preadipocytes. *Mol Cell Biol* 11:2303–2306.

Roesler WJ, Vandenbark GR, Hanson RW (1989): Identification of multiple protein binding domains in the promoter-regulatory region of the phosphoenolpyruvate carboxykinase (GTP) gene. *J Biol Chem* 264:9657–9664.

Roman C, Platero JS, Shuman J, Calame K (1990): Ig/EBP-1: A ubiquitously expressed immunoglobulin enhancer binding protein that is similar to C/EBP and heterodimerizes with C/EBP. *Genes & Dev* 4:1404–1415.

Ron D, Habener JF (1992): CHOP, a novel developmentally regulated nuclear protein that dimerizes with transcription factors C/EBP and LAP and functions as a dominant-negative inhibitor of gene transcription. *Genes & Dev* 6:439–453.

Ron D, Brasier AR, Wright KA, Habener JF (1990): The permissive role of glucocorticoids on interleukin-1 stimulation of angiotensinogen gene transcription is mediated by an interaction between inducible enhancers. *Mol Cell Biol* 10:4389–4395.

Ross RS, Li AC, Hoeg JM, Schumacher UK, Demosky SJ Jr, Brewer HB Jr (1991): Apolipoprotein B upstream suppressor site: Identification of an element which can decrease apolipoprotein B transcription. *Biochem Biophys Res Commun* 176:1116–1122.

Roth S, Stein D, Nüsslein-Volhard C (1989): A gradient of nuclear localization of the *dorsal* protein determines dorsoventral pattern in the *Drosophila* embryo. *Cell* 59:1189–1202.

Rottman JN, Widom RL, Nadal-Ginard B, Mahdavi V, Karathanasis SK (1991): A retinoic acid-responsive element in the apolipoprotein AI gene distinguishes between two different retinoic acid response pathways. *Mol Cell Biol* 11:3814–3820.

Rubin EM, Ishida BY, Clift SM, Krauss RM (1991): Expression of human apolipoprotein A-I in transgenic mice results in reduced plasma levels of murine apolipoprotein A-I and the appearance of two new high density lipoprotein size subclasses. *Proc Natl Acad Sci USA* 88:434–438.

Ruiz-Bravo N, Ernest MJ (1982): Induction of tyrosine aminotransferase mRNA by glucocorticoids and cAMP in fetal rat liver. *Proc Natl Acad Sci USA* 79:365–368.

Ruppert S, Kelsey G, Scedl A, Schmid E, Thies E, Schutz G (1992): Deficiency of an enzyme of tyrosine metabolism underlies altered gene expression in newborn liver of lethal albino mice. *Genes & Dev* 6:1430–1443.

Ruppert S, Boshart M, Bosch FX, Schmid W, Fournier REK, Schuetz G (1990): Two genetically defined *trans*-acting loci coordinately regulate overlapping sets of liver-specific genes. *Cell* 61:895–904.

Rushlow CA, Han K, Manley JL, Levine M (1989): The graded distribution of the *dorsal* morphogen is initiated by selective nuclear transport in *Drosophila*. *Cell* 59:1165–1177.

Rüther U, Woodroofe C, Fattori E, Ciliberto G (1993): Inducible formation of liver tumors in transgenic mice. *Oncogene* 8:87–93.

Rüther U, Tripodi M, Cortese R, Wagner E (1987): The expression of the human alpha-1-antitrypsin gene in transgenic mice. *Nucleic Acids Res* 15:7519–7529.

Samuelsson L, Strömberg K, Vikman K, Bjursell G, Enerbäck S (1991): The CCAAT/enhancer binding protein and its role in adipocyte differentiation:

evidence for direct involvement in terminal adipocyte development. *EMBO J* 10:3787–3793.

Sandgren EP, Quaife CJ, Pinkert CA, Palmiter RD, Brinster RL (1989): Oncogene-induced liver neoplasia in transgenic mice. *Oncogene* 4:715–724.

Sawadaishi K, Morinaga T, Tamaoki T (1988): Interaction of a hepatoma-specific nuclear factor with transcription-regulatory sequences of the human α-fetoprotein and albumin genes. *Mol Cell Biol* 8:5179–5187.

Schmid E, Schmid W, Mayer D, Jastorff B, Schuetz G (1987): Transcriptional activation of the tyrosine aminotransferase gene by glucocorticoids and cAMP in primary hepatocytes. *Eur J Biochem* 165:499–506.

Schneider JA, Weiss MC (1971): Expression of differentiated functions in hepatoma cell hybrids. I. Tyrosine aminotransferase in hepatoma-fibroblast hybrids. *Proc Natl Acad Sci USA* 68:127–131.

Seal SN, Davis DL, Burch JBE (1991): Mutational studies reveal a complex set of positive and negative control elements within the chicken vitellogenin II promoter. *Mol Cell Biol* 11:2704–2717.

Sen R, Baltimore D (1986): Multiple nuclear factors interact with the immunoglobulin enhancer sequences. *Cell* 46:705–716.

Sepulveda AR, Finegold MJ, Smith B, Slagle BL, DeMayo JL, Shen R-F, Woo SLC, Butel JS (1989): Development of a transgenic mouse system for the analysis of stages in liver carcinogenesis using tissue-specific expression of SV40 large T-antigen controlled by regulatory elements of the human α_1-antitrypsin gene. *Cancer Res* 49:6108–6117.

Seto E, Mitchell PJ, Yen TSB (1990): Transactivation by the hepatitis B virus X protein depends on AP-2 and other transcription factors. *Nature* (Lond) 344:72–74.

Shaul Y, Rutter WJ, Laub O (1985): A human hepatitis B viral enhancer element. *EMBO J* 4:427–430.

Shen R-F, Li Y, Sifers RN, Wang HH, Jardick C, Tsai SY, Woo SLC (1987): Tissue-specific expression of the human α_1-antitrypsin gene is controlled by multiple *cis*-regulatory elements. *Nucleic Acids Res* 15:8399–8415.

Shimada T, Noda T, Tashiro M, Murakami T, Takiguchi M, Mori M, Yamamura K-I, Saheki T (1991): Correction of ornithine transcarbamylase (OTC) deficiency in spf-ash mice by introduction of rat OTC gene. *FEBS Lett* 279:198–200.

Shirakata Y, Kawada M, Fujkik Y, Sano H, Oda M, Yaginuma K, Kobayashi M, Koike K (1989): The X gene of hepatitis B virus induced growth stimulation and tumorigenic transformation of mouse NIH3T3 cells. *Jpn J Cancer Res* 80:617–621.

Shoshani T, Benvenisty N, Trus M, Reshef L (1991): *cis*-Regulatory elements that confer differential expression upon the rat gene encoding phosphoenolpyruvate carboxykinase in kidney and liver. *Gene* 101:279–283.

Siddiqui A, Jameel S, Mapoles J (1987): Expression of the hepatitis B virus X gene in mammalian cells. *Proc Natl Acad Sci USA* 84:2513–2517.

Siddiqui A, Gaynor R, Srinivasan A, Mapoles J, Farr RW (1989): *trans*-Activation of viral enhancers including long terminal repeat of the human immunodeficiency virus by the hepatitis B virus X protein. *Virology* 169:479–484.

Sifers RN, Finegold MJ, Woo SLC (1989): Alpha-1-antitrypsin deficiency: Accumulation or degradation of mutant variants within the hepatic endoplasmic reticulum. *Am J Respir Cell Mol Biol* 1:341–345.

Simek S, Rice NR (1988): p59 v-*rel*, The transforming protein of reticuloendotheliosis virus, is complexed with at least four other proteins in transformed chicken lymphoid cells. *J Virol* 62:4730–4736.

Sladek FM, Zhong WM, Lai E, Darnell JE Jr (1990): Liver-enriched transcription factor HNF-4 is a novel member of the steroid hormone receptor superfamily. *Genes & Dev* 4:2353–2365.

Slagle BL, Zhou Y-Z, Butel JS (1991): Hepatitis B virus integration event in human chromosome 17p near the p53 gene identifies the region of the chromosome commonly deleted in virus-positive hepatocellular carcinomas. *Cancer Res* 51:49–54.

Spandau DF, Lee C-H (1988): *trans*-Activation of viral enhancers by the hepatitis B virus X protein. *J Virol* 62(2):427–434.

Thayer MJ, Fournier REK (1989): Hormonal regulation of TSE-1 repressed genes: Evidence for multiple genetic controls in extinction. *Mol Cell Biol* 9:2837–2846.

Theiler (1989): *The House Mouse*. New York: Springer-Verlag.

Theill LE, Castrillo JL, Wu D, Karin M (1989): Dissection of functional domains of the pituitary-specific transcription factor GHF-1. *Nature* (Lond) 342:945–948.

Tian J-M, Schibler U (1991): Tissue-specific expression of the gene encoding hepatocyte nuclear factor 1 may involve hepatocyte nuclear factor 4. *Genes & Dev* 5:2225–2234.

Tiollais P, Pourcel C, Dejean A (1985): The hepatitis B virus. *Nature* (Lond) 317:489–495.

Tognoni A, Cattaneo R, Serfling E, Schaffner W (1985): A novel expression selection approach allows precise mapping of the hepatitis B virus enhancer. *Nucleic Acids Res* 13:7457–7464.

Tomei L, Piaggio G, Toniatti C, De Francesco R, Gerstner J, Cortese R (1993): LFB1/HNF1 Acts as repressor of its own transcription. Submitted.

Tomei L, Cortese R, De Francesco R (1992): A POU-A related region dictates DNA binding specificity of LFB1/HNF1 by orienting the two XL-homeodomains in the dimer. *EMBO J* 11:4119–4129.

Toniatti C, Monaci P, Nicosia A, Cortese R, Ciliberto G (1993): A bipartite activation domian is responsible for the activity of transcription factor HNF1/LFB1 in cells of hepatic and non-hepatic origin. *DNA and Cell Biol*, in press.

Toniatti C, Monaci P, Nicosia A, Cortese R, Ciliberto G: A bipartite transactivation domain in transcriptional activator LFB1/HNF-1α functions in cells of non-hepatic origin (in manuscript).

Toniatti C, Arcone R, Majello B, Ganter U, Arpaia G, Ciliberto G (1990a): Regulation of the human C-reactive protein gene, a major marker of inflammation and cancer. *Mol Biol Med* 7:199–212.

Toniatti C, Demartis A, Monaci P, Nicosia A, Ciliberto G (1990b): Synergistic *trans*-activation of the human C-reactive protein promoter by transcription factor HNF-1 binding at two distinct sites. *EMBO J* 9:4467–4475.

Tripodi M, Abbott C, Vivian N, Cortese R, Lovell-Badge R (1991): Disruption of the LF-A1 and LF-B1 binding sites in the human alpha-1-antitrypsin gene has a differential effect during development in transgenic mice. *EMBO J* 10:3177–3182.

Tronche F, Rollier A, Bach I, Weiss MC, Yaniv M (1989): The rat albumin promoter: Cooperation with upstream elements is required when binding of APF/HNF1 to the proximal element is partially impaired by mutation or bacterial methylation. *Mol Cell Biol* 9:4759–4766.

Tronche F, Rollier A, Herbomel P, Bach I, Cereghini S, Weiss M, Yaniv M (1990): Anatomy of the rat albumin promoter. *Mol Biol Med* 7:173–185.

Trujillo MA, Letovsky J, Maguire HF, Lòpez-Cabrera M, Siddiqui A (1991): Functional analysis of a liver-specific enhancer of the hepatitis B virus. *Proc Natl Acad Sci USA* 88:3797–3801.

Trus M, Benvenisty N, Cohen H, Reshef L (1990): Developmentally regulated interactions of liver nuclear factors with the rat phosphoenolpyruvate carboxykinase promoter. *Mol Cell Biol* 10:2418–2422.

Tsutsumi K-I, Ito K, Ishikawa K (1989): Developmental appearance of transcription factors that regulate liver-specific expression of the aldolase B gene. *Mol Cell Biol* 9:4923–4931.

Tung HYL, Bargmann WJ, Lim M-Y, Bose HR Jr (1988): The v-*rel* oncogene product is complexed to a 40-kDa phosphoprotein in transformed lymphoid cells. *Proc Natl Acad Sci USA* 85:2479–2483.

Tur-Kaspa R, Burk RD, Shaul Y, Shafritz DA (1986): Hepatitis B virus DNA contains a glucocorticoid-responsive element. *Proc Natl Acad Sci USA* 83:1627–1631.

Tur-Kaspa R, Shaul Y, Moore DD, Burk RD, Okret S, Poellinger R, Shafritz DA (1988): The glucocorticoid receptor recognizes a specific nucleotide sequence in hepatitis B virus DNA causing increased activity of the HBV enhancer. *Virology* 167:630–633.

Twu JS, Robinson WS (1989): Hepatitis B virus X gene can transactivate heterologous viral sequences. *Proc Natl Acad Sci USA* 86:2046–2050.

Twu JS, Schloemer RH (1987): Transcriptional trans-activating function of hepatitis B virus. *J Virol* 61(11):3448–3453.

Twu JS, Chu K, Robinson WS (1989): Hepatitis B virus X gene activates κB-like enhancer sequences in the long terminal repeat of human immunodeficiency virus 1. *Proc Natl Acad Sci USA* 86:5168–5172.

Umek RM, Friedman AD, McKnight SL (1991): CCAAT-enhancer binding protein: A component of a differentiation switch. *Science* 251:288–293.

Vaccaro M, Pawlak A, Jost J-P (1990): Positive and negative regulatory elements of chicken vitellogenin II gene characterized by *in vitro* transcription competition assays in a homologous system. *Proc Natl Acad Sci USA* 87:3047–3051.

Van Snick J (1990): Interleukin-6: An overview. *Annu Rev Immunol* 8:253–278.

Vaulont S, Puzenat N, Kahn A, Raymondjean M (1989a): Analysis by cell-free transcription of the liver-specific pyruvate kinase gene promoter. *Mol Cell Biol* 9:4409–4415.

Vaulont S, Puzenat N, Leyrat F, Cognet M, Kahn A, Raymondjean M (1989b): Proteins binding to the liver-specific pyruvate kinase gene promoter. A unique combination of known factors. *J Mol Biol* 209:205–219.

Verrijzer CP, Kal A-J, van der Vliet PC (1990): The oct-1 homeo domain contacts only part of the octamer sequence and full oct-1 DNA-binding activity requires the POU-specific domain. *Genes & Dev* 4:1964–1974.

Walsh A, Ito Y, Breslow JL (1989): High levels of human A-I in transgenic mice result in increased plasma levels of small high density lipoprotein (HDL) particles comparable to human HDL3. *J Biol Chem* 264:6488–6494.

Wang Y, Chen P, Wu X, Sun A-L, Wang H, Zhu Y-A, Li Z-P (1990): A new enhancer element, ENII, identified in the X gene of hepatitis B virus. *J Virol* 64:3977–3981.

Weigel D, Jäckle H (1990): Forkhead: A new eukaryotic DNA binding motif? *Cell* 63:455–456.

Weih F, Stewart AF, Boshart M, Nitsch D, Schuetz G (1990): *In vivo* monitoring of a cAMP-stimulated DNA-binding activity. *Genes & Dev* 4:1437–1449.

Widom RL, Ladias JAA, Kouidou S, Karathanasis SK (1991): Synergistic interactions between transcription factors control expression of the apolipoprotein AI gene in liver cells. *Mol Cell Biol* 11:677–687.

Wijnholds J, Muller E, Ab G (1991): Oestrogen facilitates the binding of ubiquitous and liver-enriched nuclear proteins to the apoVLDL II promoter *in vivo*. *Nucleic Acids Res* 19:33–41.

Wijnholds, J, Philipsen JNF, Ab G (1988): Tissue-specific and steroid-dependent interaction of transcription factors with the oestrogen-inducible apolVLDL II promoter *in vivo*. *EMBO J* 7:2757–2763.

Williams PM, Ratajczak T, Lee SC, Ringold GM (1991): AGP/EBP (LAP) expressed in rat hepatoma cells interacts with multiple promoter sites and is necessary for maximal glucocorticoid induction of the rat alpha-1 acid glycoprotein gene. *Mol Cell Biol* 11:4959–4965.

Williams SC, Cantwell CA, Johnson PF (1991): A family of C/EBP-related proteins capable of forming covalently linked leucine zipper dimers *in vitro*. *Genes & Dev* 5:1553–1567.

Wilson DR, Juan TSC, Wilde MD, Fey GH, Darlington GJ (1990): A 58-base-pair region of the human C3 gene confers synergistic inducibility by interleukin-1 and interleukin-6. *Mol Cell Biol* 10:6181–6191.

Wilson JM, Johnston DE, Jefferson DM, Mulligan RC (1988): Correction of the genetic defect in hepatocytes from the Watanabe heritable hyperlipidemic rabbit. *Proc Natl Acad Sci USA* 85:4421–4425.

Wilson JM, Chowdhury NR, Grossman M, Wajsman R, Epstein A, Mulligan RC, Chowdhury JR (1990): Temporary amelioration of hyperlipidemia in low density lipoprotein receptor-deficient rabbits transplanted with genetically modified hepatocytes. *Proc Natl Acad Sci USA* 87:8437–8441.

Wolff JA, Yee J-K, Skelly HF, Moores JC, Respess JG, Friedmann T, Leffert H (1987): Expression of retrovirally transduced genes in primary cultures of adult rat hepatocytes. *Proc Natl Acad Sci USA* 84:3344–3348.

Wollersheim M, Debelka U, Hoschneider PH (1988): A transactivating function encoded in the hepatitis B virus X gene is conserved in the integrated state. *Oncogene* 3(5):545–552.

Won K-A, Baumann H (1991): NF-AB, a liver-specific and cytokine-inducible nuclear factor that interacts with the interleukin-1 response element of the rat α_1-acid glycoprotein gene. *Mol Cell Biol* 11:3001–3008.

Woo P, Sipe J, Dinarello CA, Colten HR (1987): Structure of a human serum amyloid A gene and modulation of its expression in transfected L cells. *J Biol Chem* 262:15790–15795.

Wu GY, Wilson JM, Shalaby F, Grossman M, Shafritz DA, Wu CH (1991): Receptor-mediated gene delivery in vivo. Partial correction of genetic analbuminemia in Nagase rats. *J Biol Chem* 266(22):14338–14342.

Wu CH, Wilson JM, Wu GY (1989): Targeting genes: Delivery and persistent expression of a foreign gene driven by mammalian regulatory elements *in vivo*. *J Biol Chem* 264:16985–16987.

Wu GY, Wu CH (1988): Receptor-mediated gene delivery and expression *in vivo*. *J Biol Chem* 263:14621–14624.

Wu J, Zhou ZY, Judd A, Cartwright CA, Robinson WS (1990): The hepatitis B virus-encoded transcriptional trans-activator hbx appears to be a novel protein serine/thereonine kinase. *Cell* 63(4):687–695.

Wuarin J, Schibler U (1990): Expression of the liver-enriched transcriptional activator protein DBP follows a stringent circadian rhythm. *Cell* 63:1257–1266.

Wuarin J, Mueller C, Schibler U (1990): A ubiquitous CCAAT factor is required for efficient *in vitro* transcription from the mouse albumin promoter. *J Mol Biol* 214:865–874.

Xanthopoulos KG, Mirkovitch J, Decker T, Kuo CF, Darnell JE Jr (1989): Cell-specific transcriptional control of the mouse DNA-binding protein mC/EBP. *Proc Natl Acad Sci USA* 86:4117–4121.

Xanthopoulos KG, Prezioso VR, Chen WS, Sladek FM, Cortese R, Darnell JE Jr (1991): The different tissue transcription patterns of genes for HNF-1, C/EBP, HNF-3, and HNF-4, protein factors that govern liver-specific transcription. *Proc Natl Acad Sci USA* 88(9):3807–3811.

Yamada K, Noguchi T, Matsuda T, Monaci P, Nicosia A, Tanaka T (1990): Identification and characterization of hepatocyte-specific regulatory regions of the rat pyruvate kinase L gene. *J Biol Chem* 265:19885–19891.

Yang N-S, Burkholder J, Roberts B, Martinell B, McCabe D (1990): *In vivo* and *in vitro* gene transfer to mammalian somatic cells by particle bombardment. *Proc Natl Acad Sci USA* 87:9568–9572.

Yee J-K (1989): A liver-specific enhancer in the core promoter region of human hepatitis B virus. *Science* 246:658–661.

Yokode M, Hammer RE, Ishibashi S, Brown MS, Goldstein JL (1990): Diet-induced hypercholesterolemia in mice: Prevention by overexpression of LDL receptors. *Science* 250:1273–1275.

Yuh C-H, Ting L-P (1990): The genome of hepatitis B virus contains a second enhancer: Cooperation of two elements within this enhancer is required for its function. *J Virol* 64:4281–4287.

Yuh C-H, Ting L-P (1991): C/EBP-like proteins binding to the functional box-α and box-β of the second enhancer of hepatitis B virus. *Mol Cell Biol* 11:5044–5052.

Zabel U, Baeuerle PA (1990): Purified human IκB can rapidly dissociate the complex of the NF-κB transcription factor with its cognate DNA. *Cell* 61:255–265.

Zhang D-E, Hoyt PR, Papaconstantinou J (1990): Localization of DNA protein-binding sites in the proximal and distal promoter regions of the mouse α-fetoprotein gene. *J Biol Chem* 265:3382–3391.

Zhang X-K, Dong J-M, Chiu J-F (1991): Regulation of α-fetoprotein gene expression by antagonism between AP-1 and the glucocorticoid receptor at their overlapping binding site. *J Biol Chem* 266:8248–8254.

Zhong W, Sladek FM, Darnell J Jr. (1993): The expression pattern of a Drosophila homolog to the mouse transcription factor HNF-4 suggests a determinative role in the gut formation. *EMBO J* 12:537–544.

Zhou D-X, Yen TSB (1990): Differential regulation of the hepatitis B virus surface gene promoters by a second viral enhancer. *J Biol Chem* 265:20731–20734.

Zhou D-X, Yen TSB (1991): The ubiquitous transcription factor Oct-1 and the liver-specific factor HNF-1 are both required to activate transcription of a hepatitis B virus promoter. *Mol Cell Biol* 11:1353–1359.

Zhou D-X, Taraboulos A, Ou JH, Yen TSB (1990): Activation of class I histocompatibility complex gene expression by hepatitis B virus. *J Virol* 64(8):4025–4028.

zur Hausen H (1991): Viruses in human cancers. *Science* 254:1167–1173.

Chapter 8

Transcriptional Control of Pituitary Gene Expression

Lars Eyde Theill

Cascades of interacting regulatory genes controlling developmental pathways have been defined in organisms such as *Drosophila* (Chalfie and Au, 1989; Cohen and Jurgens, 1991; Gehring, 1987; Ingham, 1988; Ingham and Martinez Arias, 1992; Moses, 1991; Nusslein-Volhard and Wieschaus 1980; Nusslein-Volhard et al., 1987; Perkins and Perrimon, 1991; St. Johnston and Nusslein-Volhard, 1992) and the nematode *Caenorhabditis elegans* (Davidson, 1990; Sternberg and Horvitz, 1991). The developmental regulators include transcription factors, kinases, phosphatases, growth factors, receptors, and cell-adhesion molecules. In *Drosophila*, products of maternally expressed genes control expression of zygotic segmentation genes, the products of which regulate homeotic gene expression; regulation is also exerted between genes of the same class. As a result the embryo is divided into a meshwork of metameric units, each expressing a unique combination of homeotic genes. These genes contain a conserved homeobox, encoding a 60-amino-acid homeodomain that functions in DNA binding. By acting as transcription factors, the homeodomain proteins orchestrate activation of a unique combination of target genes that causes the cells to enter a specific morphogenetic pathway.

The molecular mechanisms underlying regulation of target gene expression are poorly understood but include competition among homeoproteins for binding to recognition elements, posttranslational modifications affecting DNA binding, and transactivation properties and cooperation with specific cofactors. Further elucidation of regulatory mechanisms has been hampered because only a few target genes in

GENE EXPRESSION: GENERAL AND CELL-TYPE-SPECIFIC
Michael Karin, Editor
© 1993 Birkhäuser Boston

Drosophila regulated by homeogenes have so far been identified (Gould et al., 1990; Graba et al., 1992). Defining developmental pathways in mammalian systems has proven even more difficult. Cloning by similarity has been used to identify genes that are potentially mammalian counterparts of well-characterized *Drosophila* transcriptional regulators (Deschamps and Meijlink, 1992; Fienberg et al., 1987; Goulding and Gruss, 1989; Kessel and Gruss, 1990; McGinnis and Krumlauf, 1992). Mammalian HOX genes are homologous to the antennapedia class of homeotic genes in *Drosophila* and appear to be involved in determination of regional identity along the anterior–posterior axis (Melton, 1991; Reid, 1990). The drawback of cloning by the similarity method is the difficulty of identifying physiological target genes. In a different approach, cell-type-specific regulatory elements identified in well-characterized genes have been the key entry point for purification and cloning (Dynlacht et al., 1989; Paonessa et al., 1988) or direct cloning (Singh et al., 1988) of transcriptional regulators.

Elucidation of prominent transcriptional control and intercellular signaling mechanisms underlying cell commitment, organ development, differentiation, and the continued fine tuning of gene expression is still at an early stage. Much of our understanding of these processes in mammals originates from in-depth studies of a relatively small number of selected model systems as for instance: specification of skeletal muscle cell lineage (Weintraub et al., 1991; Blau, 1992); globin gene expression (Evans et al., 1990; Felsenfeld, 1992); liver-specific gene expression (De-Simone and Cortese, 1992) (see also preceding chapters of this volume) and systems which include viral genomes (Martin, 1991; Nevins, 1991).

The anterior pituitary receives signals from the brain in the form of neurotransmitters and polypeptide hormones directed to it from the hypothalamus and elsewhere. In turn, well-defined cell types in the anterior pituitary respond differentially to these messengers, releasing polypeptide hormones that are major regulators of metabolism, growth, and reproduction. The specification of five distinct anterior pituitary cell types from a seemingly common progenitor is a useful paradigm for the study of regulatory mechanisms controlling cell commitment and differentiation in vertebrates (Karin et al., 1990; Rosenfeld, 1991).

Growth hormone (GH), the polypeptide hormone released by one of the anterior pituitary cell types, the somatotrophs, is a major regulator of linear growth and participates in the regulation of protein and fat metabolism (Davidson, 1987; Isaksson et al., 1987; Wheeler and Styne,

1988). Recombinant human GH is used to treat children with GH deficiency. Reduced availability of GH in late adulthood may contribute to the reduction in muscle mass and increase in adipose tissue that occur with aging (Christiansen and Jorgensen, 1991; Rudman et al., 1990). The potential benefits of maintaining the availability of GH in the elderly, as well as the possibility that GH administration could benefit cancer or autoimmunodeficiency syndrome (AIDS) patients suffering from waste syndrome, underscore the importance of understanding the regulation of GH expression.

To illustrate the regulation of gene expression in the pituitary, this chapter describes the molecular mechanisms underlying somatotroph cell specification and transcriptional control of GH gene expression. The identification of transcriptional regulators binding to the GH promoter and the characterization of their individual and synergistic modes of action are also discussed. Further, this chapter reviews how the activities of these factors are regulated by environmental and cell autonomous cues during development as well as in the adult pituitary. Special emphasis is placed on the pituitary specific POU-domain protein GHF-1, also called Pit1, which functions in both specification and proliferation of three of the five anterior pituitary cell types.

Development of the Anterior Pituitary

During anterior pituitary ontogenesis, five distinct cell types that specialize in production of polypeptide hormones are believed to arise from one common lineage (Chatelain et al., 1979; Gash et al., 1982; Hoeffler et al., 1985; Lugo et al., 1989; Oliver et al., 1980; Watanabe and Daikoku, 1979). These cell types, in order of their appearance, are corticotrophs, which produce proopiomelanocortin (POMC); thyrotrophs, which produce thyroid-stimulating hormone (TSH); gonadotrophs, which produce luteinizing hormone (LH) and follicle-stimulating hormone (FSH); somatotrophs, which produce growth hormone (GH); and lactotrophs, which produce prolactin (Prl). POMC, GH, and Prl are composed of a single polypeptide, but the glycoprotein hormones TSH, LH, and FSH are heterodimers, composed of a common α subunit and distinct β subunits. Synthesis of the α-subunit of the glycoprotein hormones (αGSu) appears to be restricted to the thyrotrophs and the gonadotrophs (Simmons et al., 1990).

The posterior lobe of the pituitary develops from the neuroectoderm. In contrast, all the cell types of the anterior pituitary and the intermediate lobes of the pituitary originate from a placode of somatic ectoderm at the roof of the embryonic mouth (Schwind, 1928). This region of the oral ectoderm is in contact with the neuroectodermal plate, which forms the floor of the developing brain. This first step in primordial pituitary cell commitment in the rat is marked by the appearance of mRNA for αGSu on embryonic day 11 (ED11) (Simmons et al., 1990). By ED12 in the rat, Rathke's pouch, the anlagen for the anterior and the intermediate pituitary, emerges as an outpocketing of the oral ectoderm (Schwind, 1928). A signal from the developing diencephalic floor may be required for normal development of Rathke's pouch (Watanabe, 1982). By ED14 (ED12 in the mouse), Rathke's pouch becomes an independent structure (Schwind, 1928), and two more committed precursor cell types called acidophils and basophils can be distinguished. At this time, concomitant with increased cellular proliferation, βTSH and POMC mRNAs become detectable in cells in the ventral portion of Rathke's pouch (Lugo et al., 1989; Simmons et al., 1990). Further cytodifferentiation of basophils in this region of the anterior pituitary anlagen generates gonadotrophs expressing βLH (ED16) and βFSH (ED17) mRNA (Simmons et al., 1990). The acidophil progenitors differentiate into somatotrophs secreting GH, mammosomatotrophs secreting GH and Prl, and lactotrophs secreting Prl. Somatotrophs expressing GH mRNA appear on ED18 in the rat and ED15 in the mouse, and a few lactotroph cells (also called mammotrophs) expressing Prl mRNA appear on ED18 or ED19 (ED15 in the mouse). However, lactotrophs remain very rare in murine and rat fetal pituitaries until 3–4 days after birth, at which time an explosive increase occurs in the number of Prl-secreting cells (Chatelain et al., 1979; Dollé et al., 1990; Hoeffler et al., 1985; Leong et al., 1985; Nogami et al., 1989; Setalo and Nakane, 1976; Simmons et al., 1990; Slabaugh et al., 1982); Watanabe and Daikoku, 1979).

In transgenic mice, the GH promoter-driven expression of toxin genes gives rise to dwarf mice that are nearly devoid of GH-producing cells (Behringer et al., 1988; Borrelli et al., 1989). Interestingly, the ablation of somatotrophs also leads to either a strong reduction (Behringer et al., 1988) or the virtual absence (Borrelli et al., 1989) of Prl-secreting cells, indicating that during pituitary development somatotrophs give rise to the large majority of Prl secretors (lactotrophs). This notion supports the observation that almost all the initial Prl-secreting cells in the rat and in humans also release GH (Hoeffler et al., 1985). These dual hormone producers, called mammosomatotrophs, have been detected in different species and may account for 5%—15% of total cells in the anterior pituitary (Frawley, 1989; Frawley and Boockfor, 1991; Frawley et al., 1985a); however, a minor fraction of lactotrophs may arise directly from the acidophilic precursor. Fluctuations in the proportions of somatotrophs, lactotrophs, and mammosomatotrophs occuring independent of cell proliferation, has been correlated with changes in the *in vivo* and *ex vivo* steroid environment, suggesting that GH and Prl cells are functionally interconvertible in the adult (Frawley and Boockfor, 1991; Kineman et al., 1992). The molecular mechanisms underlying the cytodifferentiation and possibly the transdifferentiation of acidophilic cell types are discussed next.

The GH Gene

GH and *Prl* genes are present in all vertebrates. They originate from a common progenitor that underwent duplication approximately 350 million years ago (Barsh et al., 1983; Miller and Eberhardt, 1983); today, the genes are located on different chromosomes (Owerbach et al., 1980). Subsequent duplication of the *GH* gene in primates or the *Prl* gene in rodents gave rise to the placental lactogen (*PL*) genes (Chen et al., 1989; Talamantes, 1990; Walker et al., 1991). Of the five genes in the *hGH/PL* gene cluster, only the *hGH-N* gene is expressed in the pituitary; the remaining four genes, *PL-1* (*CS-L* for chorionic somatomammotropin, which is synonymous with PL), *PL-4* (*CS-A*), *GH-V*, and *PL-3* (*CS-B*), are expressed in the placenta during pregnancy. PL functions to stimulate fetal growth and alter maternal metabolism (Walker et al., 1991). The strict differential expression of these genes is striking when one considers that the first 464-bp 5′-flanking sequences of the *hGH* and *hPL* genes share 93.8% sequence identity (Selby et al., 1984). The 5′-regulatory regions of *GH* and *Prl* genes also exhibit some homology. There is indeed an overlap in populations of transcriptional regulators binding to *GH, PL,* and *Prl* gene promoters (see following). In addition to promoter sequences, a 3′-flanking enhancer is required for placenta-specific expression of *PL* genes (Walker et al., 1991), while lactotroph-specific expression of Prl depends on a 5′-flanking upstream enhancer (Crenshaw et al., 1989). For the *GH gene*, however, a short promoter region suffices for somatotroph-specific gene expression: DNA sequences located within 235 bp of the rat or 289 bp of the human *GH* promoter direct efficient somatotrophic expression of linked genes when transfected into GH-producing cell lines (Casanova et al., 1985; Crew and Spindler, 1986; Nelson et al., 1986) or after introduction into the germline of mice (Behringer et al., 1988; Borrelli et al., 1989; Burton et al., 1991; Lira et al., 1988; Struthers et al., 1991).

The Anterior Pituitary Transcriptional Regulator GHF-1

Within the *GH* promoter, two *cis* -acting elements, centered around −80 and −122 and required for somatotroph-specific expression, were found to bind a pituitary-specific factor GHF-1 (also called Pit1 or GC1) (Catanzaro et al., 1987; Lefevre et al., 1987; Nelson et al., 1988; West et al., 1987; Ye and Samuels, 1987; Ye et al., 1988). Addition of GHF-1-containing fractions to HeLa extracts stimulated hGH promoter activity *in vitro* and provided the first conclusive evidence for the importance of this transcription factor (Bodner and Karin, 1987). This activity together with a GH promoter footprint assay served as bioassays for purification of GHF-1 to homogeneity (Castrillo et al., 1989).

A partial amino acid sequence was obtained from purified GHF-1, and the information was used to isolate cDNA clones encoding rat and bovine GHF-1 (Bodner et al., 1988). GHF-1 (Pit1) was also cloned from an expression library using a GHF-1-binding element as a probe (Ingraham et al., 1988). As determined by RNA and immunohistochemical analysis of mouse and rat, GHF-1 is only expressed in the anterior pituitary (Bodner et al., 1988; Dollé et al., 1990; Ingraham et al., 1988). GHF-1 appears to be the most restricted of all cell-type-specific regulators known to date. Subsequently, the *GHF1* gene was also cloned from mouse and human (Lew and Elsholtz, 1991; Li et al., 1990; Tatsumi et al., 1992). The 291-residue amino acid sequence of GHF-1 is highly conserved; sequence identity is 96% at the amino acid level and 90% at the DNA level (Tatsumi et al., 1992), indicating that GHF-1 structure and function are conserved between rodents and primates. GHF-1 is a member of the POU-domain family of transcriptional regulators (Herr et al., 1988; Karin et al., 1990; Rosenfeld, 1991; Ruvkun and Finney, 1991; Schöler, 1991), which is characterized by conservation of a 60-amino-acid homeodomain (POU_{HD}) and a second region of about 75 amino acids located N terminal to the homeodomain that is called the POU-specific domain (POU_s) (Herr et al., 1988); POU-domain genes constitute a subclass of the homeobox genes. The binding to and transactivation of the *GH* promoter by GHF-1 was one of the first demonstrations that homeodomain proteins are indeed cell-type-specific transcriptional regulators. POU-domain proteins exert critical developmental and transcriptional functions (Finney and Ruvkun, 1990; Finney et al., 1988; Johnson and Hirsh, 1990; Rosner et al., 1990; Scholer, 1991; Verrijzer et al., 1990b).

GHF-1 DNA Binding. The GHF-1 homeodomain, which exhibits a 73% match to a consensus homeodomain sequence (Bodner et al., 1988), is essential for GHF-1 binding to DNA (Theill et al., 1989). Of the residues in the GHF-1 homeodomain, 22 are invariant among the POU homeodomains. Most of the conserved residues are located in three potential α-helices. Most notably, the invariant hexamer RVWFCN is present in the recognition helix which, in classical homeoproteins, fits into the major grove of the DNA and contacts specific bases in the binding site (Kissinger et al., 1990). The cysteine (C) in this sequence is unique to POU proteins, but its function is not understood because nonconservative amino acid substitution at this position does not affect DNA binding (Ingraham et al., 1990). However, interruption of the presumed α-helix structure by inclusion of a proline abolishes DNA binding completely (Ingraham et al., 1990). The tryptophan (W) residue in the recognition helix is also crucial for GHF-1 binding to target elements (Castrillo et al., 1991; Li et al., 1990). Together with the α-helixes, basic residues located both N- and C terminal in the homeodomain are required for GHF-1 binding to recognition elements (Ingraham et al., 1990), and they may also function as nuclear transport signals (Theill et al., 1989). The functional similarity between the POU homeodomain and the classical homeodomain was demonstrated by the efficient binding of both the GHF-1 and the Oct-1 homeodomains to classical homeodomain target elements (Ingraham et al., 1990; Verrijzer et al., 1990b). Segments of the POU-specific domain are required, however (Dixon et al., in manuscript; Ingraham et al., 1990; Theill et al., 1989), for high-affinity and high-specificity binding of GHF-1 to natural target sites (Figure 1).

Some 27 of about 81 amino acid residues in the POU-specific domain (POU$_s$) are invariant among POU proteins, indicating conserved structure and function of this domain also. Two regions, each containing a cluster of basic residues and a potential α-helix termed A and B, are separated by a conserved region of polar amino acids. Deletions of, or amino acid substitutions within, either A or B render GHF-1 inactive in DNA binding (Ingraham et al., 1990; Theill et al., 1989). DNA-binding analysis of chimeric POU domains suggested that both the homeobox and the POU_s A and B regions as well as the POU linker influence DNA-binding specificity (Aurora and Herr, 1992; Kristie and Sharp, 1990; Sturm and Herr, 1988; Verrijzer et al., 1990b). The POU-domain proteins, including GHF-1, thus appear to have a bipartite DNA-binding domain with both the homeobox and the POU-specific region contacting DNA.

DNA-binding proteins can make contact either with the DNA sugar-phosphate backbone or with individual bases within the binding site. DNA backbone contacts can be determined from hydroxyl radical footprinting experiments (Tullius and Dombroski, 1986), while base contacts can be determined from missing nucleotide experiments (Hayes and Tullius, 1989). Binding energy results mainly from electrostatic forces between the protein and the DNA backbone (von Hippel and Berg, 1986). In the POU-domain protein Oct-1, the homeodomain is responsible for most of the DNA-backbone contacts (Kristie and Sharp, 1990; Verritzer et al., 1990b). By comparing the existing structural and DNA-binding data for several homeodomains with the hydroxyl radical footprints and missing nucleoside data for GHF-1 bound to the GH promoter, Dixon and coworkers have determined which base contacts and DNA backbone contacts can be attributed to the homeodomain (Dixon et al., in manuscript). Similar to the findings for Oct-1, basic residues in the GHF-1 homeodomain were found responsible for the majority of the DNA-backbone contacts (Lys13, Lys25, Arg46, Arg53, Arg55, Lys57, Arg58), indicating that the homeodomain is important for high-affinity binding. In addition, residues in the recognition helix contact five bases in the major grove and residues in the N terminal of the homeodomain contact two bases in the minor grove (Figure 1A) (Dixon et al., in manuscript).

Mutation of the first three nucleotides (ATG) of the octamer motif affected binding by the complete Oct-1 POU domain but not binding by the Oct-1 homeodomain alone, indicating that the POU-specific domain is responsible for contacts located at one side of the octamer (Verrijzer et al., 1990b). In agreement with this finding, the first four nucleotides on the lower strand of the GHF-1 recognition element on the GH promoter (see Figure 1A) produced a missing nucleotide signal. This region appears not to be contacted by the homeodomain and therefore it is thought to be recognized by the POU_s domain of GHF-1 (see Figure 1) (Dixon et al., in manuscript).

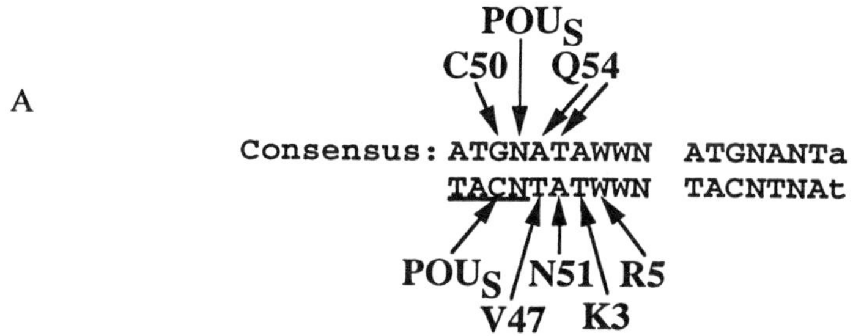

B   **GHF-1 Binding Sites:**

Source	Site	Sequence		Spacing
Consensus:		ATGNATAWN	ATGNANTa	10
GHF1	GHF1-I	ATGTATAA	ATGGATTT	8
	GHF1-II	ATGTATAT	ATGCAATA	8
GH	pGHF-1	ATGCATAA	ATGTACAC	9
	dGHF-1	ATGGATAAT	TTAGAAGC	9
PRL	P1P Pal	ATGAATATTC	ATGTAATC	10
	P3D	GTGAATATTC	ATGTAATC	10
	P1P	ATGAATATAT	ATATAATC	10
	P1D	ATGCATTTTTA	ATGCACTC	12

Figure 1   Model for GHF-1 binding to DNA. A: Consensus sequence for GHF-1 binding site with proposed GHF-1 protein contacts indicated above and below each strand. Four bases on left of lower strand produced a missing nucleotide signal, but it is not contacted by the homeodomain; instead, these bases could be contacted by the GHF-1 POU$_s$ domain. B: High-affinity GHF-1- binding sites present in GHF-1 (Chen et al., 1990; McCormick et al., 1990), GH (Lefevre et al., 1987), and prolactin (Mangalam et al., 1989; Nelson et al., 1988) genes plus a high-affinity synthetic binding element (Ingraham et al., 1988). (Figure modified and reprinted with permission from W. Dixon, L. Theill, M. Karin, and T. Tullius, in manuscript.)

The majority of GHF-1-binding sites contain a core recognition element and a Hawkins degenerate recognition element (see Figure 1) (Dixon et al., in manuscript). A complete GHF-1-binding site may therefore be composed of two half-sites positioned one helical turn apart in a head-to-tail manner. Several observations support this notion: DNase I and hydroxyl radical footprints extend beyond the core recognition element encompassing the degenerative recognition element (Dixon et al., in manuscript; Lefevre et al., 1987), and although most of GHF-1 is a monomer in solution, two GHF-1 molecules bind cooperatively to a GHF-1-binding site from the *GH* gene promoter (Dixon et al., in manuscript; Ingraham et al., 1990). Both the POU-specific domain and the homeodomain are required for cooperativity. DNA-dependent GHF-1GHF-1 interaction was observed in cross-linking experiments (Ingraham et al., 1990), indicating that stabilizing proteinprotein interactions may contribute to cooperative binding. GHF-1 and Oct-1 can also form a heterodimeric complex on DNA (Verrijzer et al., 1992a; Voss et al., 1991). It is not clear, however, whether such a complex forms *in vivo* and if so what physiological role it would play.

GHF-1 Protein–Protein Interactions. A small fraction of both GHF-1 and Oct-1 transiently forms homodimers in solution (Verrijzer et al., 1992a; Voss et al., 1991) and heterodimerization in solution between the two proteins was also observed (Voss et al., 1991). When DNA-dependent protein-protein interactions were inhibited by including ethidium bromide in the reaction weak GHF-1 homodimerization was still evident; but GHF-1-Oct-1 heterodimerization was undetectable, indicating that, while GHF-1 homodimerization in solution is real, the heterodimerization depended on binding to contaminating DNA (Lai and Herr, 1992). It is likely that these weak and transient interactions are stabilized by the presence of a compatible dimeric recognition element. Cooperative interactions between GHF-1 and heterologous POU-domain proteins on specific promoters may thus depend on the strength of their association in solution and the degree of stabilization provided by specific dimeric binding sites. With such a mechanism, complicated differential gene expression patterns can be mediated by a small number of POU-domain transcription factors. There may even exist POU-domain proteins that, as a heterodimer with GHF-1, inhibit GHF-1 binding to target elements.

A precedent for such negative interaction between POU-domain proteins has been demonstrated (Treacy et al., 1991a). Weak but specific DNA-independent interaction has also been observed between GHF-1 and members of the thyroid hormone receptor family (F. Schaufele, personal communication). GHF-1 and the thyroid hormone receptor both bind to the GH gene promoter and synergistically activate its transcription (Schaufele et al., 1992) (see following). Synergistic induction of transcription has also been observed between hormone receptors and Oct-1

(Bruggemeier et al., 1991), although in this case a direct interaction between the proteins remains to be proven. Although both the POU_{HD} and the POU_s domain are essential for protein–protein interactions involving GHF-1, the interaction interface is yet to be identified. The best characterized interactions are the interaction of I-POU with Cf1-a (Treacy et al., 1991a, 1992) and Oct-1 interaction with the herpes simplex virus transactivator VP16 (Kristie and Sharp, 1990; Pomerantz et al., 1992; Stern and Herr, 1991; Stern et al., 1989). The interaction interface was mapped to the homeodomain in both these cases.

The GHF-1 POU_s domain, together with the homeodomain, is also required for DNA bending that can be observed when GHF-1 binds to its site (Verrijzer et al., 1991). Bending of target DNA around recognition elements was also observed after binding of other POU-domain proteins (Verrijzer et al., 1991). Binding of GHF-1 to the distal (dGHF-1) and proximal (pGHF-1) sites on the GH promoter results in the creation of a strong hypersensitive site in the region between the binding sites. Interestingly, GHF-1 binds to opposite sides of the DNA helix at the dGHF-1 site relative to the pGHF-1 site (W. Dixon, unpublished data). Interaction between distally and proximally bound GHF-1 would imply under- or overwinding the DNA (creating a hypersensitive site) for the proteins to be positioned in the same spatial orientation. Proximally and distally bound GHF-1 could also serve to bring GHF-1 and other transcriptional regulators binding to the upstream enhancer of the Prl gene closer to the transcription start site. Such multimerization has been observed for the transcription factor Sp1 (Courey et al., 1989).

GHF-1 Isoforms with Promoter-Selective Transactivation Domains. The major transactivating function of GHF-1 resides in the N-terminal region (Ingraham et al., 1990; Theill et al., 1989). This 75-amino-acid region, which functions as an independent activation domain when connected to a heterologous DNA-binding domain, contains about 30% hydroxylated amino acid residues and has therefore been named STA (for Ser, Thr, Tyr activation domain). Interestingly, it is similar to the C-terminal repeat of the large subunit of RNA polymerase II. Both these regions also contain several prolines, indicating that they do not form α-helices. The mechanism by which the STA domain functions in transactivation is unknown. Major transcriptional activation domains of Oct-1 (OTF-1) and Oct-2 (OTF-2) are also rich in serine and threonine and contains proline (Tanaka and Herr, 1990).

GHF-1 functions in part by binding to and activating the promoters of both *GH* and *PRL* and that of the *GHF1* gene itself (Bodner and Karin, 1987; Bodner et al., 1988; Fox et al., 1990; Ingraham et al., 1988; Mangalam et al., 1989; McCormick et al., 1990; Nelson et al., 1988; Theill et al., 1992). A naturally occurring and functionally different isoform of GHF-1 has been identified (Konzak and Moore, 1992; Morris et al., 1992; Theill et al., 1992). In this variant GHF-2 (also called Pit-1a or Pit-1 beta), 26 additional amino acids are inserted into the activation domain of the protein as a result of alternative splicing. The 26 amino acid insert is conserved also in turkey and salmon GHF1 genes (Wong et al., 1992; Ono and Takayama, 1992). GHF-2 has DNA-binding activity identical to that of GHF-1 (Theill et al., 1992) and can activate the *GH* promoter but not the *Prl* or *GHF1* promoters (Morris et al., 1992; Theill et al., 1992).

The promoter selectivity of the GHF-2 transcription activating domain does not depend on the number of binding sites. One possibility is that GHF-2 is an effective activator only in collaboration with other sequence-specific transcription factors that bind to the *GH* promoter, but not the *Prl* and *GHF1* promoters. The insert in GHF-2, like the STA domain that it interrupts, is quite rich in hydroxylated amino acid residues. Structural constraints rather than a high content of a particular amino acid may therefore cause the transactivation domains of GHF-1 and GHF-2 to function differently. A putative protein kinase C phosphorylation site at Thr 57 of GHF-2 sequence could also play a role (Morris et al., 1992). It is possible that the insert within the GHF-2 activation domain alter its conformation and thereby disturbs the ability of GHF-2 to interact with target proteins that are either other sequence-specific activators or components of the basic transcriptional machinery. A similar change in promoter selectivity by alternative splicing and without affecting the DNA-binding domain was demonstrated for the Oct-2 proteins by Tanaka et al. (1992). The GHF-2 mRNA, which arises from the use of an alternative splice acceptor site, is expressed specifically in the anterior pituitary at a level about one-seventh that of GHF-1 mRNA (Theill et al., 1992). It is possible that changes in the levels of GHF-1 and GHF-2 production may contribute to the selective activation of GHF-1 target genes during anterior pituitary development. The octamer factor variants may also selectively activate different target genes (Müller-Immerglück et al., 1990; Tanaka et al., 1992).

In addition to the N-terminal STA domain, some transactivation is mediated by downstream domains of GHF-1 (Ingraham et al., 1990). The invaraint Arg at amino acid position 271 may form part of this downstream transactivation domain since when this residue is exchanged for a Trp residue the mutant GHF-1 acts as a dominant inhibitor of basal and wt GHF-1 mediated trans-activation (Radovick et al., 1992) (see below).

Correlation Between Genomic Organization and Functional Domains. In comparing the structure of the *GHF1* gene with that of the GHF-1 protein, a relatively good correlation is found between the exon structure and the distribution of functional domains within the protein (Theill et al., 1992). The GHF-1 transactivation domain is a composite of two subdomains that can function independently (Ingraham et al., 1990; Theill et al., 1989). These subdomains, which in the GHF-2 isoform are separated by the 26-amino-acid insert, are encoded by exon 1 and exon 2, respectively (Theill et al., 1992). A segment important for GHF-1 and thyroid hormone receptor synergism in activation of the GH gene promoter (F. Schaufele, personal communication) is located downstream of the transactivation domain. This region is encoded by exon 3, which also codes for the N-terminal portion of the POU-specific domain. Notably, this module, which is highly conserved among all POU-specific domains (Herr et al., 1988), is also present at the 3′ end of an exon in both in the *Oct-2* and *Oct-3* genes (Hatzopoulos et al., 1990; Okazawa et al., 1991).

The remainder of the POU-specific domain is located entirely within exon 4 of GHF1. Interestingly, this region of Oct-2 and Oct-3 is also encoded by a separate small exon (Hatzopoulos et al., 1990; Okazawa et al., 1991). The basic region at the N terminus of the GHF-1 homeodomain may function both in nuclear transport (Theill et al., 1989) and in contacting the minor groove of the binding site (Kissinger et al., 1990; Otting et al., 1988). This region is encoded by exon 5, whereas the helix-turn-helix motif of the homeodomain (Otting et al., 1988; Kissinger et al., 1990), is encoded by exon 6. The exon/intron organization therefore appears to reflect functional domains of the GHF-1 protein. The coding region outside the homeodomain is split by introns in phase I of the reading frame. This coincidence of phasing would permit exon shuffling. Taken together, these findings support the hypothesis that gene segment duplication and exon shuffling contributed to the evolution of the POU subfamily of regulatory proteins (Kirchgessner et al., 1989; Rogers, 1985; Südhof et al., 1985).

Distribution and Function of GHF-1 During Development. During development of the mouse, transcription of the *GHF1* gene is first detected in the anterior pituitary anlagen during ED13. Accumulation of GHF-1 protein, on the other hand, is not observed until ED15 (ED17 in the rat), a time that coincides with initiation of GH gene transcription and expansion of the somatotrophic lineage (Dollé et al., 1990; Simmons et al., 1990). Expression of GHF-1 protein in mature pituitaries is restricted to somatotrophs, lactotrophs, and thyrotrophs. The correlation between the initial accumulation of GHF-1 protein and the activation of *GH* gene expression suggests that GHF-1 is indeed a major determinator of somatotroph cell specialization. Low and transient expression of Prl was also observed following the initial appearance of GHF-1 (Crenshaw et al., 1989; Dollé et al., 1990; Simmons et al., 1990). Considerable levels of *Prl* gene expression are not reached, however, until several days after birth, indicating that other regulatory factors in addition to GHF-1 restrict or activate this gene.

A close correlation between GHF-1 and GH or Prl expression has also been observed in somatic cell hybrids between pituitary and fibroblast cells. Extinction of GH and Prl gene transcription, often observed in such hybrids, is accompanied by repression of GHF-1 synthesis, indicating that GHF-1 is required for endogenous GH and Prl expression (McCormick et al., 1988, 1991; Supowit et al., 1991; Tripputi et al., 1988).

Two well-established mouse mutant strains have defects in the *GHF1* gene. The Snell and Jackson dwarf mouse strains, which both carry allelic recessive mutations on chromosome 16, exhibit anterior pituitary hypoplasia and produce no detectable GH, Prl, or TSH (Eicher and Beamer, 1980; Roux et al., 1982; Snell, 1929; Wilson and Wyatt, 1986; Yashiro et al., 1988). The Jackson dwarf has a gross rearrangement of the *GHF1* gene (Li et al., 1990), while the Snell dwarf contains a single point mutation resulting in the substitution of the invariant tryptophan residue in the third helix of the homeodomain for a cysteine residue, creating a GHF-1 variant that is highly deficient in DNA binding (Castrillo et al., 1991; Li et al., 1990). The total cell number in the anterior lobe of Snell dwarfs is greatly reduced compared to the cell number in their normal counterparts, whereas the developmentally related intermediate lobe is unaffected. The Snell dwarf pituitary is almost totally deficient in GHF-1 mRNA and protein (Castrillo et al., 1991; Li et al., 1990). This is the combined result of the nonfunctional GHF-1 variant causing decreased transcription of the positively autoregulated *GHF1* gene and the absence of cell populations that would normally express GHF-1. Dwarf mouse embryos already exhibit anterior pituitary abnormalities, a lack of GH immunoreactivity, at day 16 of gestation confirming that the dwarf phenotype is caused by a lack of formation rather than disappearance of GH- and Prl-producing cells (Wilson and Wyatt, 1986). Several human homologues to the dwarf phenotype has been described. A subclass of cretinism, marked by irreversible mental and growth retardation, results from combined GH, TSH, and Prl deficiency. One patient was homozygous for a nonsense mutation at codon 172 in the GHF1 gene that would generate variant GHF-1 lacking the entire POU homeodomain (Tatsumi et al., 1992). Other patients with combined pituitary hormone deficiency (CPHD) are homozygous for point mutations in the POU_S region changing amino acid 143 in a basic domain from Arg to Gln (Ohta et al., 1992). The mutation at position 143 causes anterior pituitary hypoplasia (Ohta et al., 1992). The position 158 mutant GHF-1 reportedly exhibits a modest reduction in DNA binding affinity and a more pronounced decrease in transactivation of target promoters while apparently retaining properties needed to support development of a normal size anterior pituitary (Pfaffle et al., 1992). The Snell, Jackson, 143 and 158 mutant GHF1's are all affected in the POU domain. These mutations are fully or partly impaired in DNA-binding and since they are recessive may also be impaired in protein protein interaction. This supports the notion that once bound to DNA GHF-1 functions as a dimer and both the POU_S region and the homeodomain is required for dimerization on DNA. These mutants are contrasted by two very different and dominant acting mutations causing CHPD without affecting DNA-binding: One patient with very severe CPHD and anterior pituitary hypoplasia is heterozygous for a point mutation which changes a conserved Arg at amino acid position 271 in the c-terminal basic region of the homeodomain into a Trp (Ohta et al., 1992; Radovick et al., 1992). This mutant protein was found to not only abrogate wt GHF-1 trans-activation but also to inhibit basal activity from GHF-1 target promoters (Radovick et al., 1992). How the c-terminal basic region of the homeodomain functions in trans-activation is unknown. Finally a point mutation changing amino acid 24 from Pro to Leu also causes CPHD in the heterozygous state (Ohta et al., 1992).

In this patient GH and Prl was undetectable (Ohta et al., 1992), which verifies the importance of the N-terminally located trans-activation domain (Theill et al., 1989). On the other hand significant although redcued levels of TSH were detected, indicating that a part of GHF-1's activity remained. It is possible that the STA transactivation domain is dispensible for some GHF-1 functions. In this respect it would be interesting to determine whether the size of the anterior pituitary is affected by the position 24 mutant GHF-1. The dominant nature of mutations outside the DNA-binding domain also supports the idea that GHF-1 functions as a dimer. In conclusion, the human and mouse dwarf phenotypes suggest that GHF-1 functions in specification, expansion, and survival of somatotrophs, lactotrophs, and thyrotrophs.

Using antisense technology, a state of GHF-1 deficiency was created in pituitary cell lines. Specific inhibition of GHF-1 synthesis by complementary oligonucleotides led to a marked decrease in endogenous GH and Prl mRNA levels and to a significant decrease in cell proliferation (Castrillo et al., 1991). The correlation between decreased GHF-1 expression and inhibition of pituitary cell proliferation suggests that GHF-1 may function in proliferation of somatotrophic precursors. GHF-1 could influence proliferation by one of three mechanisms: (1) functioning directly as a replication factor, binding near the chromosomal origins of replication; (2) making functional heterodimers with factors involved in replication; or (3) functioning indirectly by regulating the expression of cell cycle regulatory or growth factor receptor genes. Other POU-domain proteins may also function in cell proliferation. In developing Schwan cells, SCIP/tst-1/Oct-6 is expressed to a high level only during the highly proliferative phase; indicating that this POU protein could function in both cell division and differentiation (Monuki et al., 1989, 1990). An isoform of Oct-2 called mini Oct-2 is also specifically expressed in proliferating cells of the olfactory system (Schöler, 1991). Oct-1 and Oct-2 stimulate replication of adenovirus *in vitro* (Verrijzer et al., 1990a), and Oct-1 binds to an auxiliary region next to the core origin of adenovirus replication (Coenjaerts et al., 1991).

Interestingly GHF-1 and other POU-domain proteins also stimulates replication in this system (Verrijzer et al., 1992b). The POU-proteins acts subsequent to the formation of a preinitiation complex and may facilitate replication through a conformational change in the DNA structure (Verrijzer et al., 1992). Whether GHF-1 or other POU-dmmain proteins can function directly at cellular origin of replication sites remains to be determined. Most recently GHF-1 involvement in activation of the growth hormone releasing factor receptor (GRFR) was suggested by the absence of GRFR in the Snell dwarf anterior pituitary, indicating that GHF-1 also functions in cell proliferation/survival by directly or indirectly activating growth factor receptors (Lin et al., 1992a) (see below).

Although GHF-1 is required for both specification and proliferation of somatotrophs and lactotrophs, the factor may have a more limited function in thyrotrophs because this cell type (as identified by its expression of TSHβ) is present in the anterior lobe 2 days before the initial appearance of the GHF-1 protein (Simmons et al., 1990). Other regulatory proteins must therefore direct the initial specification of thyrotrophs. The temporal and spatial expression pattern of the leucine-zipper protein thyrotroph embryonic factor (TEF) has been reported to match that of TSHβ in the embryo, and TEF can bind and activate an exogenous TSHβ promoter (Drolet et al., 1991). TEF may therefore be required for the initial activation of the TSHβ gene. However, once present in thyrotrophic cells, GHF-1 may also bind to and participate in TSHβ gene regulation (Shupnik et al., 1992; Steinfelder et al., 1992a,b) and function in cell maintenance or proliferation (Li et al., 1990).

Regulation of GHF1 Gene Expression

The regulation of GHF-1 expression and activity has attracted much interest with the establishment of GHF-1 as a key determinator for specification proliferation and survival of several anterior pituitary cell types.

Autoregulation. *GHF1* transcription is positively autoregulated via a GHF-1 response element at position −60 to −45 (Chen et al., 1990; McCormick et al., 1990). Because *GHF1* is involved in terminal differentiation, autoregulation can ensure that it will remain active for the life of the cell after it is turned on. A second GHF-1 recognition element located directly downstream of the start site of transcription may be an inhibitory element (Chen et al., 1990). The two elements may work together to control precisely the basal level of GHF-1 expression. Autoregulation may be widespread among transcription factors involved in terminal differentiation (Bienz and Tremml, 1988; Hiromi and Gehring, 1987; Thayer et al., 1989).

Regulation by cAMP. GHF1 gene transcription is also regulated by environmental cues that affect the intracellular level of cyclic adenosine monophosphate (cAMP) (McCormick et al., 1990). This regulation is mediated at least in part by the cAMP-responding element binding protein (CREB) or related family members such as CREM (Brindle and Montminy, 1992) that bind to the GHF1 promoter at two sites around positions -200 and -155 (McCormick et al., 1990). In response to elevation of intracellular cAMP levels, the catalytic subunit of Protein Kinase A (PKA) dissociates from its regulatory unit, resulting in migration of the catalytic subunit to the nucleus where it phosphorylates CREB. The phosphorylation of CREB by PKA on serine 133, which resides in its transactivating domain, strongly increases its ability to activate transcription (Gonzalez and Montminy, 1989; Gonzalez et al., 1989). It is possible that *GHF1* transcription is also activated by calcium, because CREB activity in some cell lines is thought to be induced after elevation of intracellular Ca^{2+}. In this case the same residue in CREB becomes phosphorylated, but the kinase is calmodulin kinase (CAM kinase), rather than PKA, or other kinases activated by Ca^{2+} (Sheng et al., 1991). Whether CREB in pituitary cells can be activated in response to Ca^{2+} remains to be tested. A gene encoding protein highly homologous to CREB has recently been cloned from a pituitary cDNA library. Isoforms of this protein (CREM) also bind at least a subset of the cAMP response elements, but in contrast to CREB, CREM appears to be a transcriptional antagonist (Foulkes et al., 1991). Whether CREM isoforms can bind to CREs in the *GHF1* promoter and downregulate cAMP-induced transcription is not known. It is also possible that CREM will be found to act as a positive regulator of transcription in collaboration with other factors.

GH-Releasing Factor and Somatostatin. The hypothalamic peptide growth hormone-releasing factor (GRF) (Gelato and Merriam, 1986; Rivier et al., 1982; Spies et al., 1983) binds to a receptor on somatotrophs, causing elevation of intracellular cAMP levels (Ikuyama et al., 1988; Mayo, 1985; Seifert et al., 1985). Long before the GRF receptor was actually isolated or cloned, it was known to belong to the family of Gs linket cell-surface receptors that, on binding the ligand, activate adenylate cyclase and causes increased cAMP levels and PKA activation (Bilezikjian and Vale, 1983; Struthers et al., 1989). Molecular cloning of the GH-releasing factor receptor (GRFR) gene confirmed that GRFR is homologous to other G-protein-coupled receptors containing seven potential membrane-spanning domains (Lin et al., 1992; Mayo, 1992). Binding of GRF to kidney cells transfected with GRFR stimulated intracellular cAMP production. *GRFR* expression is pituitary specific (Lin et al., 1992; Mayo, 1992). Binding of GRF to its receptor on pituitary cells results in increased *GH* gene transcription by a mechanism that is independent of the intracellular pool of GH (Barinaga et al., 1985a, 1985b). The responsiveness to cAMP has been mapped to sequences within the first 100 bp of the GH promoter, which includes its proximal GHF-1-binding site (Copp and Samuels, 1989; Dana and Karin, 1989). It is possible therefore that GRF induction of GH is mediated via GHF-1.

In addition to its effect on GH expression and secretion, GRF is also a specific mitogen for somatotrophs (Billestrup et al., 1986, 1987). GRF was actually isolated from human pancreatic tumors from patients with agromegaly and somatotrophic hyperplasia (Frohman and Szabo, 1981; Guillemin et al., 1982). Overexpression of GRF in transgenic mice leads to specific somatotrophic hyperplasia (Mayo et al., 1988), and many human GH-secreting pituitary adenomas have mutations that constitutively activate the stimulatory G protein that couples the GRF receptor to adenylate cyclase (Landis et al., 1989; Lyons et al., 1990). Expression of a GH-cholera toxin fusion gene also leads to pituitary hyperplasia and gigantism in transgenic mice, an effect caused by chronic activation of adenylate cyclase (Burton et al., 1991).

It has been suggested, by the observation that overexpression of a nonphosphorylatable (noninducible dominant negative mutant) variant of CREB driven by the GH promoter in transgenic mice leads to pituitary hypoplasia, that this signaling system for expansion of the somatotrophic lineage may indeed work via the PKA-mediated activation of CREB (Struthers et al., 1991). Although this experiment suggests that CREB is a crucial link between the GRF receptor and somatotroph proliferation, it is also possible that the dominant negative CREB mutant was acting by inhibiting the binding of a related transcription factor such as CREM. Regardless, the *GHF1* gene is likely to be the next link in this signaling chain because it is required for somatotroph proliferation (Castrillo et al., 1991).

Alternatively, the proliferative effect of GRF may be mediated by the c-*fos* gene, which is also induced by GRF (Billestrup et al., 1987). GHF-1 activity could also be regulated by posttranslational modifications. Elucidation of the GFR signal transduction pathway from the cell membrane to specific gene activation in the nucleus (Figure 2) has been hampered by the lack of GRF receptors in established pituitary cell lines (Zeytin et al., 1984) and by the limited availability of pure populations of somatotrophs. Freshly dispersed GRF- responsive cells from the transplantable rat somatotrophic tumor MtTW15 (Judd and MacLeod, 1984) or new pituitary cell lines reported to secrete GH in response to GRF (Chomczynski et al., 1988; Inoue et al., 1990) may be valuable tools for such investigation. Alternatively, GRFR signal transduction can now be studied in GH3 or GC cells expressing recombinant wild-type or mutant GRFR using isolated cDNA (Mayo, 1992).

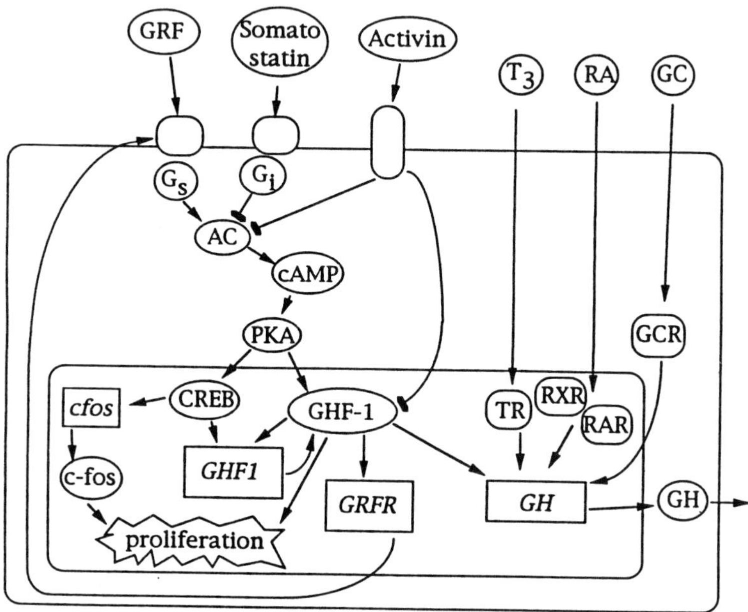

Figure 2   Schematic diagram of signal transduction and somatotroph-specific gene expression. Arrow, activation; bar, inhibition; box and italics, gene; AC, adenylate cyclase; cAMP, cyclic adenosine monophosphate; CREB, cAMP- response element-binding protein; Gs, Gi, G, proteins; GC, glucocorticoid; GCR, GC receptor; GH, growth hormone; GHF-1, GH factor 1; GRF, GH-releasing factor; GRFR, GRF receptor; RA, retinoic acid; RAR, RA receptor; RXR, retinoid X receptor; TH, thyroid hormone; TR, thyroid hormone receptor.

GRF was first detected on ED18 in the rat in hypothalamic nerve terminals reaching the median eminence (Ishikawa et al., 1986). GRF is released into special capillaries that are connected to capillaries in the anterior pituitary via a hypophysal portal vessel system (Dearden and Holmes, 1976; Fink and Smith, 1971). Because of this direct connection, GRF (like other releasing factors) reaches the pituitary before being diluted in the general bloodstream. The hypophysial-portal vessels do not appear until ED21. However, a direct vascular connection that exists between the brain and the anterior pituitaries several days earlier could allow GRF to reach the pituitary. The appearance of GRF on ED18 in the rat coincides with activation of GH gene expression and proliferation of somatotrophic cells (Dollé et al., 1990; Frawley et al., 1985b; Simmons et al., 1990). It is quite possible, therefore, that during development GRF stimulates the final step in both differentiation and proliferation of somatotrophs. Starting on ED16 in the mouse (equivalent to ED18 in the rat), slight differences can be detected in anterior pituitary development between normal mice and the Snell dwarf that evolve to marked postnatal differences in the size and morphology of the gland (Wilson and Wyatt, 1986). The correlation of high-level GHF-1 protein expression and somatotroph proliferation with GRF expression and the requirement for functional GHF-1 for somatotroph proliferation (Castrillo et al., 1991) suggest that GHF-1 is a mediator of the signal originated by binding of GRF to its receptor. GRF may induce high-level expression of GHF-1 via activation of CREB or may stimulate GHF-1 accumulation or activity post-trasncriptionally. GRF is clearly not the initial activator of GHF-1 transcription because considerable levels of GHF-1 mRNA accumulate before establishment of the neuroendocrine connection and GRF expression. On the contrary, GHF-1 is expected to be required for activation of GRF-receptor gene expression in the somatotrophic precursor cell, a view supported by the pituitary-specific distribution of GRFR mRNA (Mayo, 1992). Most recently Lin and coworkers verified that GRFR expression which in the normal mouse is first detected at ED 16 is absent in the *dw* mouse and thereby provided evidence that GHF-1 is indeed required for the initial appearance of GRFR (Lin et al., 1992a). The hypoplastic nature of the Snell dwarf is therefore explainable in part by the lack of one or more growth factor receptors. Conversely these receptors downstream signaling is likely to go through GHF-1.

The hypothalamic hormone somatostatin has effects on somatotroph function opposite to those of GFR and is an important negative regulator of GH secretion (Brazeau et al., 1973; Lamberts, 1988; Patel and Srikant, 1986; Reichlin, 1987; Schonbrunn and Koch, 1987). Somatostatin binds to a cell-surface receptor that is coupled to G_i, the inhibitory guanine nucleotide binding protein; G_i in turn mediates inhibition of adenylate cyclase. Several somatostatin receptors have been cloned (Yamada et al., 1992). It is possible that somatostatin and its receptor play a role in restricting GHF-1 expression and somatotrophic lineage expansion during development.

GRF immunoreactivity is markedly reduced in the median eminence of old rats as compared to young rats, whereas no remarkable difference could be detected between SRIF immunoreactivities in the same animals (Morimoto et al., 1988). A decrease in the synthesis/release of GRF may thus be responsible for the reduction in GH production that occurs with age. Whether GHF-1 expression is decreased in aged pituitaries remains to be determined.

Activin and Inhibin. Somatotroph activities are also modulated by activin and inhibin through a mechanism that involves modulation of GHF-1 expression or activity. Activin and inhibin belong to the TGFβ family of growth and differentiation factors (Massague, 1987; Ying, 1988). Activins are dimers of two β subunits ($\beta_A\beta_A$, $\beta_A\beta_B$, or $\beta_B\beta_B$), whereas inhibins are heterodimers containing an α and a β subunit ($\alpha\beta_A$ $\alpha\beta_B$). Activin was first characterized from gonadal extracts as an activity that stimulates pituitary FSH secretion, but α and β subunits are expressed in several other tissues including brain and adrenal gland. Other prominent activin functions include the promotion of nerve cell survival (Schubert et al., 1990) and the induction of mesoderm formation (Thomsen et al., 1990). Inhibin competes for the same classes of activin-binding sites and in general antagonizes the effects of activin.

Within the pituitary gland, a and β subunits are synthesized by the gonadotrophs, indicating an autocrine function (Corrigan et al., 1991). Activin also exerts paracrine actions on other pituitary cell types including somatotrophs; activin inhibits basal as well as glucocorticoid- (GC-) and thyroid hormone- (TH-) stimulated biosynthesis and secretion of GH from somatotrophic cell lines or cultures of pituitary cells (Bilezikjian et al., 1990; Billestrup et al., 1990; Kitaoka et al., 1991). Pretreatment with activin also inhibits GRF-stimulated cAMP synthesis, GH biosynthesis and secretion, and somatotroph proliferation *in vitro* (Bilezikjian et al., 1990; Billestrup et al., 1990; Kitaoka et al., 1991). Activin represses GH mRNA levels and decreases expression of a GH promoter-CAT reporter construct in MtTW15 somatotrophic tumor cells. The inhibition of transcription appears to be mediated via the distal GHF-1-binding site and correlates with reduced GHF-1 binding to GHF-1 recognition elements (Struthers et al., 1992). Full effects of activin were only observed after prolonged incubation (24–72 h), and the effects were blocked by actinomycin D and cycloheximide. Unfortunately, it was not established whether the change in binding was accompanied by a decrease in GHF-1 mRNA and protein. It is suspected, however, that because GHF-1 binding is required for the expression of *GHF1* itself the reduction in binding is primarily caused by reduction in synthesis of GHF-1 mRNA and protein. What causes this reduction in GHF-1 expression or binding affinity is not known. The inhibitory mechanism could be entirely at the level of GHF-1 transcription/mRNA accumulation but post-translational modification of GHF-1 may also be a factor.

Activin and somatostatin act additively to inhibit GH secretion. and inhibition by activin is only partially prevented by pertussis toxin, indicating that there is an additional signal transduction pathway independent of adenylate cyclase (Bilezikjian et al., 1990). Somatotrophs from normal rat pituitaries appear to respond directly to activin by increasing intracellular calcium (Tasaka et al., 1992). These results suggest that the inhibitory effects of activin on somatotrophs are direct and are mediated by two different pathways, only one of which involves inhibition of G_s activation of adenylate cyclase. Sequence comparison of cloned activin receptors predicts that they are transmembrane serine or threonine kinases (Attisano et al., 1992; Mathews and Vale, 1991; Mathews et al., 1992). Indeed, a related TGF-β receptor is capable of autophosphorylation (Lin et al., 1992). So far, however, kinase activity for the activin receptor has not been demonstrated and the signaling pathway downstream of the receptor remains undetermined.

Activin and its receptors may be involved in controlling the temporal and spatial expansion of the somatotrophic lineage. It is of interest that this control is mediated at least in part by restricting the expression and activity of GHF-1.

Early Activators of GHF1 Transcription. The minimal GHF1 promoter fragment required for maximum cell-specific expression in differentiated somatotroph cell lines (GC or GH3) consists of two CREB-binding sites, a GHF-1-binding site, and a 15-bp region centered around the TATA box, which also exhibits pituitary-specific activity (McCormick et al., 1990, 1991). GHF-1 of course cannot be responsible for initial activation of its own gene, and the ubiquitous factor CREB is rather unlikely to mediate initial *GHF1* gene activation because CREB-binding sites are not required for basal *GHF1* expression and because GHF-1 mRNA is detected before GRF expression and establishment of the hypothalamic hypophysal connection. The exact nature of the somatotroph-specific activity mediated via the *GHF1* TATA region and its role in initial *GHF1* activation have yet to be established but are likely to involve a pituitary-specific cofactor that interacts with the TATA box binding protein (TBP).

Identification of the mechanism responsible for the initial activation of GHF1 is essential for understanding the regulatory hierarchies that control development of the anterior pituitary. Analysis in transgenic mice has revealed that a region located far upstream of the *GHF1* proximal promoter is crucial for the initial activation of the *GHF1* gene in somatotrophic precursors. Although 2.5 kbp of the *GHF1* promoter was ineffective in directing expression to the anterior pituitary, a 15-kbp GHF1 promoter fragment directed expression of the SV40 large T-antigen to the anterior pituitary in transgenic mice, causing developmental entrapment of somatotrophic progenitor cells. The immortalized progenitor cells expressed high levels of GHF-1 mRNA but rather low levels of protein and failed to express either the *GH* or *Prl* gene. As a result, these transgenic mice exhibit dramatic dwarfism (Lew et al., 1993). The existence of such a progenitor was predicted from the observation that GHF-1 mRNA appears 2 days before initiation of GH gene expression (Dollé et al., 1990). Cell lines derived from a GHF-*Tag* transgenic anterior pituitary tumor maintain the characteristics of the somatotroph/lactotroph progenitor. Using these cells, an enhancer element was identified upstream of the GHF1 promoter that activates its transcription at this early stage of development but is inactive in cells representing later developmental stages or in other cell types (Lew et al., 1993). This stage-specific enhancer, which is located between −3.1 kbp and −5.3 kbp upstream from the start of GHF-1 transcription, is likely to serve as a target for a transcriptional regulator that is active in the somatotrophic progenitor but not in its more differentiated derivatives. Identification of this enhancer element is an important step toward the identification of additional genes that control both the cell-type specificity and the timing of GHF-1 expression.

Model for Development Regulation of *GHF1* Expression

The available data are consistent with the following model for activation of the *GHF1* gene during development: At or before day ED13 (in the mouse), a stage-specific factor(s) (or a specific network of factors) that interacts with the stage-specific enhancer of the *GHF1* gene appears or becomes activated. This factor(s) may be induced by a signal originating from the neighboring diencephalon. The stage-specific factor works in conjunction with the pituitary- specific TBP cofactor to activate the *GHF1* transcription. This event may signal the divergence of the somatotrophic progenitor from cells destined to express glycoprotein hormones. In the somatotrophic progenitor, GHF-1, protein levels are initially low because of a posttranscriptional inhibitory mechanism (Dollé et al., 1990; Lew et al., in press). GHF-1 accumulation eventually causes activation of *GH* and *GRFR* genes. Once the GRF receptor is expressed, the emerging somatotrophs respond to GRF, which is released from hypothalamic neurons around ED16 (ED18 in the rat). This leads to the activation of CREB and to a faster increase in *GHF1* transcription. The GRF receptor signal may also mediate the attenuation of GHF-1 posttranscriptional repression and/or activation of GHF-1 protein by posttranslational modification. As a result, an autoregulatory loop is activated ensuring permanent activation of the *GHF1* promoter by the GHF-1 protein. The increased level of GHF-1 may also provide a signal leading to inactivation or shutdown of the factor(s) that act on the stage-specific enhancer element (Lew et al., in press).

Identification of the regulatory factors binding the stage-specific enhancer will be an important step toward an elucidation of the signal(s) regulating initial *GHF1* activation. The retinoic acid receptor family is one example of transcription factors involved in developmental regulation of other POU genes. Expression of the embryonic octamer-binding transcription factor Oct3 is rapidly repressed by retinoic acid (RA) acting through a stem cell-specific, RA-repressible enhancer element in an upstream region of the gene (Okazawa et al., 1991). Conversely, the Oct-6 gene is transiently upregulated during RA-induced differentiation of embryonic stem cells (Meijer et al., 1990).

Transcriptional Regulators Interacting with the GH Promoter.

GHF-1 . GHF-1 was first identified because of its binding to and transactivation of the *GH* gene promoter (Bodner and Karin, 1987; Catanzaro et al., 1987; Lefevre et al., 1987; West et al., 1987; Ye and Samuels, 1987) and was subsequently found to also target the *Prl* gene (Nelson et al., 1988). GHF-1 was subsequently shown to be crucial for endogenous GH and PRL gene expression (Castrillo et al., 1991; Li et al., 1990). Thus, the lack of *GH* and *Prl* expression in the somatotrophic progenitors may be explained by the low level of GHF-1 present in these cells compared to GH-producing pituitary tumor cell lines (Lew et al., 1993).

As summarized previously, the hypothalamic regulatory peptides GRF and somatostatin respectively upregulate and downregulate intracellular cAMP levels and thereby affect *GH* expression. This response is likely to be mediated, at least in part, via the GHF-1 binding sites (Copp and Samuels, 1989; Dana and Karin, 1989; Peers et al., 1991). Likewise, thyrotrophic GHF-1, or a factor with similar DNA-binding specificity, could mediate cAMP stimulation of the TSH-β gene (Steinfelder et al., 1992a,b).

Increased biosynthesis of GHF-1 may not be the primary event in cAMP stimulation of GH transcription because GRF stimulation of GH transcription is already observed 10 min after GRF infusion into live rats and within 30 min of GRF treatment of cultured pituitary cells (Barinaga et al., 1985a), a timeframe that suggests posttranslational modification of GHF-1 itself or of a factor that synergizes with it. GHF-1 phosphorylated *in vitro* by PKA has unchanged or decreased binding affinity to GHF-1 response elements in the GH-promoter (Kapiloff et al., 1991). In contrast, phosphorylation enhanced GHF-1 binding to recognition elements in the TSHβ gene (Steinfelder et al., 1992b). Whether cAMP stimulation of GH transcription can be mediated by phosphorylation of GHF-1 remains to be determined. The lack of GH and PRL expression in somatotrophic progenitors could actually be caused by the absence of either additional transcription factors or posttranslational processes required for developmental activation of these promoters. Additional *cis-* and *trans-*acting elements involved in regulating GH expression are discussed next.

Thyroid hormone and retinoic acid receptors. Thyroid hormone (TH) stimulates *GH* gene transcription (Brent et al., 1988; Crew and Spindler, 1986; Diamond and Goodman, 1985; Dobner et al., 1981; Evans et al., 1982; Flug et al., 1987; Martial et al., 1977; Spindler et al., 1982; Yaffe and Samuels, 1984). The effect of TH is mediated by its receptors (TRs), which belong to the nuclear receptor superfamily (Brent et al., 1991a). These receptors contains two zinc fingers necessary for contacting TH-response elements (TREs) and a C-terminal ligand-binding domain. In the rat, *GH* promoter sequences between -190 to -167 confer TH induction (Brent et al., 1989a, 1989b; Glass et al., 1987; Norman et al., 1989). Using crude or partially purified nuclear extracts, TR-DNA contacts were reported at $-188/-174$ (Koenig et al., 1987), $-180/-165$

(Glass et al., 1987), or −190/−170 (Norman et al., 1989). Mutagenesis and analysis of this region revealed that both the imperfect direct repeat −189′- AGGTAAGATCAGGGACG-172′ and the overlapping imperfect palindrome - 180′-CAGGGACGTGACCGCA-164′ can function as TR-response elements (Brent et al., 1989a, 1989b; Glass et al., 1987, 1988; Norman et al., 1989). The complete TRE in the rGH promoter consists of three hexamers related to the consensus (A/G)GGT(C/A)A (Brent et al., 1989a, 1989b, 1991b, 1992). An additional high-affinity TR-binding site was identified in the third intron of the rat GH gene (Sap et al., 1990).

In mammals, two distinct but related genes termed c-*erb*Aα and c-*erb*Aβ encode TRs, and isoforms of both TRα and TRβ are produced by alternative splicing (Brent et al., 1991a). TRβ isoforms were identified in rat pituitary as well as in other tissues (Bradley et al., 1989; Koenig et al., 1988; Murray et al., 1988). A pituitary-specific β isoform termed TRβ2 was identified by Hodin et al. (1989). The TRβ2 form differs from TRβ1 in the N-terminal transactivating domain, but the functional significance is unknown. Isoforms of both TRβ and TRβ are expressed in GH-producing pituitary cell lines (Davis and Lazar, 1992; Koenig et al., 1988).

In transient cotransfection experiments, both TRa1 and TRb receptors transactivated promoters containing the rat GH TRE (Forman et al., 1988; Koenig et al., 1989; Thompson and Evans, 1989) whereas the TRα2 variant, which cannot bind TH, functions as an inhibitor (Koenig et al., 1989). In the absence of TH, TRα1 and TRβ also repress transcription from the GH promoter via the TRE (Brent et al., 1989c). A possible TH switch changing the receptors from repressors to activators is likely to be an important mechanism in the regulation of GH gene expression. TRs can also bind to the estrogen receptor (ER) recognition elements (EREs). On these sites, the TRs function as repressors in both the absence and presence of T3 (Glass et al., 1988). It is possible therefore that TRs could also function as constitutive inhibitors of Prl gene expression via interaction with EREs present in the *Prl* upstream enhancer.

The observation that 235-1 prolactinoma cells express lower levels of TR than GC cells (Forman et al., 1988) raises the interesting possibility that GH gene inactivity in these GHF-1-positive cells could be caused by TR deficiency. However, the *GH* gene expression in 235-1 cells was not rescued by the ectopic expression of chicken TRα or human TRβ (Forman et al., 1988). It remains to be tested whether other TR isoforms can mediate activation. 235-1 cells could also be deficient in posttranslational modification of the TRs, because their activity may be controlled by phosphorylation (Glineur et al., 1989; Goldberg et al., 1988).

In the presence of TH, TRs and GHF-1 synergize in activation of the GH promoter (Schaufele et al., 1992). This response is increased by stimulation of both PKA and Protein Kinase C (PKC). The molecular basis for this synergistic response may involve weak but specific inter-actions between GHF-1 and TR (Schaufele, personal communication). A detailed temporal and spatial analysis of the TR mRNA and protein isoform expression in the developing pituitary should help determinate whether there is a correlation between the TR expression and the GH gene activation during development.

Vitamin A function in normal growth and development is mediated via its metabolite retinoic acid (RA). RA activates *GH* gene expression in pituitary cells via the RA receptors (RARs) (DeLuca, 1991; Giguere et al., 1987), and it acts synergistically with both TH and GC (Bedo et al., 1989; Morita et al., 1989). The RARs can bind to the TREs identified in the rGH promoter (Umesono et al., 1988; Williams et al., 1991). Both RARs and TRs also form heterodimers with RXR (Kliewer et al., 1992; Zhang et al., 1992); the ligand for RXR is 9-cis-retinoic acid (Heyman et al., 1992). It is possible that RAR/RXR as well as TR/RXR heterodimers can mediate strong ligand-dependent activation of the *GH* promoter. Conversely, the COUP protein, another member of the nuclear receptor family, represses RAR activity (Tran et al., 1992).

Glucocorticoid receptor. GH gene expression is stimulated by glucocorti-coids (GCs) (Dobner et al., 1981; Evans et al., 1982; Martial et al., 1977; Robins et al., 1982; Seo, 1985; Spindler et al., 1982). The immediate 5'- flanking region of the rat GH gene was shown to function as a GC-inducible enhancer when associated with a heterologous promoter (RSV); a GC- responsive element was mapped to position $-97/-111$ (Treacy et al., 1991b). On the other hand, the intact rat GH promoter including 1.8 kbp of 5'-sequences does not mediate induction by glucocorticoids (Flug et al., 1987). In a separate study, GC stimulation of GH mRNA in GH3 cells was found to require ongoing protein synthesis and did not involve GH gene transcription (Strobl et al., 1989). The effect of GC may thus be mediated by both transcriptional and posttranscriptional mechanisms in the rat.

GC also increases GH mRNA levels in human pituitary cells (Isaacs et al., 1987). GC-responsive elements were identified in the 5'-flanking region and in the first intron of the *h*GH gene (Cattini and Eberhardt, 1987; Robins et al., 1982; Slater, 1985), and binding of the GCR to these regions has been demonstrated (Eliard et al., 1985; Rousseau et al., 1987; Slater, 1985). Potential GCR recognition elements are located in the -245 to -206 region of the *h*GH promoter and $+97$ to $+111$ of the first intron. *h*GH mRNA stability is also stimulated by GC (Paek and Axel, 1987). A more detailed analysis is required to assess the precise role of GCR in regulation of rat and human GH expression.

Ubiquitous transcription factors. Sp1 binds a site that is conserved between the rat and the human GH promoters. This site overlaps the distal GHF-1 binding site, and it has been suggested that GHF-1 and Sp1 binding are mutually exclusive (Lemaigre et al., 1990; Schaufele et al., 1990b; Tansey and Catanzaro, 1991). Sp1 appears to act positively on the GH promoter even though it can inhibit GHF-1 binding. It is not understood how these mutually exclusive binding proteins can cooperate in the activation of transcription. The activator protein AP2 can bind to two regions on the *h*GH promoter: -142 to -168 and -262 to -290 (Imagawa et al., 1987). AP2 can mediate cAMP and phorbol ester responses of some promoters, but it is not known whether transcriptional regulation of the GH gene involves AP2. A third ubiquitous factor upstream stimulatory factor (USF) (Sawadogo and Roeder, 1985) or a similar factor appears to bind to the -267 to -257 region of the *h*GH gene (Lemaigre et al., 1989); its role is also unknown.

A different ubiquitous factor termed GHF3 was shown to bind between nucleotides -239 and -219 in the rat GH promoter (Schaufele et al., 1990a). Deletion or mutagenesis of the GHF3-binding site reduces the activity of the GH promoter to 30% of wild-type activity, both in transient transfection of pituitary cells and in *in vitro* transcription assay (Nelson et al., 1986; Schaufele et al., 1990a; West et al., 1987). Several different complexes were found to interact with the GHF3 site; however, this protein has not been identified (Schaufele et al., 1990a). In a separate study, GHF3-binding activity was found to be present in GH-producing pituitary cell lines but absent in a fibroblast cell line and extinguished in a pituitary/fibroblast hybrid cell line (Tripputi et al., 1988). However, GHF3-binding activity was later identified in several tissues that do not express GH (Schaufele et al., 1990a), suggesting that GHF3 merely functions to adjust GH expression levels.

Silencer element. A negative regulatory element called silencer 1 is located between −325 and −280 bp upstream of the start of rGH transcription (Larsen et al., 1986; Pan et al., 1990; Tripputi et al., 1988). This element was found to be more inhibitory in nonpituitary cells than in GH-producing cells (Tripputi et al., 1988). The silencing function appears to be mediated by the binding of a 45-kDa protein identified in crude extracts of cells that do not express GH and is undetectable in pituitary extracts (Roy et al., 1992). The function of this element remains uncertain, because it exhibits a strong repressing function both in non-GH- and GH-producing cells when inserted upstream of a heterologous promoter (Roy et al., 1992). An additional proximal repressor element (PRE) has been identified in the −169 to −152 region (Pan et al., 1990). The function of this inhibitory element is also unclear because it is functional in GH-producing cells.

In Vivo Footprinting of the GH Promoter

GHF-1 is required for activation of GH or Prl genes in somatotrophs and lactotrophs, respectively. The level of GHF-1 and its DNA-binding activities appears to be similar in somatotroph- and lactotroph-derived cell lines (Klausing, in manuscript). It is not well understood why expression of GHF-1 does not activate the GH gene in lactotroph-derived cells, such as 235-1. Genomic footprinting experiments indicate that the GHF-1 binding sites on the GH promoter are unoccupied in 235-1 cells, while high occupancy is observed in GH3 and GC cells that express GH (Klausing, in manuscript). On the other hand, GHF-1-binding sites in the GHF1 gene itself are equally occupied in both 235-1 and GH-producing cell lines. In the lactotrophic 235-1 cell line, the GH gene 5′-flanking region is also less sensitive to digestion by micrococcal nuclease, indicating that the GH promoter in these cells is in a compact chromatin structure (Klausing, in manuscript). In GH-expressing cells, however, the GH promoter is sensitive to micrococcal nuclease digestion. It is possible therefore that the failure of GHF-1 to bind and activate the GH promoter in 235-1 cells and in lactotrophs is caused in part by a chromatin structure which prevents GHF-1 binding. This may explain why exogenous expression of GHF-1 does not activate the GH gene in heterologous cells (L. E. Theill and J.-L. Castrillo, unpublished data). Similar differential transcription factor binding *in vivo* was previously observed on the MMTV LTR (Archer et al., 1992; Cordingley et al., 1987; Pina et al., 1990) and the tyrosine aminotransferase gene promoter (Rigaud et al., 1991).

Model for Transcriptional Control of GH Gene Expression

The binding of transcriptional regulators to the human and rat GH gene is summarized in Figure 3. Four categories of factors interact with the promoter. The first category includes basic transcriptional machinery factors (TFIID) (Kao et al., 1990) and ubiquitous sequence-specific factors such as CTF/NF-1, USF/MCTF, and SP1 (Lemaigre et al., 1989). Binding of these factors to the GH promoter *in vivo* may depend on an "opening" of the chromatin structure mediated by other transcriptional regulators (Klausing, in manuscript). Lack of promoter access may help explain why these ubiquitous factors do not activate the GH gene in nonpituitary cells. These factors are required for high-level activity of the GH promoter in pituitary cells as demonstrated for TFIID, Sp1, and GHF3 (Lemaigre et al. 1990; Schaufele et al., 1990a, 1990b).

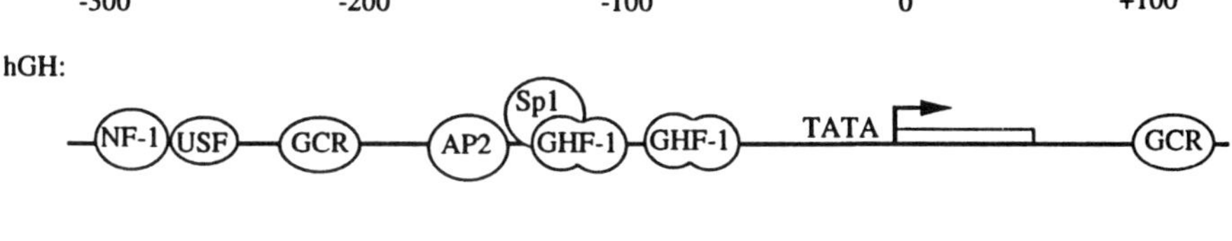

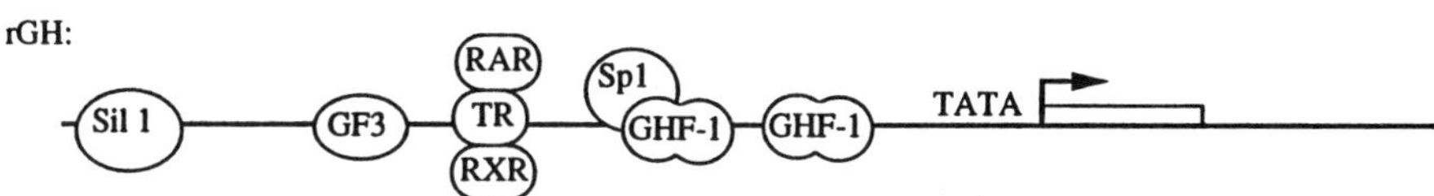

Figure 3   Transcription factors binding to the rat and human growth hormone gene promoters. Scale refers to base pairs relative to transcription initiation site. (References in text.)

Factors of the second category are inhibitory proteins that bind to silencer elements (Roy et al., 1992). Neither the silencer elements nor the inhibitory factors are well characterized. The elements do not overlap with positive elements, indicating that the inhibitory factors do not work by hindering access of positive factors. The inhibitory factors are ubiquitously expressed and are also active in pituitary cells. The silencer proteins thus appear to function to attenuate GH gene transcription. Part of the function of positive factors acting on the GH promoter in pituitary cells must therefore be to neutralize the silencer proteins.

The strict pituitary-specific GHF-1 and its isoforms encoded by the GHF1 gene represent a third category of factors (Bodner et al., 1988; Ingraham et al., 1988). Variant forms of GHF-1 may preferably activate one target promoter and not the other, as evidenced for GHF-2 (Theill et al., 1992).

The level of GH transcription is highly regulated by the hypothalamic peptides GRF and somatostatin (see Figure 2). The response elements in the GH promoter have not been clearly identified but the responsive region includes one of the GHF-1-binding sites (Copp and Samuels, 1989; Dana and Karin, 1989). Asynchronous oscillations in the levels of GRF and somatostatin translate via cAMP and PKA into oscillating levels of transcriptionally active CREB. Because CREB stimulates GHF-1 transcription. Part of the signal from GRF and somatostatin to the GH gene is likely to be mediated via regulated transcription of GHF-1. Conversely, because GRF can induce GH transcription quickly and independently of protein synthesis, GRF signaling may also activate GHF-1 by a post-translational mechanism. On the GH promoter, GHF-1 can also form a transcriptionally active heterodimer with the ubiquitous POU-domain protein Oct-1 (Verrijzer et al., 1992a; Voss et al., 1991). A cAMP signal could therefore also be mediated via modification of Oct-1 or unidentified factors interacting with GHF-1. The transcriptional regulator AP-2 that mediates induction by cAMP and PKC in other systems binds to the GH promoter upstream of the cAMP-responsive region and is therefore unlikely to mediate the signals from GRF and somatostatin (Imagawa et al., 1987).

GHF-1 is absolutely required for transcription of the GH gene *in vivo* (Castrillo et al., 1991; Li et al., 1990). However, as demonstrated by *in vivo* footprinting, in lactotrophs GHF-1, despite being fully active, cannot bind to the GH promoter *in vivo* (Klausing, in manuscript). Under the same set of conditions, however, it does bind to the GHF1 promoter. Other positive-acting factors must therefore be responsible for making the GH promoter available for GHF-1 binding. The Sp1- and dGHF-1-binding sites in the GH promoter partly overlap, and binding of the factors is mutually exclusive (Lemaigre et al., 1990; Schaufele et al., 1990b), indicating that Sp1 may function via a hit-and-run mechanism: Sp1 could assist in loading of GHF-1 before leaving its own binding site.

The opening of the GH chromatin structure could be mediated by a member of the fourth category of factors, which consists of the hormone-activated nuclear receptors GCR, TR, RAR, and RXR. It is reasonable to assume that only a specific subset of nuclear receptor homodimers and heterodimers can function in the context of the GH promoter. The spatial distribution of these functional units within the anterior pituitary is likely to be a major determinant for whether a GHF-1-producing cell can indeed express GH. Binding of the activated nuclear receptors may cause rearrangement of nucleosomes over the GH promoter, thereby allowing access for GHF-1 and other factors. According to this model, a specific subset of ligand-bound nuclear receptors forms a master switch for activation of GH gene expression in cells that already express GHF-1. The fact that exogenous expression of GHF-1 does not suffice to activate GH outside the pituitary indicates that only a pituitary-specific combination of receptor isoforms suffices to remodel the chromatin structure of the GH promoter. The finding of a pituitary-specific isoform of a TR is consistent with this hypothesis (Hodin et al., 1989). Nuclear receptor-mediated nucleosome rearrangement resulting in differential transcription factor binding is well documented in other systems (Archer et al., 1992; Cordingley et al., 1987; Pina et al., 1990; Rigaud et al., 1991). In addition to their proposed role at the chromatin level, nuclear receptors TR, RAR, and GCR also synergize with GHF-1 to activate presumably chromatin-free GH promoter constructs (Schaufele et al., 1992; Theill, unpublished data), suggesting that the receptors also cooperate more directly with GHF-1 in transcription complex.

GHF-1 is positioned on its recognition site as a dimer (Dixon et al., in manuscript; Ingraham et al., 1990), and the binding causes bending of the DNA (Verrijzer et al., 1991). In addition, when GHF-1 binds to the two recognition elements present in the GH promoter, a strong DNaseI-hypersensitive site appears. This hypersensitivity, located in the area between the binding sites, could result from marked bending of DNA in response to contact between GHF- 1 dimers on the proximal site and GHF-1 dimers on the distal site. In this scheme, the two levels of DNA topological changes induced by GHF-1 binding and GHF-1 homodimerization/multimerization contribute to the final positioning of GH promoter binding factors into an active transcriptional complex.

This model, in which binding of activated nuclear receptor(s) allows access of GHF-1, which in turn modifies the GH promoter DNA topology and facilitates formation of a transcription complex, is illustrated

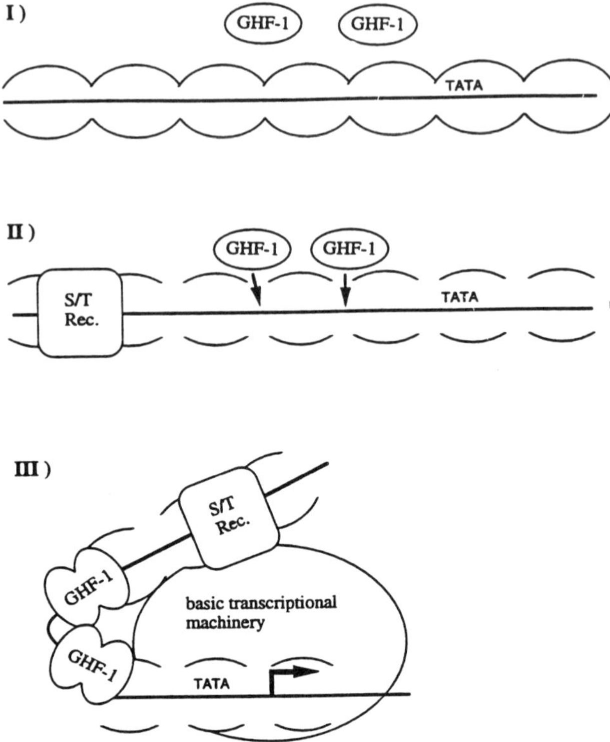

Figure 4   Model for transcriptional control of GH gene expression *in vivo*. I: Closed arches illustrate closed chromatin structure over GH promoter in all cell types except somatotrophs and mammosomatotrophs. GHF-1 is excluded from its binding site. II: Somatotroph-specific steroid or thyroid hormone receptor activity (S/T Rec.) binds to GH promoter and causes nucleosome rearrangement, thereby providing GHF-1 access to GHF-1-binding sites. III: Binding and multimerization of GHF-1 results in bending of DNA, which facilitates relative positioning of upstream binding factors and assembly of active transcriptional complex, causing transcription of the gene.

in Figure 4. In this paradigm, the composition of the nuclear receptor population and the availability of their ligands become the determining factors for the development of the somatotrophic progenitor into a somatotroph, a lactotroph, or a mammosomatotroph cell. In the simplest scenario, activated TR and ER would be required for somatotroph and lactotroph differentiation, respectively. The mammosomatotroph population would express both TRs and ERs and, depending on the pool of ligands, exhibit either the mammosomatotroph, somatotroph, or lactotroph phenotype. Expression of the nuclear receptor genes may be controlled by the local steroid and other hormone environment. In essence, the GH- and Prl-producing cells could be functionally interconvertible and controlled by the hormonal state. Evidence for such transdifferentiation is accumulating (Frawley and Boockfor, 1991).

Acknowledgments. The author wishes to thank Dr. Michael Karin for his critical reading of the manuscript. The discussions and contributions of Wendy Dixon, Kay Klausing, Katsuyuki Taginuma, Tom Tullius, Denise Lew, and Helen Brady are gratefully acknowledged. I am indebted to Donna Caruso for her help and encouragement in preparing this manuscript. This study was supported by a Senior Researcher Fellowship from the Danish Natural Science Research Council.

References

Archer TK, Lefebvre P, Wolford RG, Hager GL (1992): Transcription factor loading on the MMTV promoter: A bimodal mechanism for promoter activation. *Science* 255(5051):1573–1576.

Attisano L, Wrana JL, Cheifetz S, Massague J (1992): Novel activin receptors: Distinct genes and alternative mRNA splicing generate a repertoire of serine/threonine kinase receptors. *Cell* 68(1):97–108.

Aurora R, Herr W (1992): Segments of the POU domain influence one another's DNA-binding specificity. *Mol Cell Biol* 12(2):455–467.

Barinaga M, Yamonoto G, Rivier C, Vale W, Evans R, Rosenfeld MG (1985a): Transcriptional regulation of growth hormone gene expression by growth hormone-releasing factor. *Nature* (Lond) 306:84–85.

Barinaga M, Bilezikjian LM, Vale WW, Rosenfeld MG (1985b): Independent effects of growth hormone releasing factor on growth hormone release and gene transcription. *Nature* (Lond) 314:279–281.

Barsh GS, Seeburg PH, Gelinas RE (1983): The human growth hormone gene family: structure and evolution of the chromosomal locus. *Nucleic Acids Res* 11:3939.

Bedo G, Santisteban P, Aranda A (1989): Retinoic acid regulates growth hormone gene expression. *Nature* (Lond) 339(6221):231–234.

Behringer RR, Mathews LS, Palmiter RD, Brinster RL (1988): Dwarf mice produced by genetic ablation of growth hormone-expressing cells. *Genes & Dev* 2:453–461.

Bienz M, Tremml G (1988): Domain of Ultrabithorax expression in *Drosophila* visceral mesoderm from autoregulation and exclusion. *Nature* (Lond) 333(6173): 576–578.

Bilezikjian LM, Vale W (1983): Stimulation of adenosine 3′,5′-monophosphate production by growth hormone-releasing factor and its inhibition by somatostatin in anterior pituitary cells *in vitro* . *Endocrinology* 113:1726–1731.

Bilezikjian LM, Corrigan AZ, Vale W (1990): Activin-A modulates growth hormone secretion from cultures of rat anterior pituitary cells. *Endocrinology* 126(5):2369–2376.

Billestrup N, Swanson LW, Vale W (1986): Growth hormone-releasing factor stimulates proliferation of somatotrophs *in vitro*. *Proc Natl Acad Sci USA* 83:6854–6857.

Billestrup N, Gonzalez Manchon C, Potter E, Vale W (1990): Inhibition of somatotroph growth and growth hormone biosynthesis by activin *in vitro*. *Mol Endocrinol* 4(2):356–362.

Billestrup N, Mitchell RL, Vale W, Verma IM (1987): Growth hormone-releasing factor induces c-fos expression in cultured primary pituitary cells. *Mol Endocrinol* 1(4):300–305.

Blau HM (1992): Differentiation requires continuous active control. *Annu Rev Biochem* 61:1213–1230.

Bodner M, Karin M (1987): A pituitary-specific trans-acting factor can stimulate transcription from the growth hormone promoter in extracts of nonexpressing cells. *Cell* 50(2):267–275.

Bodner M, Castrillo J-L, Theill LE, Deerinck T, Ellisman M, Karin M (1988): The pituitary-specific transcription factor GHF-1 is a homeobox-containing protein. *Cell* 55(3):505–518.

Borrelli E, Hayman RA, Arias C, Sawchenko PE, Evans RM (1989): Transgenic mice with inducible dwarfism. *Nature (Lond)* 339(6225):538–541.

Bradley DJ, Young WSE, Weinberger C (1989): Differential expression of alpha and beta thyroid hormone receptor genes in rat brain and pituitary. *Proc Natl Acad Sci USA* 86(18):7250–7254.

Brazeau P, Vale W, Burgus R, Ling N, Butcher M, Rivier J, Guillemin R (1973): Hypothalamic polypeptide that inhibits the secretion of immunoreactive pituitary growth hormone. *Science* 179:77–79.

Brent GA, Harney JW, Moore DD, Larsen PR (1988): Multihormonal regulation of the human, rat, and bovine growth hormone promoters: Differential effects of 3′,5′-cyclic adenosine monophosphate, thyroid hormone, and glucocorticoids. *Mol Endocrinol* 2(9):792–798.

Brent GA, Larsen PR, Harney JW, Koenig RJ, Moore DD (1989a): Functional characterization of the rat growth hormone promoter elements required for induction by thyroid hormone with and without a co-transfected beta type thyroid hormone receptor. *J Biol Chem* 264(1):178–182.

Brent GA, Harney JW, Chen Y, Warne RL, Moore DD, Larsen PR (1989b): Mutations of the rat growth hormone promoter which increase and decrease response to thyroid hormone define a consensus thyroid hormone response element. *Mol Endocrinol* 3(12):1996–2004.

Brent GA, Dunn MK, Harney JW, Gulick T, Larsen PR, Moore DD (1989c): Thyroid hormone aporeceptor represses T3-inducible promoters and blocks activity of the retinoic acid receptor. *New Biol* 1(3):329–136.

Brent GA, Moore DD, Larsen PR (1991a): Thyroid hormone regulation of gene expression. *Annu Rev Physiol* 53:17–35.

Brent GA, Williams GR, Harney JW, Forman BM, Samuels HH, Moore DD, Larsen PR (1991b): Effects of varying the position of thyroid hormone response elements within the rat growth hormone promoter: Implications for positive and negative regulation by 3,5,3'-triiodothyronine. *Mol Endocrinol* 5(4):542–548.

Brent GA, Williams GR, Harney JW, Forman BM, Samuels HH, Moore DD, Larsen PR (1992): Capacity for cooperative binding of thyroid hormone (T3) receptor dimers defines wild type T3 response elements. *Mol Endocrinol* 6(4):502–514.

Brindle PK, Montminy MR (1992): The CREB family of transcription activators. *Curr Opin Genet Dev* 2(2):199–204.

Bruggemeier U, Kalff M, Franke S, Scheidereit C, Beato M (1991): Ubiquitous transcription factor OTF-1 mediates induction of the MMTV promoter through synergistic interaction with hormone receptors. *Cell* 64(3):565–572.

Burton FH, Hasel KW, Bloom FE, Sutcliffe JG (1991): Pituitary hyperplasia and gigantism in mice caused by a cholera toxin transgene. *Nature* (Lond) 350(6313):74–77.

Casanova J, Copp RP, Janocko L, Samuels HH (1985): 5'-Flanking DNA of the rat growth hormone gene mediates regulated expression by thyroid hormone. *J Biol Chem* 260:11744–11748.

Castrillo J-L, Bodner M, Karin M (1989): Purification of growth hormone-specific transcription factor GHF-1 containing homeobox. *Science* 243(4892):814–817.

Castrillo J-L, Theill LE, Karin M (1991): Function of the homeodomain protein GHF-1 in pituitary cell proliferation. *Science* 253(5016):197–199.

Catanzaro DF, West BL, Baxter JD, Reudelhuber TL (1987): A pituitary-specific factor interacts with an upstream promoter element in the rat growth hormone gene. *Mol Endocrinol* 1(1):90–96.

Cattini PA, Eberhardt NL (1987): Regulated expression of chimaeric genes containing the 5'-flanking regions of human growth hormone-related genes in transiently transfected rat anterior pituitary tumor cells. *Nucleic Acids Res* 15:1297–1309.

Chalfie M, Au M (1989): Genetic control of differentiation of the Caenorhabditis elegans touch receptor. *Science* 243(4894 Pt 1):1027–1033.

Chatelain A, Dupuoy JP, Dubois MP (1979): Ontogenesis of cells producing polypeptide hormones (ACTH, MSH, LPH, GH, prolactin) in the fetal hypophysis of the rat: Influence of the hypothalamus. *Cell Tissue Res* 196:409–427.

Chen EY, Liao YC, Smith DH, Barrera-Saldaña HA, Gelinas RE, Seeburg PH (1989): The human growth hormone locus: Nucleotide sequence, biology, and evolution. *Genomics* 4(4):479–497.

Chen RP, Ingraham HA, Treacy MN, Albert VR, Wilson L, Rosenfeld MG (1990): Autoregulation of pit-1 gene expression mediated by two cis-active promoter elements. *Nature* (Lond) 346(6284):583–586.

Chomczynski P, Brar A, Frohman LA (1988): Growth hormone synthesis and secretion by a somatomammotroph cell line derived from normal adult pituitary of the rat. *Endocrinology* 123(5):2276–2283.

Christiansen JS, Jorgensen JO (1991): Beneficial effects of GH replacement therapy in adults. *Acta-Endocrinol* (Copenh) 125(1)7–13.

Coenjaerts FE, De Vries E, Pruijn GJ, Van Driel W, Bloemers SM, Van der Lugt NM, Van der Vliet PC (1991): Enhancement of DNA replication by transcription factors NFI and NFIII/Oct-1 depends critically on the positions of their binding sites in the adenovirus origin of replication. *Biochim Biophys Acta* 1090(1):61–69.

Cohen S, Jurgens G (1991): *Drosophila* headlines. *Trends Genet* 7(8):267–272.

Copp RP, Samuels HH (1989): Identification of an adenosine 3′,5′-monophosphate (cAMP)-responsive region in the rat growth hormone gene: Evidence for independent and synergistic effects of cAMP and thyroid hormone on gene expression. *Mol Endocrinol* 3(5):790–796.

Cordingley MG, Riegel AT, Hager GL (1987): Steroid-dependent interaction of transcription factors with the inducible promoter of mouse mammary tumor virus *in vivo*. *Cell* 48(2):261–270.

Corrigan AZ, Bilezikjian LM, Carroll RS, Bald LN, Schmelzer CH, Fendly BM, Mason AJ, Chin WW, Schwall RH, Vale W (1991): Evidence for an autocrine role of activin B within rat anterior pituitary cultures. *Endocrinology* 128(3):1682–1684.

Courey AJ, Holzman DA, Jackson SP, Tjian R (1989): Synergistic activation by the glutamine-rich domains of human transcription factor Sp1. *Cell* 59(5):827–836.

Crenshaw EB III, Kalla K, Simmons DM, Swanson LW, Rosenfeld MG (1989): Cell-specific expression of the prolactin gene in transgenic mice is controlled by synergistic interactions between promoter and enhancer elements. *Genes & Dev* 3(7):959–972.

Crew MD, Spindler SR (1986): Thyroid hormone regulation of the transfected rat growth hormone promoter. *J Biol Chem* 261(11):5018–5022.

Dana S, Karin M (1989): Induction of human growth hormone promoter activity by the adenosine 3′,5′-monophosphate pathway involves a novel responsive element. *Mol Endocrinol* 3(5):815–821.

Davidson EH (1990): How embryos work: A comparative view of diverse modes of cell fate specification. *Development* (Camb) 108(3):365–389.

Davidson MB (1987): Effect of growth hormone on carbohydrate and lipid metabolism. *Endocr Rev* 8(2):115–131.

Davis KD, Lazar MA (1992): Selective antagonism of thyroid hormone action by retinoic acid. *J Biol Chem* 267(5):3185–3189.

Dearden NM, Holmes RL (1976): Cyto-differentiation and portal vascular development in the mouse adenohypophysis. *J Anat* 121(3):551–569.

DeLuca LM (1991): Retinoids and their receptors in differentiation, embryogenesis, and neoplasia. *FASEB J* 5(14):2924–2933.

Deschamps J, Meijlink F (1992): Mammalian homeobox genes in normal development and neoplasia. *Crit Rev Oncog* 3(1-2):117–173.

De-Simone V, Cortese R (1992): Transcription factors and liver-specific genes. *Biochim Biophys Acta* 1132(2):119–126.

Diamond DJ, Goodman HM (1985): Regulation of growth hormone messenger RNA synthesis by dexamethasone and triiodothyronine. *J Mol Biol* 181:41–62.

Dixon WJ, Theill LE, Karin M, Tullius TD: Structural dissection of the DNA-binding interactions of GHF-1 with the GH promoter (in manuscript).

Dobner PR, Kawasaki ES, Yu L-Y, Bancroft FC (1981): Thyroid or glucocorticoid hormone induces pre-growth-hormone mRNA and its probable nuclear precursor in rat pituitary cells. *Proc Natl Acad Sci USA* 78(4):2230–2234.

Dollé P, Castrillo J-L, Theill LE, Deerinck T, Ellisman M, Karin M (1990): Expression of GHF-1 protein in mouse pituitaries correlates both temporally and spatially with the onset of growth hormone gene activity. *Cell* 60(5):809–820.

Drolet DW, Scully KM, Simmons DM, Wegner M, Chu KT, Swanson LW, Rosenfeld MG (1991): TEF, a transcription factor expressed specifically in the anterior pituitary during embryogenesis, defines a new class of leucine zipper proteins. *Genes & Dev* 5(10):1739–1753.

Dynlacht BD, Attardi LD, Admon A, Freeman M, Tjian R (1989): Functional analysis of NTF-1, a developmentally regulated *Drosophila* transcription factor that binds neuronal cis elements. *Genes & Dev* 3(11):1677–1688.

Eicher EM, Beamer WG (1980): New mouse dw allele: Genetic location and effects on lifespan and growth hormone levels. *J Hered* 71:187–190.

Eliard PH, Marchand MJ, Rousseau GG, Formstecher P, Mathy-Hartert M, Belayew A, Martial JA (1985): Binding of the human glucocorticoid receptor to defined regions in the human growth hormone and placental lactogen genes. *DNA* (NY) 4:409–417.

Evans T, Frelsenfeld G, Reitman M (1990): Control of globin gene transcription. *Annu Rev Cell Biol* 6:95–124.

Evans RM, Birnberg NC, Rosenfeld MG (1982): Glucocorticoid and thyroid hormones transcriptionally regulate growth hormone gene expression. *Proc Natl Acad Sci USA* 79:7659–7663.

Felsenfeld G (1992): Chromatin as an essential part of the transcriptional mechanism. *Nature* 355(6357):219–224.

Fienberg AA, Utset MF, Bogarad LD, Hart CP, Awgulewitsch A, Ferguson Smith A, Fainsod A, Rabin M, Ruddle FH (1987): Homeobox genes in murine development. *Curr Top Dev Biol* 23:233–256.

Fink G, Smith GC (1971): Ultrastructural features of the developing hypothalamo-hypophysial axis in the rat. *Z Zellforsch* 119:208–226.

282   Lars Eyde Theill

Finney M, Ruvkun G (1990): The unc-86 gene product couples cell lineage and cell identity in C. elegans. *Cell* 63(5):895–905.

Finney M, Ruvkun G, Horvitz HR (1988): The C. elegans cell lineage and differentiation gene unc-86 encodes a protein with a homeodomain and extended similarity to transcription factors. *Cell* 55(5):757–769.

Flug F, Copp RP, Casanova J, Horowitz ZD, Janocko L, Plotnick M, Samuels HH (1987): cis-Acting elements of the rat growth hormone gene which mediate basal and regulated expression by thyroid hormone. *J Biol Chem* 262(13):6373–6382.

Forman BM, Yang CR, Stanley F, Casanova J, Samuels HH (1988): c-erbA protooncogenes mediate thyroid hormone-dependent and independent regulation of the rat growth hormone and prolactin genes. *Mol Endocrinol* 2(10):902–911.

Foulkes NS, Borrelli E, Sassone Corsi P (1991): CREM gene: Use of alternative DNA-binding domains generates multiple antagonists of cAMP-induced transcription. *Cell* 64(4):739–749.

Fox SR, Jong MT, Casanova J, Ye ZS, Stanley F, Samuels HH (1990): The homeodomain protein, Pit-1/GHF-1, is capable of binding to and activating cell-specific elements of both the growth hormone and prolactin gene promoters. *Mol Endocrinol* 4(7):1069–1080.

Frawley LS (1989): Mammosomatotrophs: current status and possible functions. *Trends Endocrinol Metabol* 1:31–34.

Frawley LS, Boockfor FR (1991): Mammosomatotrophs: Presence and functions in normal and neoplastic pituitary tissue. *Endocr Rev* 12(4):337–355.

Frawley LS, Boockfor FR, Hoeffler JP (1985a): Functional maturation of somatotropes in fetal rat pituitaries: analysis by reverse hemolytic plaques assay. *Endocrinology* 116:2355–2360.

Frawley LS, Boockfor FR, Hoeffler JP (1985b): Identification by plaque assays of a pituitary cell type that secretes both growth hormone and prolactin. *Endocrinology* 116:734–737.

Frohman LA, Szabo M (1981): Ectopic production of growth hormone-releasing factor by carcinoid and pancreatic islet tumors associated with acromegaly. *Prog Clin Biol Res* 74:259–271.

Gash D, Ahmad N, Schechter J (1982): Comparison of gonadotroph, thyrotroph and mammotroph development in situ, in transplants and in organ culture. *Neuroendocrinology* 34:222–228.

Gehring WJ (1987): Homeo boxes in the study of development. *Science* 236(4806):1245–1252.

Gelato MC, Merriam GR (1986): Growth hormone releasing hormone. *Annu Rev Physiol* 48:569–91.

Giguere V, Ong ES, Segui P, Evans RM (1987): Identification of a receptor for the morphogen retinoic acid. *Nature* (Lond) 330(6149):624–629.

Glass CK, Franco R, Weinberger C, Albert VR, Evans RM, Rosenfeld MG (1987): A c-erb-A binding site in rat growth hormone gene mediates trans-activation by thyroid hormone. *Nature* (Lond) 329(6141):738–741.

Glass CK, Holloway JM, Devary OV, Rosenfeld MG (1988): The thyroid hormone receptor binds with opposite transcriptional effects to a common sequence motif in thyroid hormone and estrogen response elements. *Cell* 54(3):313–323.

Glineur C, Bailly M, Ghysdael J (1989): The c-erbA alpha-encoded thyroid hormone receptor is phosphorylated in its amino terminal domain by casein kinase II. *Oncogene* 4(10):1247–1254.

Goldberg Y, Glineur C, Gesquiere JC, Ricouart A, Sap J, Vennstrom B, Ghysdael J (1988): Activation of protein kinase C or cAMP-dependent protein kinase increases phosphorylation of the c-erbA-encoded thyroid hormone receptor and of the v-erbA-encoded protein. *EMBO J* (8):2425–2433.

Gonzalez GA, Montminy MR (1989): Cyclic AMP stimulates somatostatin gene transcription by phosphorylation of CREB at serine 133. *Cell* 59(4):675–680.

Gonzalez GA, Yamamoto KK, Fischer WH, Karr D, Menzel P, Biggs W III, Vale WW, Montminy MR (1989): A cluster of phosphorylation sites on the cyclic AMP-regulated nuclear factor CREB predicted by its sequence. *Nature* (Lond) 337(6209):749–752.

Gould AP, Brookman JJ, Strutt DI, White RA (1990): Targets of homeotic gene control in *Drosophila*. *Nature* (Lond) 348(6299):308–312.

Goulding MD, Gruss P (1989): The homeobox in vertebrate development. *Curr Opin Cell Biol* 1(6):1088–1093.

Graba Y, Aragnol D, Laurenti P, Garzino V, Charmot D, Berenger H, Pradel J (1992): Homeotic control in *Drosophila*; the scabrous gene is an *in vivo* target of Ultrabithorax proteins. *EMBO J* 11(9):3375–3384.

Guillemin R Brazeau P, Böhlen P, Esch F, Ling N, Wehrenberg WB (1982): Growth hormone-releasing factor from a human pancreatic tumor that caused acromegaly. *Science* 218:585–587.

Hatzopoulos AK, Stoykova AS, Erselius JR, Goulding M, Neuman T, Gruss P (1990): Structure and expression of the mouse Oct2a and Oct2b, two differentially spliced products of the same gene. *Development* 109(2):349–362.

Hayes J, Tullius TD (1989): The missing nucleoside experiment: a new technique to study recognition of DNA by protein. *Biochemistry* 28(24):9521–9527.

Herr W, Sturm RA, Clerc RG, Corcoran LM, Baltimore D, Sharp PA, Ingraham HA, Rosenfeld MG, Finney M, Ruvkun G (1988): The POU domain: A large conserved region in the mammalian pit-1, oct-1, oct-2, and Caenorhabditis elegans unc-86 gene products. *Genes & Dev* 2(12a):1513–1516.

Heyman RA, Mangelsdorf DJ, Dyck JA, Stein RB, Eichele G, Evans RM, Thaller C (1992): 9-cis Retinoic acid is a high affinity ligand for the retinoid X receptor. *Cell* 68(2)397–406.

Hiromi Y, Gehring WJ (1987): Regulation and function of the *Drosophila* segmentation gene *fushi tarazu*. *Cell* 50(6):963–974.

Hodin RA, Lazar MA, Wintman BI, Darling DS, Koenig RJ, Larsen PR, Moore

DD, Chin WW (1989): Identification of a thyroid hormone receptor that is pituitary-specific. *Science* 244(4900):76–79.

Hoeffler JP, Boockfor FR, Frawley LL (1985): Ontogeny of prolactin cells in neonatal rats: Initial prolactin secretors also release growth hormone. *Endocrinology* 117:187–195.

Ikuyama S, Natori S, Nawata H, Kato K, Ibayashi H, Kariya T, Sakai T, Rivier J, Vale W (1988): Characterization of growth hormone-releasing hormone receptors in pituitary adenomas from patients with acromegaly. *J Clin Endocrinol Metab* 66(6):1265–1271.

Imagawa M, Chiu R, Karin M (1987): Transcription factor AP-2 mediates induction by two different signal-transduction pathways: Protein kinase C and cAMP. *Cell* 51(2):251–260.

Ingham PW (1988): The molecular genetics of embryonic pattern formation in *Drosophila*. *Nature* (Lond) 335:25–34.

Ingham PW, Martinez Arias A (1992): Boundaries and fields in early embryos. *Cell* 68(2):221–235.

Ingraham HA, Chen RP, Mangalam HJ, Elsholtz HP, Flynn SE Lin CR, Simmons DM, Swanson L, Rosenfeld MG (1988): A tissue-specific transcription factor containing a homeodomain specifies a pituitary phenotype. *Cell* 55(3):519–529.

Ingraham HA, Flynn SE, Voss JW, Albert VR, Kapiloff MS, Wilson L, Rosenfeld MG (1990): The POU-specific domain of Pit-1 is essential for sequence-specific, high affinity DNA binding and DNA-dependent Pit-1Pit-1 interactions. *Cell* 61(6):1021–1033.

Inoue K, Hattori M, Sakai T, Inukai S, Fujimoto N, Ito A (1990): Establishment of a series of pituitary clonal cell lines differing in morphology, hormone secretion, and response to estrogen. *Endocrinology* 126(5):2313–2320.

Isaacs RE, Findell PR, Mellon P, Wilson CB, Baxter JD (1987): Hormonal regulation of expression of the endogenous and transfected human growth hormone gene. *Mol Endocrinol* 1:569–576.

Isaksson OG, Lindahl A, Nilsson A, Isgaard J (1987): Mechanism of the stimulatory effect of growth hormone on longitudinal bone growth. *Endocr Rev* 8(4):426–438.

Ishikawa K, Katakami H, Jansson J-O, Frohman LA (1986): Ontogenesis of growth hormone-releasing hormone neurons in the rat hypothalamus. *Neuroendocrinology* 43: 537–542.

Johnson WA, Hirsh J (1990): Binding of a *Drosophila* POU-domain protein to a sequence element regulating gene expression in specific dopaminergic neurons. *Nature* (Lond) 343(6257):467–470.

Judd AM, MacLeod RM (1984): Growth hormone releasing factor increases growth hormone from MtTW15 pituitary tumors. *Brain Res* 308:137–140.

Kao CC, Lieberman PM, Schmidt MC, Zhou Q, Pei R, Berk AJ (1990): Cloning of a transcriptionally active human TATA binding factor. *Science* 248(4963): 1646–1650.

Kapiloff MS, Farkash Y, Wegner M, Rosenfeld MG (1991): Variable effects of phosphorylation of Pit-1 dictated by the DNA response elements. *Science* 253(5021):786–789.

Karin M, Castrillo J-L, Theill LE (1990): Growth hormone gene regulation: A paradigm for cell-type-specific gene activation. *Trends Genet* 6(3):92–96.

Kessel M, Gruss P (1990): Murine developmental control genes. *Science* 249(4967):374–379.

Kineman RD, Faught WJ, Frawley LS (1992): Steroids can modulate transdifferentiation of prolactin and growth hormone cells in bovine pituitary cultures. *Endocrinology* 130(6):3289–3294.

Kirchgessner TG, Chuat JC, Heinzmann C, Etienne J, Guilhot S, Svenson K, Ameis D, Pilon C, dAuriol L, Andalibi A (1989): Organization of the human lipoprotein lipase gene and evolution of the lipase gene family. *Proc Natl Acad Sci USA* 86(24):9647–9651.

Kissinger CR, Liu BS, Martin Blanco E, Kornberg TB, Pabo CO (1990): Crystal structure of an engrailed homeodomain-DNA complex at 2.8 Åresolution: A framework for understanding homeodomain-DNA interactions. *Cell* 63(3):579–590.

Kitaoka M, Kojima I, Ogata E (1988): Activin A: A modulator of multiple types of anterior pituitary cells. *Biochem Biophys Res Commun* 157(1):48–54.

Kitaoka M, Takano K, Tanaka Y, Kojima I, Teramoto A, Ogata E (1991): Inhibition of growth hormone secretion by activin A in human growth hormone-secreting tumour cells. *Acta Endocrinol* (Copenh) 124(6):666–671.

Klausing K, Babin J, Karin M (1993): Differential growth hormone gene activation in pituitary cells is determined by promoter accessibility. In manuscript.

Kliewer SA, Umesono K, Mangelsdorf DJ, Evans RM (1992): Retinoid X receptor interacts with nuclear receptors in retinoic acid, thyroid hormone and vitamin D3 signalling. *Nature* (Lond) 355(6359)446–449.

Koenig RJ, Brent GA, Warne RL, Larsen PR, Moore DD (1987): Thyroid hormone receptor binds to a site in the rat growth hormone promoter required for induction by thyroid hormone. *Proc Natl Acad Sci USA* 84(16):5670–5674.

Koenig RJ, Warne RL, Brent GA, Harney JW, Larsen PR, Moore DD (1988): Isolation of a cDNA clone encoding a biologically active thyroid hormone receptor. *Proc Natl Acad Sci USA* 85(14):5031–5035.

Koenig RJ, Lazar MA, Hodin RA, Brent GA, Larsen PR, Chin WW, Moore DD (1989): Inhibition of thyroid hormone action by a non-hormone binding c-erbA protein generated by alternative mRNA splicing. *Nature* (Lond) 337(6208):659–661.

Konzak KE, Moore DD (1992): Functional isoforms of Pit-1 generated by alternative messenger RNA splicing. *Mol Endocrinol* 6(2):241–247.

Kristie TM, Sharp PA (1990): Interactions of the Oct-1 POU subdomains with specific DNA sequences and with the HSV a-trans-activator protein. *Genes & Dev* 4(12B):2383–2396.

Lamberts SW (1988): The role of somatostatin in the regulation of anterior pituitary hormone secretion and the use of its analogs in the treatment of human pituitary tumors. *Endocr Rev* 9(4):417–436.

Lai JS, Herr W (1992): Ethidium bromide provides a simple tool for identifying genuine DNA-independent protein associations. *Proc Natl Acad Sci USA* 89(15):6958–6962.

Landis CA, Masters SB, Spada A, Pace AM, Bourne HR, Vallar L (1989): GT-Pase inhibiting mutations activate the alpha chain of Gs and stimulate adenylyl cyclase in human pituitary tumours. *Nature* (Lond) 340(6236):692–696.

Larsen PR, Harney JW, Moore DD (1986): Repression mediates cell-type-specific expression of the rat growth hormone gene. *Proc Natl Acad Sci USA* 83(21): 8283–8287.

LeBowitz JH, Clerc RG, Brenowitz M, Sharp PA (1989): The Oct-2 protein binds cooperatively to adjacent octamer sites. *Genes & Dev* 3(10):1625–1638.

Lefevre C, Imagawa M, Dana S, Grindlay J, Bodner M, Karin M (1987): Tissue-specific expression of the human growth hormone gene is conferred in part by the binding of a specific trans-acting factor.*EMBO J* 6(4):971–981.

Lemaigre FP, Courtois SJ, Durviaux SM, Egan CJ, LaFontaine DA, Rousseau GG (1989): Analysis of cis- and trans-acting elements in the hormone-sensitive human somatotropin gene promoter. *J Steroid Biochem* 34(1-6):79–83.

Lemaigre FP, Lafontaine DA, Courtois SJ, Durviaux SM, Rousseau GG (1990): Sp1 can displace GHF-1 from its distal binding site and stimulate transcription from the growth hormone gene promoter. *Mol Cell Biol* 10(4):1811–1814.

Leong DA, Lau SK, Sinha YN, Kaiser DL, Thorner MO (1985): Enumeration of lactotropes and somatotropes among male and female pituitary cells in culture: Evidence in favor of a mammosomatotrope subpopulation in the rat. *Endocrinology* 116:1371–1378.

Lew AM, Elsholtz HP (1991): Cloning of the human cDNA for transcription factor Pit-1. *Nucleic Acids Res* 19(22):6329.

Lew D, Brady H, Klausing K, Yaginuma K, Theill LE, Stauber C, Karin M, Mellon PL: GHF-1-promoter targeted immortalization of a somatotropic progenitor cell results in dwarfism in transgenic mice. *Genes & Dev* (in press).

Li S, Crenshaw EB III, Rawson EJ, Simmons DM, Swanson LW, Rosenfeld MG (1990): Dwarf locus mutants lacking three pituitary cell types result from mutations in the POU-domain gene pit-1. *Nature* (Lond) 347(6293):528–533.

Lin C, Lin S-C, Chang C-P, Rosenfeld MG (1992a): Pit-1-dependent expression of the receptor for growth hormone releasing factor mediates pituitary cell growth. *Nature* 360:765–768.

Lin HY, Wang XF, Ng EE, Weinberg RA, Lodish HF (1992b): Expression cloning of the TGF-beta type II receptor, a functional transmembrane serine/threonine kinase. *Cell* 68(4):775–785.

Lira SA, Crenshaw EB III, Glass CK, Swanson LW, Rosenfeld MG (1988): Identification of rat growth hormone genomic sequences targeting pituitary expression in transgenic mice. *Proc Natl Acad Sci USA* 85(13):4755–4759.

Lufkin T, Bancroft C (1987): Identification by cell fusion of gene sequences that interact with positive trans-acting factors. *Science* 237(4812):283–286.

Lugo DI, Roberts JL, Pintar JE (1989): Analysis of proopiomelanocortin gene expression during prenatal development of the rat pituitary gland. *Mol Endocrinol* 3(8):1313–1324.

Lyons J, Landis CA, Harsh G, Vallar L, Grunewald K, Feichtinger H, Duh QY, Clark OH, Kawasaki E, Bourne HR (1990): Two G protein oncogenes in human endocrine tumors. *Science* 249(4969):655–659.

Mangalam HJ, Albert VR, Ingraham HA, Kapiloff M, Wilson L, Nelson C, Elsholtz H, Rosenfeld MG (1989): A pituitary POU domain protein, Pit-1, activates both growth hormone and prolactin promoters transcriptionally. *Genes & Dev* 3(7):946–958.

Martial JA, Seeburg PH, Guenzi D, Goodman HM, Baxter JD (1977): Regulation of growth hormone gene expression: Synergistic effects of thyroid and glucocorticoid hormones. *Proc Natl Acad Sci USA* 74(10):4293–4295.

Martin KJ (1991): The interactions of transcription factors and their adaptors, coactivators and accessory proteins. *Bioessays* 13(10) 499–503.

Mathews LS, Vale WW (1991): Expression cloning of an activin receptor, a predicted transmembrane serine kinase. *Cell* 65(6):973–982.

Mathews LS, Vale WW, Kintner CR (1992): Cloning of a second type of activin receptor and functional characterization in Xenopus embryos. *Science* 255(5052):1702–1705.

Massague J (1987): The TGF-β family of growth and differentiation factors. *Cell* 49(4):437–438.

Mayo KE (1985): *Proc Natl Acad Sci USA* 82:63–67.

Mayo KE (1992): Molecular cloning and expression of a pituitary-specific receptor for growth hormone-releasing hormone. *Mol Endocrinol* 6(10):1734–1744.

Mayo KE, Hammer RE, Swanson LW, Brinster RL, Rosenfeld MG, Evans RM (1988): Dramatic pituitary hyperplasia in transgenic mice expressing a human growth hormone-releasing factor gene. *Mol Endocrinol* 2(7):606–612.

McCormick A, Brady H, Fukushima J, Karin M (1991): The pituitary-specific regulatory gene GHF1 contains a minimal cell type-specific promoter centered around its TATA box. *Genes & Dev* 5(8):1490–1503.

McCormick A, Brady H, Theill LE, Karin M (1990): Regulation of the pituitary-specific homeobox gene GHF1 by cell-autonomous and environmental cues. *Nature* (Lond) 345(6278):829–832.

McCormick A, Wu D, Castrillo J-L, Dana S, Strobl J, Thompson EB, Karin M (1988): Extinction of growth hormone expression in somatic cell hybrids involves repression of the specific trans-activator GHF-1. *Cell* 55(2):379–389.

McGinnis W, Krumlauf R (1992): Homeobox genes and axial patterning. *Cell* 68(2):283–302.

Meijer D, Graus A, Kraay R, Langeveld A, Mulder MP, Grosveld G (1990): The octamer binding factor Oct6: cDNA cloning and expression in early embryonic cells. *Nucleic Acids Res* 18(24):7357–7365.

Melton DA (1991): Pattern formation during animal development. *Science* 252(5003):234–241.

Miller WL, Eberhardt NL (1983): Structure and evolution of the growth hormone gene family. *Endocr Rev* 4:97.

Monuki ES, Weinmaster G, Kuhn R, Lemke G (1989): SCIP: A glial POU domain gene regulated by cyclic AMP. *Neuron* 3(6):783–793.

Monuki ES, Kuhn R, Weinmaster G, Trapp BD, Lemke G (1990): Expression and activity of the POU transcription factor SCIP. *Science* 249(4974):1300–1303.

Morimoto N, Kawakami F, Makino S, Chihara K, Hasegawa M, Ibata Y (1988): Age-related changes in growth hormone releasing factor and somatostatin in the rat hypothalamus. *Neuroendocrinology* 47(5):459–464.

Morita S, Fernandez Mejia C, Melmed S (1989): Retinoic acid selectively stimulates growth hormone secretion and messenger ribonucleic acid levels in rat pituitary cells. *Endocrinology* 124(5):2052–2056.

Morris AE, Kloss B, McChesney RE, Bancroft C, Chasin LA (1992): An alternatively spliced Pit-1 isoform altered in its ability to trans-activate. *Nucleic Acids Res* 20(6):1355–1361.

Moses K (1991): The role of transcription factors in the developing *Drosophila* eye. *Trends Genet* 7(8):250–255.

Müller-Immergluck MM, Schaffner W, Matthias P (1990): Transcription factor Oct-2A contains functionally redundant activating domains and works selectively from a promoter but not from a remote enhancer position in non-lymphoid (HeLa) cells. *EMBO J* 9(5):1625–1634.

Murray MB, Zilz ND, McCreary NL, MacDonald MJ, Towle HC (1988): Isolation and characterization of rat cDNA clones for two distinct thyroid hormone receptors. *J Biol Chem* 263(25):12770–12777.

Nelson C, Albert VR, Elsholtz HP, Lu LI-W, Rosenfeld MG (1988): Activation of cell-specific expression of rat growth hormone and prolactin genes by a common transcription factor. *Science* 239:1400–1405.

Nelson C, Crenshaw EB III, Franco R, Lira SA, Albert VR, Evans RM, Rosenfeld MG (1986): Discrete cis-active genomic sequences dictate the pituitary cell type-specific expression of rat prolactin and growth hormone genes. *Nature* (Lond) 322:557–562.

Nevins JR (1991): Transcriptional activation by viral regulatory proteins. *Trends Biochem Sci* 16(11)435–439.

Nogami H, Suzuki K, Enomoto H, Ishikawa H (1989): Studies on the development of growth hormone and prolactin cells in the rat pituitary gland by in situ hybridization. *Cell Tissue Res* 255(1):23–28.

Norman MF, Lavin TN, Baxter JD, West BL (1989): The rat growth hormone gene contains multiple thyroid response elements. *J Biol Chem* 264:12063–12073.

Nusslein-Volhard C, Wieschaus E (1980): Mutations affecting segment number and polarity in *Drosophila*. *Nature* (Lond) 287:795–801.

Nusslein-Volhard C, Frohnhofer HG, Lehmann R (1987): Determination of anteroposterior polarity in *Drosophila*. *Science* 238(4834):1675–1681.

Ohta K, Nobukuni Y, Mitsubchi H, Fujimoto S, Matsuo N, Inagaki H, Endo F, Matsuda I (1992): Mutations in the Pit-1 gene in children with combined pituitary hormone deficiency. *Biochem Biophys Res Commun* 189(2):851–855.

Okazawa H, Okamoto K, Ishino F, Ishino-Kaneko T, Takeda S, Toyoda Y, Muramatsu M, Hamada H (1991): The oct3 gene, a gene for an embryonic transcription factor, is controlled by a retinoic acid repressible enhancer. *EMBO J* 10(10):2997–3005.

Oliver C, Eskay RL, Porter JC (1980): Developmental changes in brain TRH and in plasma and pituitary TSH and prolactin levels in the rat. *Biol Neonate* 37:145–152.

Ono M, Takayama Y (1992): Structures of cDNAs encoding chum salmon pituitary-specific transcription factor, Pit-1/GHF-1. *Gene* 226:275–279.

Otting G, Qian YQ, Muller M, Affolter M, Gehring W, Wuthrich K (1988): Secondary structure determination for the Antennapedia homeodomain by nuclear magnetic resonance and evidence for a helix-turn-helix motif. *EMBO J* 7(13):4305–4309.

Owerbach D, Rutter W, Martial J, Baxter J (1980): Genes for growth hormone, chorionic somatomammotropin, and a growth hormone-like genes on chromosome 17 in humans. *Science* 209:289.

Paek I, Axel R (1987): Glucocorticoids enhance stability of human growth hormone mRNA. *Mol Cell Biol* 7(4):1496–1507.

Paonessa G, Gounari F, Frank R, Cortese R (1988): Purification of a NF1-like DNA-binding protein from rat liver and cloning of the corresponding cDNA. *EMBO J* 7(10):3115–3123.

Pan WT, Liu QR, Bancroft C (1990): Identification of a growth hormone gene promoter repressor element and its cognate double- and single-stranded DNA-binding proteins. *J Biol Chem* 265(12):7022–7028.

Patel YC, Srikant CB (1986): Somatostatin mediation of adenohypophysial secretion. *Annu Rev Physiol* 48:551–567.

Perkins LA, Perrimon N (1991): The molecular genetics of tail development in *Drosophila* melanogaster. *in vivo* 5(5):521–531.

Peers B, Monget P, Nalda MA, Voz ML, Berwaer M, Belayew A, Martial JA (1991): Transcriptional induction of the human prolactin gene by cAMP requires two cis-acting elements and at least the pituitary-specific factor Pit-1. *J Biol Chem* 266(27):18127–18134.

Pfaffle RW, DiMattia GE, Parks JS, Brown MR, Wit JM, Jansen M, Van der Nat H, Van den Brande JL, Rosenfeld MG, Ingraham HA (1992): Mutation of the POU-specific domain of Pit-1 and hypopituitarism without pituitary hypoplasia. *Science* 257(5073):1118–1121.

Pina B, Bruggemeier U, Beato M (1990): Nucleosome positioning modulates accessibility of regulatory proteins to the mouse mammary tumor virus promoter. *Cell* 60(5):719–731.

Pomerantz JL, Kristie TM, Sharp PA (1992): Recognition of the surface of a homeodomain protein. *Genes & Dev* 6(11):2047–2057.

Radovick S, Nations M, Du Y, Berg LA, Weintraub BD, Wondisford FE (1992): A mutation in the POU-domain of Pit-1 responsible for combined pituitary hormone deficiency. *Science* 257(5073):1115–1118.

Reichlin S, ed. (1987): Somatostatin. Basic and Clinical Status, p. 1. Plenum, New York.

Reid L (1990): From gradients to axes, from morphogenesis to differentiation. *Cell* 63(5):875–882.

Rigaud G, Roux J, Pictet R, Grange T (1991): *in vivo* footprinting of rat TAT gene: Dynamic interplay between the glucocorticoid receptor and a liver-specific factor. *Cell* 67(5):977–986.

Rivier J, Spiess J, Thorner M, Vale W (1982): Characterization of a growth hormone-releasing factor from a human pancreatic islet tumour. *Nature* (Lond) 300:276–278.

Robins DM, Paek I, Seeburg PH, Axel R (1982): Regulated expression of human growth hormone genes in mouse cells. *Cell* 29:623–631.

Rosenfeld MG (1991): POU-domain transcription factors: Pou-er-ful developmental regulators. *Genes & Dev* 5(6):897–907.

Rosner MH, Vigano MA, Ozato K, Timmons PM, Poirier F, Rigby PW, Staudt LM (1990): A POU-domain transcription factor in early stem cells and germ cells of the mammalian embryo. *Nature* (Lond) 345(6277):686–692.

Rousseau GG, Eliard PH, Barlow JW, Lemaigre FP, Lafontaine DA, De Nayer P, Economidis IV, Formstecher P, Idziorek T, Mathy Hartert M (1987): Approach to the molecular mechanisms of the modulation of growth hormone gene expression by glucocorticoid and thyroid hormones. *J Steroid Biochem* 27(1-3):149–158.

Roux M, Bartke A, Dumont F, Dubois MP (1982): Immunohistological study of the anterior pituitary glandpars distalis and pars intermediain dwarf mice. *Cell Tissue Res* 223:415–420.

Roy RJ, Gosselin P, Anzivino MJ, Moore DD, Guerin SL (1992): Binding of a nuclear protein to the rat growth hormone silencer element. *Nucleic Acids Res* 20(3):401–408.

Rudman D, Feller AG, Nagraj HS, Gergans GA, Lalitha PY, Goldberg AF, Schlenker RA, Cohn L, Rudman IW, Mattson DE (1990): Effects of human growth hormone in men over 60 years old. *N Engl J Med* 323(1):1–6.

Ruvkun G, Finney M (1991): Regulation of transcription and cell identity by POU domain proteins. *Cell* 64(3):475–478.

Sap J, deMagistris L, Stunnenberg H, Vennstrom B (1990): A major thyroid hormone response element in the third intron of the rat growth hormone gene. *EMBO J* 9(3):887–896.

Sawadogo M, Roeder RG (1985): Interaction of a gene-specific transcription factor with the adenovirus major late promoter upstream of the TATA box region. *Cell* 43:165–175.

Schaufele F, Cassill JA, West BL, Reudelhuber T (1990a): Resolution by diagonal gel mobility shift assays of multisubunit complexes binding to a functionally important element of the rat growth hormone gene promoter. *J Biol Chem* 265(24):14592–14598.

Schaufele F, West BL, Reudelhuber TL (1990b): Overlapping Pit-1 and Sp1 binding are both essential to full rat growth hormone gene promoter activity despite mutually exclusive Pit-1 and Ps1 binding. *J Biol Chem* 265(28):17189–17196.

Schaufele F, West BL, Baxter JD (1992): Synergistic activation of the rat growth hormone promoter by Pit-1 and the thyroid hormone receptor. *Mol Endocrinol* 6(4):656–665.

Schöler HR (1991): Octamania: The POU factors in murine development. *Trends Genet* 7(10):323–329.

Schonbrunn A, Koch BD (1987): Mechanisms by which somatostatin inhibits pituitary hormone release. In: Somatostatin. Basic and Clinical Status, Reichlin S, ed., p. 121. Plenum, New York.

Schubert D, Kimura H, LaCorbiere M, Vaughan J, Karr D, Fischer WH (1990): Activin is a nerve cell survival molecule. *Nature* (Lond) 344(6269):868–870.

Schwind (1928): The development of the hypophysis cerebri of the albino rat. *Am J Anat* 41:295–319.

Seifert H, Perrin M, Rivier J, Vale W (1985): Binding sites for growth hormone releasing factor on rat anterior pituitary cells. *Nature* (Lond) 313:487–489.

Selby MJ, Barta A, Baxter JD, Bell GI, Eberhardt NL (1984): Analysis of a major human chorionic somatomammotropin gene. Evidence for two functional promoter elements. *J Biol Chem* 259:13131.

Seo H (1985): In: *The Pituitary Gland*, pp. 57–82. Raven Press, New York.

Setalo G, Nakane P (1976): Functional differentiation of the fetal anterior pituitary cells in the rat. *Endocrinol Exp* (Bratisl) 10:155.

Sheng M, Thompson MA, Greenberg ME (1991): CREB: A $Ca^{(2+)}$-regulated transcription factor phosphorylated by calmodulin-dependent kinases. *Science* 252(5011):1427–1430.

Shupnik MA, Rosenzweig BA, Friend KE, Mason ME (1992): Thyrotropin (TSH)-releasing hormone-responsive elements in the rat TSH beta gene have distinct biological and nuclear protein-binding properties. *Mol Endocrinol* 6(1):43–52.

Simmons DM, Voss JW, Ingraham HA, Holloway JM, Briode RS, Rosenfeld MG, Swanson LW (1990): Pituitary cell phenotypes involve cell-specific Pit-1 mRNA translation and synergistic interactions with other classes of transcription factors. *Genes & Dev* 4(5):695–711.

Singh H, LeBowitz JH, Baldwin AS, Jr., Sharp PA (1988): Molecular cloning of an enhancer binding protein: isolation by screening of an expression library with a recognition site DNA. *Cell* 52(3):415–423.

Slabaugh MD, Lieberman ME, Rutledge JJ, Gorski J (1982): Ontogeny of growth hormone and prolactin gene expression in mice. *Endocrinology* 110:1489–1497.

Slater EP, Rabenau O, Karin M, BAxter JD, Beato M (1985): Glucocorticoid receptor binding and activation of a heterologous promoter by dexamethasone by the first intron of the human growth hormone gene. *Mol Cell Biol* 5:2984–2992.

Snell GD (1929): Dwarf, a new mendelian recessive character of the house mouse. *Proc Natl Acad Sci USA* 15:733–734.

Spiess J, Rivier J, Vale W (1983): Characterization of rat hypothalamic growth hormone-releasing factor. *Nature* (Lond) 503:532–535.

Spindler SR, Mellon SH, Baxter JD (1982): Growth hormone gene transcription is regulated by thyroid and glucocorticoid hormones in cultured rat pituitary tumor cells. *J Biol Chem* 257(19):11627–11632.

St. Johnston D, Nusslein Volhard C (1992): The origin of pattern and polarity in the Drosophila embryo. *Cell* 68(2):201–219.

Steinfelder HJ, Radovick S, Mroczynski MA, Hauser P, McClaskey JH, Weintraub BD, Wondisford FE (1992a): Role of a pituitary-specific transcription factor (pit-1/GHF-1) or a closely related protein in cAMP regulation of human thyrotropin-beta subunit gene expression. *J Clin Invest* 89(2):409–419.

Steinfelder HJ, Radovick S, Wondisford FE (1992b): Hormonal regulation of the thyrotropin beta-subunit gene by phosphorylation of the pituitary-specific transcription factor Pit-1. *Proc Natl Acad Sci USA* 89(13):5942–5945.

Stern S, Herr W (1991): The herpes simplex virus trans-activator VP16 recognizes the Oct-1 homeodomain: Evidence for a homeodomain recognition subdomain. *Genes & Dev* 5(12B):2555–2566.

Stern S, Tanaka M, Herr W (1989): The Oct-1 homeodomain directs formation of a multiprotein-DNA complex with the HSV transactivator VP16. *Nature* (Lond) 341(6243):624–630.

Sternberg PW, Horvitz HR (1991): Signal transduction during *C. elegans* vulval induction. *Trends Genet* 7(11-12):366–371.

Strobl JS, van Eys GJ, Thompson EB (1989): Dexamethasone control of growth hormone mRNA levels in GH3 pituitary cells is cycloheximide-sensitive and primarily posttranscriptional. *Mol Cell Endocrinol* 66(1):71–82.

Struthers RS, Gaddy-Kurten D, Vale WW (1992): Activin inhibits binding of transcription factor Pit-1 to the growth hormone promoter. *Proc Natl Acad Sci USA* 89:11451–11455.

Struthers RS, Perrin MH, Vale W (1989): Nucleotide regulation of growth hormone-releasing factor binding to rat pituitary receptors. *Endocrinology* 124(1):24–29.

Struthers RS, Vale WW, Arias C, Sawchenko PE, Montminy MR (1991): Somatotroph hypoplasia and dwarfism in transgenic mice expressing a non-phosphorylatable CREB mutant. *Nature* (Lond) 350(6319):622–624.

Sturm RA, Herr W (1988): The POU domain is a bipartite DNA-binding structure. *Nature* (Lond) 336(6199):601–604.

Südhof TC, Russell DW, Goldstein JL, Brown MS (1985): Cassette of eight axons shared by genes for LDL receptor and EGF precursor. *Science* 228:893–895.

Supowit SC, Ramsey T, Thompson EB (1992): Extinction of prolactin gene expression in somatic cell hybrids is correlated with the repression of the pituitary-specific trans-activator GHF-1/Pit-1. *Mol Endocrinol* 6(5):786–792.

Talamantes F (1990): Structure and regulation of secretion of mouse placental lactogens. *Prog Clin Biol Res* 342:81–85.

Tanaka M, Herr W (1990): Differential transcriptional activation by Oct-1 and Oct-2: Interdependent activation domains induce Oct-2 phosphorylation. *Cell* 60(3):375–386.

Tanaka M, Lai J-S, Herr W (1992): Promoter-selective activation domains in Oct-1 and Oct-2 direct differential activation of an snRNA and mRNA promoter. *Cell* 68(4):755–767.

Tansey WP, Catanzaro DF (1991): Sp1 and thyroid hormone receptor differentially activate expression of human growth hormone and chorionic somatomammotropin genes. *J Biol Chem* 266(15):9805–9813.

Tasaka K, Kasahara K, Masumoto N, Mizuki J, Kurachi H, Miyake A, Tanizawa O (1992): Activin A increases cytosolic free calcium concentration in rat pituitary somatotropes. *Biochem Biophys Res Commun* 185(3):974–980.

Tatsumi KI, Notomi T, Amino N, Miyai K (1992): Nucleotide sequence of the complementary DNA for human Pit-1/GHF-1. *Biochim Biophys Acta* 1129(2):231–234.

Thayer MJ, Tapscott SJ, Davis RL, Wright WE, Lassar AB, Weintraub H (1989): Positive autoregulation of the myogenic determination gene MyoD1. *Cell* 58(2):241–248.

Theill LE, Castrillo J-L, Wu D, Karin M (1989): Dissection of functional domains of the pituitary-specific transcription factor GHF-1. *Nature* (Lond) 342(6252):945–948.

Theill LE, Hattori K, Lazzaro D, Castrillo J-L, Karin M (1992): Differential splicing of the GHF1 primary transcript gives rise to two functionally distinct homeodomain proteins. *EMBO J* 11(6):2261–2269.

Thompson CC, Evans RM (1989): Trans-activation by thyroid hormone receptors: Functional parallels with steroid hormone receptors. *Proc Natl Acad Sci USA* 86(10):3494–3498.

Thomsen G, Woolf T, Whitman M, Sokol S, Vaughan J, Vale W, Melton DA (1990): Activins are expressed early in Xenopus embryogenesis and can induce axial mesoderm and anterior structures. *Cell* 63(3):485–493.

Tran P, Zhang XK, Salbert G, Hermann T, Lehmann JM, Pfahl M (1992): COUP orphan receptors are negative regulators of retinoic acid response pathways. *Mol Cell Biol* 12(10):4666–4676.

Treacy MN, He X, Rosenfeld MG (1991a): I-POU: A POU-domain protein that inhibits neuron-specific gene activation *Nature* (Lond) 350(6319):577–584.

Treacy MN, Ryan F, Martin F (1991b): Functional glucocorticoid inducible enhancer activity in the 5′-flanking sequences of the rat growth hormone gene. *J Steroid Biochem Mol Biol* 38(1):1–15.

Treacy MN, Neilson LI, Turner EE, He X, Rosenfeld MG (1992): Twin of I-

POU: A two amino acid difference in the I-POU homeodomain distinguishes an activator from an inhibitor of transcription. *Cell* 68(3):491–505.

Tripputi P, Guerin SL, Moore DD (1988): Two mechanisms for the extinction of gene expression in hybrid cells. *Science* 241(4870):1205–1207.

Tullius TD, Dombroski BA (1986): Hydroxyl radical "footprinting:" High resolution information about DNA-protein contacts and the application to 1 repressor and Crp protein. *Proc Natl Acad Sci USA* 83:5469–5473.

Umesono K, Giguere V, Glass CK, Rosenfeld MG, Evans RM (1988): Retinoic acid and thyroid hormone induce gene expression through a common responsive element. *Nature* (Lond) 336(6196):262–265.

Verrijzer CP, van Oosterhout JA, van der Vliet PC (1992a): The Oct-1 POU domain mediates interactions between Oct-1 and other POU proteins. *Mol Cell Biol* 12(2):542–551.

Verrijzer CP, Strating M, Mul YM, van der Vliet PC (1992b): POU domain transcription factors from different subclasses stimulate adenovirus DNA replication. *Nuc Acids Res* 20(23):6369–6375.

Verrijzer CP, Kal AJ, van der Vliet PC (1990a): The DNA binding domain (POU domain) of transcription factor oct-1 suffices for stimulation of DNA replication. *EMBO J* 9(6):1883–1888.

Verrijzer CP, Kal AJ, van der Vliet PC (1990b): The oct-1 homeodomain contacts only part of the octamer sequence and full oct-1 DNA-binding activity requires the POU-specific domain. *Genes & Dev* 4(11):1964–1974.

Verrijzer CP, van Oosterhout JA, van Weperen WW, van der Vliet PC (1991): POU proteins bend DNA via the POU-specific domain.*EMBO J* 10(10):3007–3014.

Verrijzer CP, van Oosterhout JA, van der Vliet PC (1992): The Oct-1 POU domain mediates interactions between Oct-1 and other POU proteins. *Mol Cell Biol* 12(2):542–551.

von Hippel PH, Berg OG (1986): On the specificity of DNA-protein interactions. *Proc Natl Acad Sci USA* 83:1608–1612.

Voss JW, Wilson L, Rosenfeld MG (1991): POU-domain proteins Pit-1 and Oct-1 interact to form a heteromeric complex and can cooperate to induce expression of the prolactin promoter. *Genes & Dev* 5(7):1309–1320.

Walker WH, Fitzpatrick SL, Barrera-Saldaña HA, Reséndez-Pérez D, Saunders GF (1991): The human placental lactogen genes: Structure, function, evolution and transcriptional regulation. *Endocr Rev* 12(4):316–328.

Watanabe YG (1982): Effects of brain and mesenchyme upon the cytogenesis of rat adenohyphresis *in vitro* . *Cell Tissue Res* 227:257–266.

Watanabe YG, Daikoku S (1979): An immunohistochemical study on the cytogenesis of adenohypophysial cells in fetal rats. *Dev Biol* 68:559–567.

Weintraub H, Davis R, Tapscott S, Thayer M, Krause M, Benezra R, Blackwell TK, Turner D, Rupp R, Hollenberg S et al. (1991): The myoD gene family: nodal point during specification of the muscle cell lineage. *Science* 251(4995):761–766.

West BL, Catanzaro DF, Mello SH, Cattini PA, Baxter JD, Reudelhuber TL (1987): Interaction of a tissue-specific factor with an essential rat growth hormone gene promoter element. *Mol Cell Biol* 7(3):1193–1197.

Wheeler MD, Styne DM (1988): The nonhuman primate as a model of growth hormone physiology in the human being. *Endocr Rev* 9(2):213–246.

Williams GR, Franklyn JA, Sheppard MC (1991): Thyroid hormone and glucocorticoid regulation of receptor and target gene mRNAs in pituitary cells. *Mol Cell Endocrinol* 80(1–3):127–138.

Wilson DB, Wyatt DP (1986): Ultrastructural immunocytochemistry of somatotrophs and mammotrophs in embryos of the dwarf mutant mouse. *Anat Rec* 215:282–287.

Wong EA, Ferrin NH, Silsby JL, El Halawani ME (1992): Complementary cDNA cloning and expression of Pit1/GHF-1 from the domestic turkey. *DNA Cell Biol* 11:651–660.

Yaffe B, Samuels HH (1984): Hormonal regulation of the growth hormone gene. *J Biol Chem* 259(10):6284–6291.

Yamada Y, Post SR, Wang K, Tager HS, Bell GI, Seino S (1992): Cloning and functional characterization of a family of human and mouse somatostatin receptors expressed in brain, gastrointestinal tract, and kidney. *Proc Natl Acad Sci USA* 89(1):251–255.

Yashiro T, Arai M, Miyashita E, Yamashita K, Suzuki T (1988): Fine-structural and immunohistochemical study of anterior pituitary cells of Snell dwarf mice. *Cell Tissue Res* 251(2):249–255.

Ye Z-S, Samuels HH (1987): Cell- and sequence-specific binding of nuclear proteins to 5′-flanking DNA of the rat growth hormone gene. *J Biol Chem* 262(13):6313–6317.

Ye Z-S, Forman BM, Aranda A, Pascual A, Park H-Y, Casanova J, Samuals HH (1988): Rat growth hormone gene expression: Both cell-specific and thyroid hormone response elements are required for thyroid hormone regulation. *J Biol Chem* 263(16):7821–7829.

Ying SY (1988): Inhibins, activins, and follistatins: Gonadal proteins modulating the secretion of follicle-stimulating hormone. *Endocr Rev* 9(2):267–293.

Zeytin FN, Gick GG, Brazeau P, Ling N, McLaughlin M, Bancroft C (1984): Growth hormone (GH)-releasing factor does not regulate GH release or GH mRNA levels in GH3 cells. *Endocrinology* 114(6):2054–2059.

Zhang XK, Hoffmann B, Tran PB, Graupner G, Pfahl M (1992): Retinoid X receptor is an auxiliary protein for thyroid hormone and retinoic acid receptors. *Nature* (Lond) 355(6359):441–446.

Index